MFPT 2018

Intelligent Technologies for Equipment and Human Performance Monitoring

Virginia Beach, Virginia, USA
15-17 May 2018

ISBN: 978-1-5108-8840-1

Printed from e-media with permission by:

Curran Associates, Inc.
57 Morehouse Lane
Red Hook, NY 12571

CURRAN ASSOCIATES INC. proceedings.com

Some format issues inherent in the e-media version may also appear in this print version.

Printed by Curran Associates, Inc. (2019)

MFPT Society
715 Twining Road
Suite 135
Dresher PA 19025

Phone: (215) 346-2870
Fax: (215) 352-6902

https://mfpt.org

Additional copies of this publication are available from:

Curran Associates, Inc.
57 Morehouse Lane
Red Hook, NY 12571 USA
Phone: 845-758-0400
Fax: 845-758-2633
Email: curran@proceedings.com
Web: www.proceedings.com

MFPT

MFPT 2018

Intelligent Technologies for Equipment and Human Performance Monitoring

PROCEEDINGS

MAY 15-17, 2018

Sheraton Virginia Beach Oceanfront Hotel
Virginia Beach, VA

Table of Contents

Table of Contents

A Novel Approach for Stress Cycle Analysis based on Empirical Mode Decomposition

Ruoyu Li*, Ali Marzban*, Jing Ping*, Jae Yoon*, and Gilbert Chahine*
***National Oilwell Varco**
7909 Parkwood Circle Dr., Houston, TX 77036
ruoyu.li@nov.com, ali.marzban@nov.com, jing.ping@nov.com, JaeMyung.Yoon@nov.com, gilbert.chahine@nov.com

In real industrial stress/strain analysis applications, calculating equivalent constant amplitude cycles is important. Rainflow counting is a process to obtain equivalent constant amplitude cycles. The method is designed to count reversals in accordance with the material's stress/strain relationship including hysteresis loops. However, Rainflow counting needs to identify the peaks/valleys in the collected sensor signal and it is sensitive to noise. In this paper, a novel approach is proposed to count the fatigue cycles. The approach first uses empirical mode decomposition method to decompose the signal adaptively. Then a systematic count method is developed to calculate the cycles based on the decomposed signal components. The effectiveness and the performance of this method are compared with the Rainflow counting algorithm on simulated data with different frequencies and levels of noise.

Keywords: Rainflow count, Empirical mode decomposition, Stress cycle analysis, fatigue analysis

1. Introduction

Fatigue analysis is one of the main challenges in man-made structures health monitoring. It is more challenging these days due to excessive usage of modern technology. Advances in the field of fatigue monitoring helped to reduce the amount of failure in structures. Recent understanding of how the materials fail has been considerably increased in the recent years.

One of the main factors in fatigue analysis is calculating the number cycles (i.e. loading history). There are different methods to estimate this number. One very popular method is Rainflow counting, which was first proposed by Matsuishi and Endo[1] . It is now widely applied to estimate the fatigue cycles in different industry applications, such as, railway, aircraft, bridge, and automotive industries [2, 3, 4]. However, Rainflow counting algorithm has a complicated procedure, which makes it difficult to apply when their statistical properties are to be studied [5, 6]. Moreover, Rainflow counting algorithm is very sensitive to noise as it is depending on the local minima and local maxima of the sensor signal.

In this paper, a new stress cycle counting algorithm has been developed. The approach first uses empirical mode decomposition to decompose the signal into different narrowband signals and then estimates the stress counts based on the decomposed signals. The performance of the developed method is evaluated with simulated data with different frequency and noise levels. The reminder of the paper is organized as follow. Section 2 describes the Rainflow counting algorithm, empirical mode decomposition (EMD) and the proposed framework. Section 3 presents a case study comparing the Rainflow counting algorithm and the developed method. Section 4 summarizes the work and concludes this paper.

2. Theoretic Basis

2.1 Rainflow counting algorithm

Rainflow counting algorithm is widely used in estimating the stress cycles of the structure [7]. The procedure of conducting the Rainflow counting algorithm can be summarized in the following five steps [8],

Assume X denotes range under consideration; Y denotes the previous range adjacent to X; and S denotes the starting point in the stress history.

(a) Read next peak or trough. If out of data, go to step (f).

(b) If there are less than three points, go to step (a). Form ranges X and Y using the three most recent peaks and trough that have not been discarded.

(c) Compare the absolute values of ranges X and Y.

If $X < Y$, go to step (a).

If $X > Y$, go to step (d).

(d) If range Y contains the starting point S, go to step (e); otherwise, count range Y as one cycle; discard the peak and trough of Y and go to step (b).

(e) Count range Y as one-half cycle; discard the first point in range Y; move the starting point to the second point in range Y and go to (b).

(f) Count each range that has not been previously counted as one-half cycle.

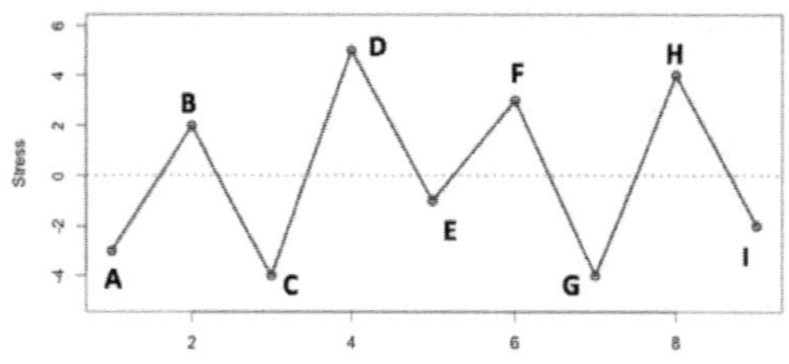

Figure 1. Example Stress vs Time Plot

A simple example is applied here to illustrate how Rainflow counting algorithm works. The stress versus time plot is shown in Figure 1. And the number of cycles corresponding to stress range illustrated in Fig. 3 is summarized in Table I.

Table I. Stess cycle count

Path	Cycles	Sress Range
A-B	0.5	5
B-C	0.5	6
C-D	0.5	9
D-G	0.5	9
E-F	1.0	4
G-H	0.5	8
H-I	0.5	6

2.2 Empirical mode decomposition

EMD was first proposed by Huang et al. as part of the Hilbert–Huang transform (HHT) [9]. Recently, EMD has been widely applied in different fields, such as acoustic emission signal processing [10], medical image processing [11], vibration analysis [12].

Assuming X represent the time series of the signal, the procedures of implementing the EMD algorithm is shown below,

(a) Find the local maxima and local minima of the signals.

(b) Construct the lower and upper envelopes of the signals by the cubic spline based on the local maxima and local minima, respectively.

(c) Calculate the mean values $m(t)$ by averaging the lower envelope and the upper envelope.

(d) Subtract the mean values from the original signals to produce the IMF candidate component $h_1(t) = X(t) - m(t)$. If it is the true IMF, go to the next step. In addition, the IMF component $C_i(t) = h_m(t)$ is saved. If it is not the IMF, repeat Steps a)–d). The stop condition for the iteration is given by

$$\sum_{t=0}^{T} \frac{[h_{m-1}(t) - h_m(t)]^2}{h_{m-1}^2(t)} \leqq SD$$

where $h_{m-1}(t)$ and $h_m(t)$ denote the IMF candidates of the $m - 1$ and m iterations, respectively, and, usually, SD is set between 0.2 and 0.3.

(e) Calculate the residual component by subtracting the IMF component obtained in Step d) from the original signals $res_i(t) = X(t) - C_i(t)$. This residual component is treated as new data and is subjected to the same processes described previously to calculate the next IMF component.

(f) Repeat Steps a)–e) until the final residual component becomes a monotonic function and no more IMF component can be extracted or the envelopes become smaller than a predetermined value. Through Steps a)–f), the original signals X.

A simulated example is used here to demonstrate how EMD process works. In the simulation, we add two sine waves (0.2 Hz and 3 Hz) together to generate a synthetic data. EMD is applied to decompose the signal into different components. From the Figure 2, we could see that the synthetic signal has been successfully separated into two components with 0.2 Hz and 3 Hz respectively.

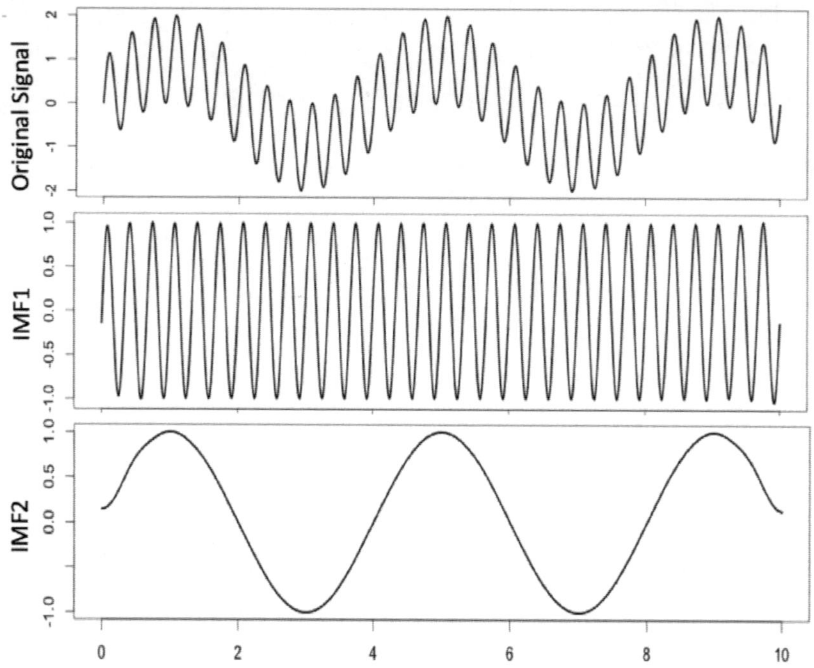

Figure 2. Simulated signal and IMF decomposition components

2.3 EMD-based Stress Cycle Counting Methodology

The framework of the developed methodology is shown in Figure 3.

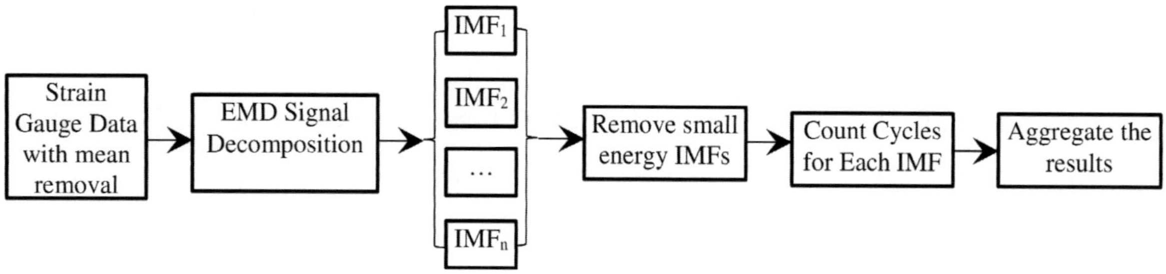

Figure 3. The framework of the developed methodology

As shown in Figure 3, the mean is first estimated and removed from the collected strain data. The estimated mean value is also used as the mean stress for fatigue analysis. Then the EMD is applied to decompose the strain data with mean value removed into multiple IMF components. To make the developed algorithm more robust to the noise, the energy ratios between each IMF component and the original signal are estimated and a threshold is set to remove any IMF component with energy ratio less than the threshold value. In this paper, 0.1 is used as the threshold to remove the IMF components. The energy ratio is calculated by using the following equation.

$$EnergyRatio_k = \frac{\sum_{i=1}^{N}(IMF_k^2(i))}{\sum_{i=1}^{N}\left(X^2(i)\right)}$$

where, IMF_k is the kth IMF component, X is the original signal, i is the ith sampling point, and N is the number of the sampling points of signal X.

For stress cycle counting, the zero crossings of each selected i^{th} IMF component are counted as fatigue cycles (N^i) and the amplitude between zero crossings are calculated as the alternating stress (σ_{ai}) to be used for N_f calculation . After the cycle counts and its corresponding alternating stress have been estimated, they are combined with the mean stress (σ_m), which is the mean value of the original signal, to calculate the fatigue damage. The following equation is used to compute the accumulative fatigue damage (AFD).

$$AFD_i = \sum_{i=1}^{K}\frac{N_i}{N_{fi}}$$

Where N_i is the number of cycles in i^{th} IMF component, k is the total number of the selected IMFs, and N_{fi} in the ith IMF is calculated as following,

$$N_{fi} = 10^{\text{intercept}} \times \sigma_0^{\text{slope}}$$

$$\sigma_0 = \frac{\sigma_{ai}}{1 - \left(\frac{\sigma_m}{\sigma_u}\right)^2}$$

where σ_u is ultimate stress, which is a material property, σ_{ai} is the alternating stress of the ith IMF component, σ_m is the mean stress, and intercept and slope are two constant related to material property.

3. Experimental Analysis

We have simulated five groups of signals. The simulated signals are composed by sine wave, random noise and a constant offset. For each group, it contains 100 simulated datasets with same level Gaussian noise. The simulated signals in group 1 have the lowest frequency content for sine wave and those in group 5 have the highest frequency content for sine wave. For each dataset, 200 Hz sampling frequency is used and 10 seconds' signal is simulated.

To obtain the accumulative fatigue value, a fatigue curve for a typical material (1050HR) was considered in this analysis. The relationship between number of cycles before failure (N_f) and corrected alternating stress (σ_0) is shown below

$$N_f = 1.298 \times 10^{38} \times \sigma_0^{-12.853}$$

Where

$$\sigma_0 = \frac{\sigma_a}{1 - \left(\frac{\sigma_m}{\sigma_u}\right)^2}$$

Where σ_a is alternating stress, σ_m is mean stress and σ_u is ultimate stress (630 MPa for 1050HR). For each cycle, corrected alternating stress (σ_0) and number of cycles before failure (N_f) need to be calculated. The accumulative damage (AFD) is introduced by N/N_f .

An example of the theoretical value for damage due to a 4.5 cycles of sine wave with 215 amplitude is summarized in table II.

Table II Theoretical damage

σ_a	σ_m	N	σ_0	N_f	AFD (N/N_f)
215	450	4.5	439.0	14,132	3.184E-04

All damage calculation for Rainflow and EMD are compared with their theoretical values. The calculated cycles are converted to be the AFD based on the SN curve approach described above. The performance of each model is measured by the percentage of the ratio difference, which is defined in the following equation.

$$AverageErrorPercentage = \sum_{i=1}^{J} \frac{\left|\frac{AFD_i}{AFD_{Theory}} - 1\right|}{J}$$

where AFD_i is the AFD for ith signal in the same group, AFD_{Theory} is the theoretical AFD for that group, J is the number of the signals in that group. The performance comparison is shown in Table III.

Table III Average of Error Percentage of different models

	Rainflow counting	EMD counting
Error percentage Group 1	9.8%	5.02%
Error percentage Group 2	13.17%	5.47%
Error percentage Group 3	9.17%	7.89%
Error percentage Group 4	14.7%	12.59%
Error percentage Group 5	19.99%	15.24%

From the results in Table III, one could see that EMD based counting approach works better than Rainflow counting algorithm, especially when the noise level is high.

4. Conclusions

In this paper, a novel robust approach was developed to count the fatigue cycles. The approach applied EMD algorithms to decompose the signal into IMFs. The stress cycles are estimated based on the IMFs. The performance of the developed method was evaluated with simulated data with different level of noise added to the signal. The experimental results had shown that the developed EMD based stress cycle counting algorithm outperformed the Rainflow counting algorithm and was more robust to the signal with noise.

Reference:

[1] Matsuishi, M. end Endo, T. Fatigue of metals subjected to varying stress, Proceedings of Japan Society of Mechanical Engineering, Jukvoka, Japan, 1968

[2] Baek, S.H., Cho, S.S., and Joo, W.S., Fatigue life prediction based on the Rainflow cycle counting method for the end beam of a freight car bogie, International Journal of Automotive Technology, Vol. 9, No. 1, pp 95 – 101, 2008

[3] Chan, T., Li, Z., and Ko, J., Fatigue analysis and life prediction of bridges with structural health monitoring data – Part II: application, Vol. 23, No. 1, pp 55 – 64, 2001

[4] Kermanidis, A., and Pantelakis, S., Fatigue crack growth analysis of 2024 T3 aluminium specimens under aircraft service spectra, Vol. 24, No. 10, pp. 699 – 710, 2001

[5] Rychlik, I., A new definition of the rainflow cycle counting method, International Journal of Fatigue, Vol. 9 , No2., pp. 119 -121, 1987

[6] Downing, S.D., Socie, D.F., Simple rainflow counting algorithms. International Journal of Fatigue, Vol. 4, No. 1, pp. 31-40, 1982

[7] Tovo, R., Cycle distribution and fatigue damage under broad-band random loading, International Journal of Fatigue, Vol. 24, No. 11, pp. 1137-1147, 2002

[8] Lee, J., and Palazzolo, A., Catcher bearing life prediction using a rainflow counting approach, Journal of Tribology, Vol. 134, No. 3, pp. 100-115, 2012

[9] N. Huang, Z. Shen, S. Long, M. Wu, H. Shih, Q. Zheng, N. Yen, C. Tung, and H. Liu, The empirical mode decomposition and the Hilbert spectrum for nonlinear and non-stationary time series analysis, Proc. Roy. Soc. London A, vol. 454, no. 1971, pp. 903–995, 1998.

[10] Li, R., and He D., "Rotational machine health monitoring and fault detecting using EMD-based acoustic emission feature quantification", IEEE Transactions on Instrumentation and Measurement, Vol. 61, No. 4, pp. 990 – 1001, 2012

[11] Nunes JC., Bouaoune Y., Delechelle E., Niang O, Bunel Ph, Image analysis by bidimensional empirical mode decomposition , Image and Vision Computing, Vol. 21, No. 12, pp. 1019 – 1026, 2003

[12] Liu, B., Riemenscheider, S., and Xu, Y., Gearbox Fault Diagnosis using Empirical Mode Decomposition and Hilbert Spectrum, Mechanical Systems and Signal Processing, Vol. 20, No. 3, pp. 718 – 734, 2016

TOWARDS POINT-OF-NEED MANUFACTURING TO REPLACE FAILED COMPONENTS ON THE BATTLEFIELD

Marc Pepi
Nicole Zander
US Army Research Laboratory
Weapons and Materials Research Directorate
Aberdeen Proving Ground, Maryland

Ian Jaqua
Bill Gleason
Courtney Young
Montana Tech
Butte, Montana

Abstract
The DoD is intrigued with "point of need" manufacturing as a means of being able to produce parts "on-demand" in extreme environments such as on a forward operating base, or on a ship, to increase our operational readiness, and reduce our huge military logistics tail. Additionally, research is being performed to determine whether recycled, reclaimed and/or indigenous materials can be utilized as feedstock for these components. However, there are technical challenges that need to be overcome to fully achieve this capability in the future. One such challenge is part quality, and whether such parts can provide a true replacement for military components.

Introduction

The vision for "point of need" manufacturing is to be able to replace or repair failed components at or near the theater of operations. The point of need could be the battlefield itself, a contingency base, or a reach-back CONUS facility. Readiness is a key concept for the military, and having this capability will not only improve operational readiness and materiel availability, it will provide critical repair parts with reduced resupply transportation, as well as responsiveness and flexibility, management of demand uncertainty and reduction in required inventory [1,2,3]. "Manufacturing" in the context of this paper is referring to additive manufacturing (AM) in most cases, although traditional manufacturing (i.e. metal casting, etc.) could be performed on the battlefield with enough resources. AM for the lay person is simply referring to a set of manufacturing processes that build a part "additively" in layers, versus the traditional means of manufacturing, which "subtractively" machines a part from bulk material.

Point-of-Need Manufacturing

The Army has been using 3D-printers in forward deployed areas in Afghanistan since 2012 [4]. These machines have made various plastic value-added components. In general, the US military is interested in the premise of AM for point-of-use manufacturing (the ability to produce spare parts, in the field, for immediate repair to support a mission) [5,6,7]. As

Dr. Thomas Russell, former Director of the US Army Research Laboratory, points out with respect to having the capability of having AM capabilities in-theater [8], "Logistically there are benefits. One of our biggest challenges in the Army is that there is a huge logistics burden. If we could forward-deploy manufacturing capabilities, we would have the opportunity to manufacture parts in-theater, or repair parts. This is not just about manufacturing a new part, it's often about how we can repair something that has been damaged. We have the opportunity to do that in-theater and use local materials. It's an exciting area. I don't think we've realized its full potential." Military officials see a lot of potential for on-demand 3D-printing on the battlefield. And while there's still a lot to be hashed out, commanders and decision-makers want to get it into the hands of troops sooner rather than later [9]. This new way of thinking will have ripple effects on the traditional military manufacturing supply chain and the resultant reduction in the logistics footprint.

Supply Chain

It is no secret that the time to resupply spare parts to the tip of the spear on the battlefield can be onerous. A 2008 GAO report indicates that managing spare parts for military weapon systems is a complicated, time-consuming and expensive task involving large inventories [10]. Spare parts are indeed important to keep weapon systems operational, but it can be difficult to assess which spare parts to bring to the battlefield. Which parts are likely to fail? How many should be carried in? One reference indicates that it can take weeks to get certain spare parts to the weapon system in-theater [11]. With the capability of making or repairing parts in-theater, this timeframe could be drastically reduced, perhaps to a day or two [4,12]. The making of spare parts through AM at the point-of-need is enticing, and these parts could be prime candidates for 3D-printing, especially if they have become obsolete or have long lead times, or are backordered due to diminishing sources and/or contractors not bidding on their production [13,14]. Whether it be the ability to additively manufacture on the battlefield, or on a ship, point-of-need manufacturing could provide the warfighter with parts that are good enough for the next mission, while the original part is ordered through the traditional channels. Or in other words, this technology, at first (until culture dictates acceptance, and safety critical parts are certified and qualified in-situ), may not replace existing supply chains, it may augment existing capabilities [13,15].

Logistics Footprint

It is often said that additive manufacturing is a "disruptive" technology, as it offers many advantages over traditional manufacturing. Although the technology for AM with metal powders is not mature enough for the battlefield as of yet, one can envision how it could dramatically reshape sustainment, improve system availability and affordability through reduced cost and logistics footprint associated with distribution, stowage and management of spare part inventories [16]. This streamlined process would go a long way in improving operational readiness, as weapons systems would not be idle for long, as a result of failed long-lead time or obsolete parts.

Challenges

Many challenges are currently restricting metal AM use in place of traditional manufacturing, never mind the challenges associated with battlefield manufacturing. These challenges include part certification and qualification (process control), legal concerns (intellectual property (IP) issues), security concerns (data storage and cyber threats), materials development, build volume limitations, lack of standards, AM workforce development [2,3,7,16]. When it comes to using this equipment on a battlefield, there are also issues of the surrounding environment (shock, vibration, sand, dust, temperature, and humidity, etc.), as well as the equipment footprint, power consumption, and part post-processing needs, etc. The two very important challenges of certification/qualification and intellectual property are discussed below.

AM Certification / Qualification

It is anticipated that AM on the battlefield will initially be used in a maintenance capacity for the repair of parts. As the technology matures, it is foreseen that spare parts may subsequently be produced. It is the author's belief that if and when this does occur, parts that are manufactured will only be required to be "good enough" to get the materiel back up and running and into the next battle, while simultaneously, the original equipment manufacturer (OEM) spare part would be ordered through the traditional supply chain. This "Band-Aid" approach would meet mission needs, and provide a strategic and tactical advantage on the battlefield. This may be the most logical route, since currently, no guidelines, neither documented procedures nor certification standards, exist on the approach for qualification of aeronautical parts manufactured through AM route [17]. It is anticipated that the inherent differences in AM compared to traditional manufacturing techniques will likely necessitate modifications to standard validation, verification and certification procedures [11], if safety critical parts were to be produced in-theater. This idea of temporary replacement could negate the need for qualification/certification [18], especially for non-critical parts.

Intellectual Property Concerns

Prior to the manufacturing of parts on the battlefield, decisions will need to be made regarding whether intellectual property rights (for example, patents, copyrights and perhaps trademarks) exist for those parts. Ignoring established protections could expose DoD organizations to the risk of litigation, and possibly jeopardize their relationships with defense industry partners [7]. According to [7], reverse engineering of a failed part in order to generate a CAD file would not be protected, but could be a patent infringement. However, it is imagined that in an emergency on a legacy systems, where the OEM is no longer in business, this may not be a problem (unless ownership of the part has transferred to another company) [3]. If the Government owns the technical data packages (TDPs) of these parts, it would not be an issue, but in most cases, these TDPs are cost prohibitive and ownership falls to the OEM. Liability of a failed AM part in service, that was fabricated on the battlefield, and nullified warranties of the weapon system it was emplaced, are also concerns, but beyond the scope of this paper.

Current Research

The goal of current research in this area by the US Army Research Laboratory (ARL) is technology development for point-of-need manufacturing, incorporating the use of indigenous, recycled, reclaimed and/or scrap resources for in-theatre additive or traditional manufacturing, to provide functional products of use to the warfighter. This paper will briefly describe research being performed by ARL in the following areas: the production of (AM) grade metal powder in a shipping container (intended to supply an operating base with feedstock for future metal AM operations at the point of need), and the utilization of indigenous desert sand for 3D-printed casting molds.

Use of Indigenous Materials on the Battlefield

Historically, the warfighter has used indigenous materials on the battlefield; from sticks and earth to form gabions and fascines, to sand and rock for sandbags, expeditionary earth-filled protective-barriers and Hesco-barriers. According to MIL-PRF-32277 [9], this latter family of earth-filled barriers is intended to provide protection from visual detection, small arms fire, indirect fire, and perimeter intrusion. It should be noted that all of these products are more utilitarian in nature, rather than technological innovations.

Metal Casting of Recycled Aluminum

Traditional casting has been in existence since at least 3,200 BC, and the earliest known sand casting dates back to 645 BC [19]. Research is being conducted to show a proof of concept of being able to cast parts on the battlefield with indigenous sand, and recycled and reclaimed scrap materials. This work focused on recycled soda cans, although there is not very much aluminum in them. It was found that salt fluxing was needed to help remove oxides and dross during the melting process. Sand found local to Montana Tech, Butte, Montana was screened to -50 mesh, and a non-dairy creamer was added as a binder, at approximately 2-3% by mass. The creamer used is the same as found in Meals-Ready-to-Eat (MRE) bags found on the battlefield. These dry components were mixed thoroughly. Subsequently, water was mixed a little at a time until the sand reached a relatively stable and compactable consistency. Since the amount of water needed seemed to vary, this is one area where further research and experimentation may be needed to provide the optimal mixture. Next, the part that will be cast is placed on a plastic plate within a mold frame (see Figure 1), and the casting sand is packed into the frame. For this experiment, a round section of tube approximately 2-inches in height was used. A mold release was used to keep sand from sticking to the part as it was packed into the ring firmly, and in multiple layers. Sand was leveled off at the top. The mold was then flipped, and the part removed carefully (see Figure 2), so as not to damage the imprint of the wrench. It was found that the non-dairy creamer mixture was much less stable as a binder than other mixtures, so this may also be an area for further exploration. Once this was completed, the recycled cans were melted in a Lindberg/Blue M Model 1100 box furnace (1.8kW) electric furnace at approximately 850°C. The furnace used was a traditional tabletop variety. A higher temperature was found to facilitate easier pouring and rapid melting. The melted aluminum was poured into the mold and allowed to cool (See Figure 3). Because oxide formation is so rapid, the aluminum flow was viscous and difficult to control in an open mold. Once cooled, the part was removed and allowed to finish cooling. It was found that the burned sand was non-reusable, but preliminary work with Bentonite clay showed this material could be reused. The cast wrench was subjected to milling and filing (tools available on the battlefield), with before and after photos shown in Figures 4 and 5. Although there were imperfections in the tool, and it most likely would not be strong enough to hold up to high stresses, it showed what could be accomplished towards casting of functional items on the battlefield.

Figure 1 Materials used to produce a casting of a wrench.

Figure 2 Wrench is removed from sand mold.

Figure 3 Melted soda cans being poured into a sand mold.

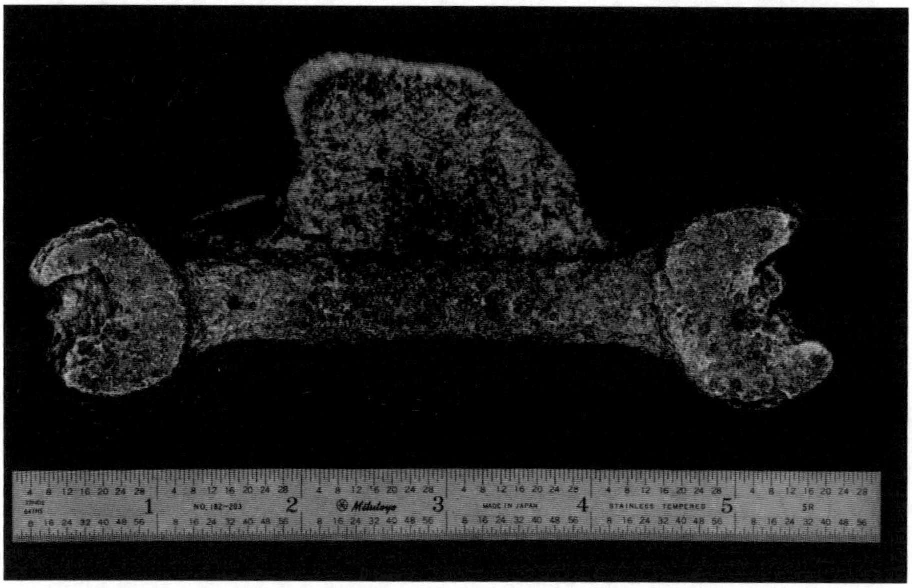

Figure 4 Crude "wrench" cast from melted soda cans.

Figure 5 Milled and filed end of wrench made from melted soda cans.

Metal Casting of Reclaimed Brass Cartridges

In similar research, attention was focused on melting and casting spent brass cartridges knowing there will probably be much of this material on the battlefield. The brass was also melted in a Lindberg/Blue M Model 1100 box furnace (1.8kW) at 1100ºC, and poured with no fluxing agent produced poor quality plates (see Figure 6). Samples that were cast with 7.5% by weight professional grade flux produced a viscous liquid metal upon melting that failed to fill the mold completely. This problem was likely the result of quick cooling due to the water in the sand cast mold as part of the binder. In an attempt to counter this, pouring from both ends of the mold was performed, but this also failed to yield a complete plate. It was determined that this problem could possibly be alleviated by taking advantage of gravity. An open top casting mold (as that shown in Figure 7) was built to improve liquid metal flow, and allow the resulting gas to escape the pour. The results of using this mold yielded a better plate, although it did have some imperfections (see Figure 8). It is anticipated that these brass plates could be used on the battlefield to produce functional items through subtractive manufacturing, either with a traditional CNC mill, or by employing an "Othermill" or similar such equipment. This unit can subtractively produce items from plate material using a CAD drawing. Items such as washers, fittings and brackets, etc. could be machined for use on the battlefield.

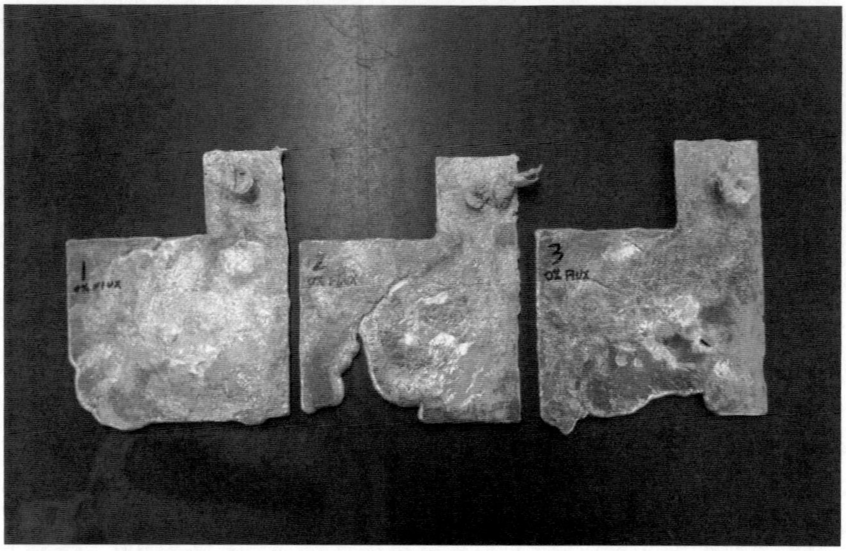

Figure 6 Castings of melted cartridge brass with no fluxing (each casting was approximately 5-inches wide).

Figure 7 Open top mold used to improve metal flow and allow for the gas to escape.

Figure 8 Plate made with melted spent brass cartridges using an open top mold.

3D-Printing Casting Molds with Desert Sand
The ability exists to 3D-print with sand using commercial off-the-shelf printers. These printers require the use of OEM provided sand, as well as a binder to keep the build together. The benefits of using 3D-printed sand molds versus traditional sand molds include the following:

- Molds can be made in a shorter time, without complex and expensive tooling
- Molds are generated from CAD models
- Complex geometries can be accommodated, with faster design modifications

The benefit of being able to use indigenous sand means one less item that would need to be shipped to the battlefield. The concept is to 3D-print casting molds based on a CAD drawing of a part with desert sand in a shipping container at the point-of-need, and melt scrap, recycled and reclaimed materials for casting of metal spare or replacement parts. This process would yield a part that was not 3D-printed, so it may be more readily accepted by the end user.

ARL teamed with the University of Northern Iowa (UNI) on a Defense Logistics Agency (DLA) funded project, to determine whether indigenous sand (such as that found in a desert or beach) could be used with existing 3D-printers with an appropriate binder, for the purpose of manufacturing appropriate long lead time parts in-theater. As a proof-of-concept, sand from the Defense Training Center at Ft. Irwin, CA (Mojave Desert) was sent to UNI to research its potential use with a 3D-printer. The original equipment manufacturer (OEM) sand that is typically used with these printers is silicon dioxide, shown in Figure 9. For comparison, Mojave Desert sand is shown in Figure 10 at the same magnification. Noticeable differences exist, including size distribution (which can be countered through sieving), composition, and the fact that the OEM sand is washed (no dust). The OEM sand actually looks closer to beach sand than desert sand based on the lack of dust (see Figure 11 for comparison). A CAD drawing of a mock component (Figure 12) was utilized to determine if it could be cast in A356 aluminum using Mojave Desert sand. A mold was made using the OEM powder for comparison. UNI screened the material to eliminate oversize and undersize particles and bond strength tested the sand using conventional foundry resins. Because desert sands may contain materials that effect the curing of conventional resins, various chemical hardeners were investigated. Once the molds were 3D-printed with desert sand, the aluminum was poured, and the parts allowed to cool. Figures 13 and 14 show the parts made with OEM, and desert sand, respectively. The part made with desert sand appeared to have a rougher surface finish, and would most likely need more post-processing than the part made with the OEM sand.

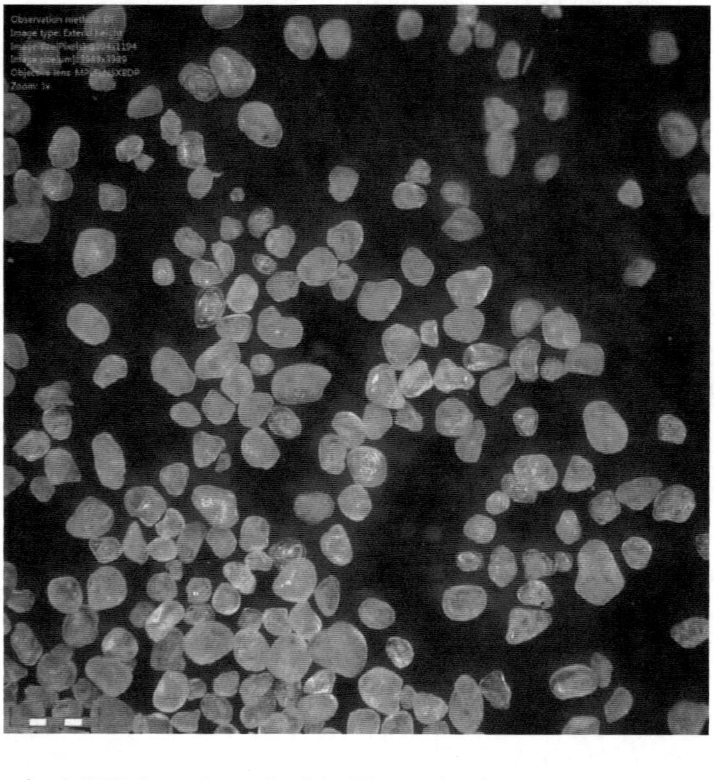

Figure 9 Typical OEM sand used with 3D sand printers. Micron marker = 500μm.

Figure 10 Sand from the Defense Training Center, Ft. Irwin, CA (Mojave Desert). Micron marker = 500μm.

Figure 11 Sand from Clam Pass, Naples, FL. Aside from the bits of shells, etc., the beach sand compares favorably to the OEM sand used with 3D sand printers. Micron marker = 500μm.

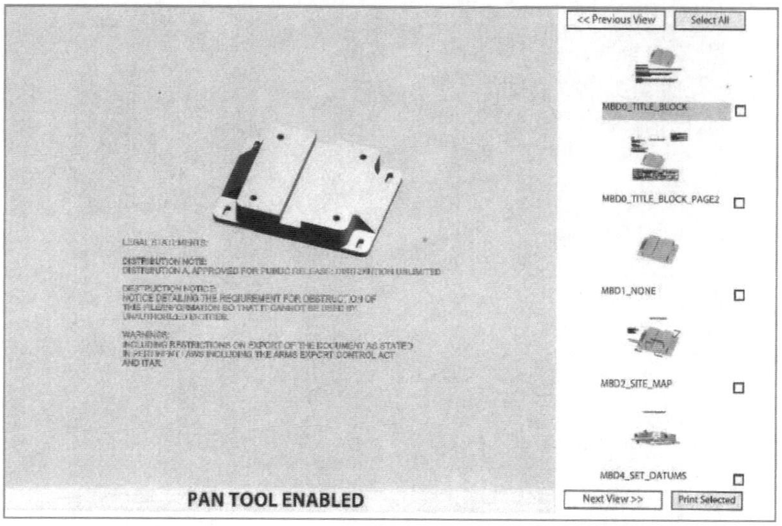

Figure 12 CAD drawing of a mock component to be aluminum cast using 3D-printed sand molds.

Figure 13 Cast aluminum A356 part using a 3D-printed mold with OEM sand.

Figure 14 Cast aluminum A356 part using a 3D-printed mold with Mojave Desert sand. Note what appears to be a rougher surface as compared to the part made with the OEM sand (Figure 13).

Towards Production of Metal Powder on the Battlefield

One of the challenges associated with future metal AM on a forward operating base is the transport of flammable metal powders for use with these processes. To counter this, ARL has teamed with MolyWorks Corp in an effort to produce AM-grade metal powder in a shipping container. The problem with traditional metal powder production in-theater (such as gas or water atomization), is that too much infrastructure would be needed, including the equipment, utilities (electric power, water, and inert gas), post-processing of the powder, need for cleanliness, etc. [20]. This report concluded that it would be more practical to stock an inventory of alloy powder for anticipated battlefield needs. Although these are legitimate concerns, the goal of this research was to determine whether the operations could be optimized to reduce the burden caused by these hindrances.

MolyWorks used their existing mobile foundry, contained within a shipping container to produce metallic powder from certified alloy bar stock, including AISI 4130 steel, titanium 6Al-4V, copper (Cu-101), 316 stainless steel (Figure 15) and 6061 aluminum alloy (Figure 16). With further research and development, it is anticipated that the ancillary equipment (controller, power supply, gas supply, etc.) as well as the mobile foundry could also all be contained in a shipping container. Within the mobile foundry, the metal is placed into the crucible, melted, and poured over flowing argon gas. The metal powder is formed and collected in the cyclones at the far end of the equipment.

To determine the feasibility of using scrap in this process, ARL furnished MolyWorks with a piece of actual battlefield scrap steel and battlefield scrap aluminum, to be added to certified steel and aluminum alloys, respectively. MolyWorks was also able to produce aluminum powder made solely from aluminum battlefield scrap (Figure 17). It is clear that the particles are, for the most part, spherical, and contain some satellites.

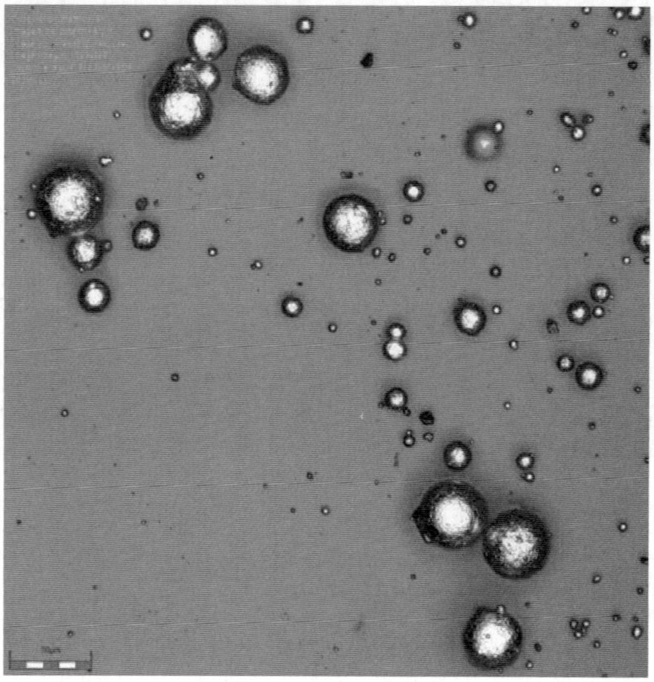

Figure 15 316 stainless steel powder (-325 mesh) made in the MolyWorks mobile foundry using certified alloy starter material. Micron marker = 50μm.

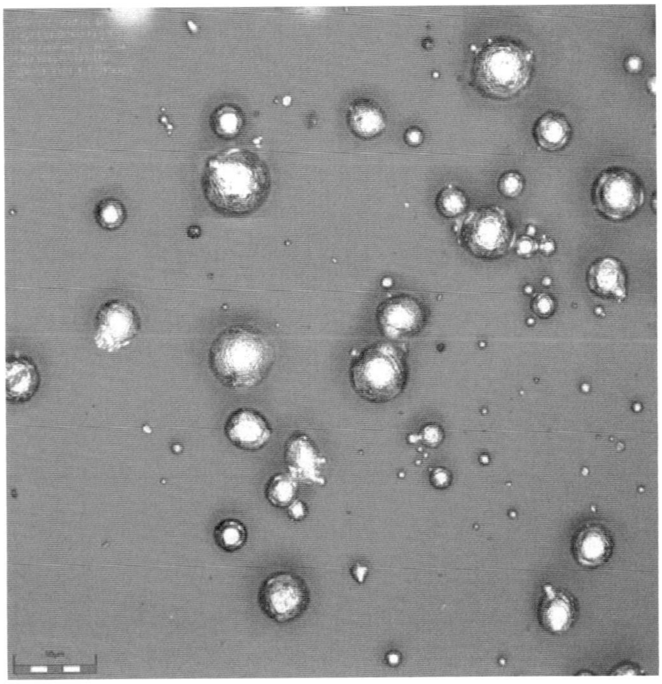

Figure 16 6061 aluminum alloy powder (-325 mesh) made in the MolyWorks mobile foundry using certified alloy starter material. Micron marker = 50μm.

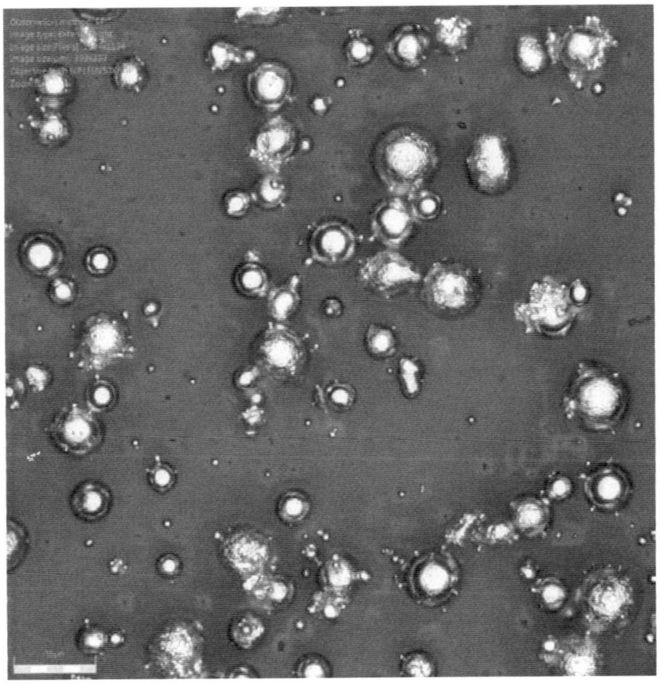

Figure 17 Aluminum powder (-325 mesh) made entirely from aluminum battlefield scrap within the MolyWorks mobile foundry. Micron marker = 50μm.

ARL also wanted to determine whether the powder produced in the mobile foundry could be used with the cold gas dynamic spray (cold spray) process. This is important, in that it would show that AM-grade metallic powder produced on the battlefield could potentially be used with a portable cold spray machine for the repair of parts in-situ, extending the life cycle of these components, and reducing the logistics needed to get a spare part back to the theater of operations. Both the 316 stainless steel powder and aluminum powder made entirely from battlefield scrap was placed within the powder feeder of the VRC Gen II portable cold spray system, and sprayed onto 316L stainless steel substrate panels, and AA 6061 panels, respectively. For comparison to the 316 stainless steel powder made by MolyWorks, a sample of Praxair FE-101 316 stainless steel powder was also cold sprayed onto the 316 stainless steel substrate. It took 35 passes to build up 0.10-inch of MolyWorks powder, as opposed to only 25 passes for the Praxair powder. In addition, the surface finish of the cold spray build using the MolyWorks powder was rougher than that of the Praxair powder (see Figure 18). To determine the reason for this difference, the panel was sectioned and metallographically prepared. Figure 19 shows a comparison of the cross sections of the two cold sprayed powders. The MolyWorks powder appeared to have less porosity, but more microcracks, indicative of higher residual forces within the build. This process would need to be optimized, but it did show that the powder could be cold sprayed with a portable unit. The Praxair powder was analyzed to determine how it compared to the MolyWorks powder. Figure 20 shows that the Praxair powder was water atomized, making "splat" shaped particles, versus the spherical particles produced from gas atomization.

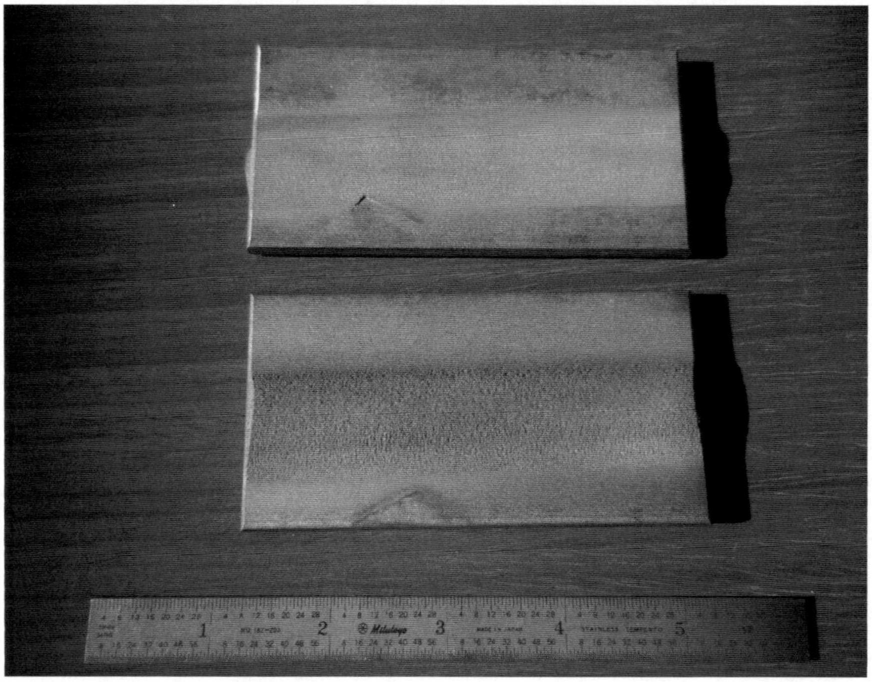

Figure 18 Oblique lighting photograph of cold sprayed 316 stainless steel powder (-325 mesh) made by Praxair (top) and the MolyWorks mobile foundry (bottom). Note the rougher surface finish of the cold spray build using the MolyWorks powder.

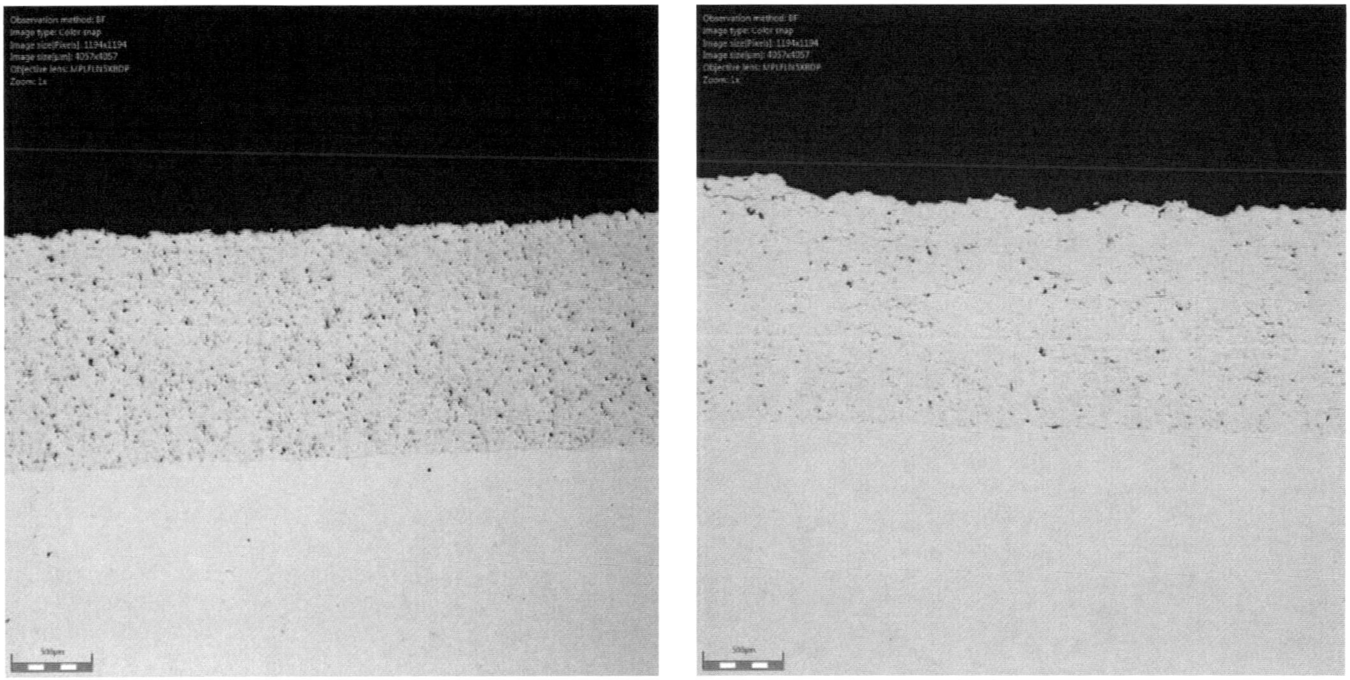

Figure 19 Micrographs of cold sprayed 316 stainless steel powder (-325 mesh) made by Praxair (left) and the MolyWorks mobile foundry (right). Although the MolyWorks deposit appeared to have less porosity, it contained microcracking. Micron marker = 500μm.

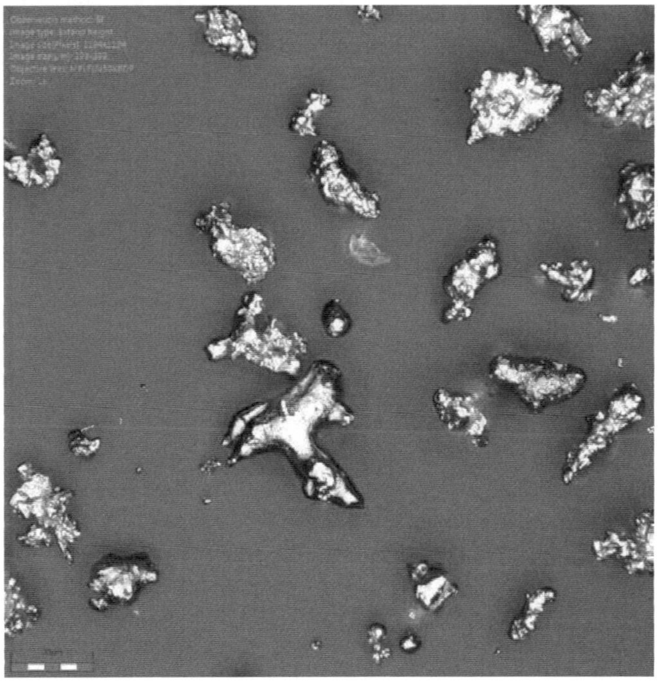

Figure 20 Praxair FE-101 -325 mesh 316 stainless steel powder Praxair that has been water atomized. Contrast to 316 stainless steel powder made by MolyWorks that was gas atomized (Figure 15).

Particles of 316 stainless steel were mounted and polished, and subjected to scanning electron microscopy. Figure 21 confirms that the powder produced by the mobile foundry is mostly spherical, with a smaller amount of spheroidal shapes, and very few angular shapes. Some porosity is seen in the cross sections, which is likely a result of atomization gas becoming trapped in the molten particles during solidification. The microstructure appears to vary from particle to particle for the 316 stainless steel, perhaps due to variability in the manufacturing process.

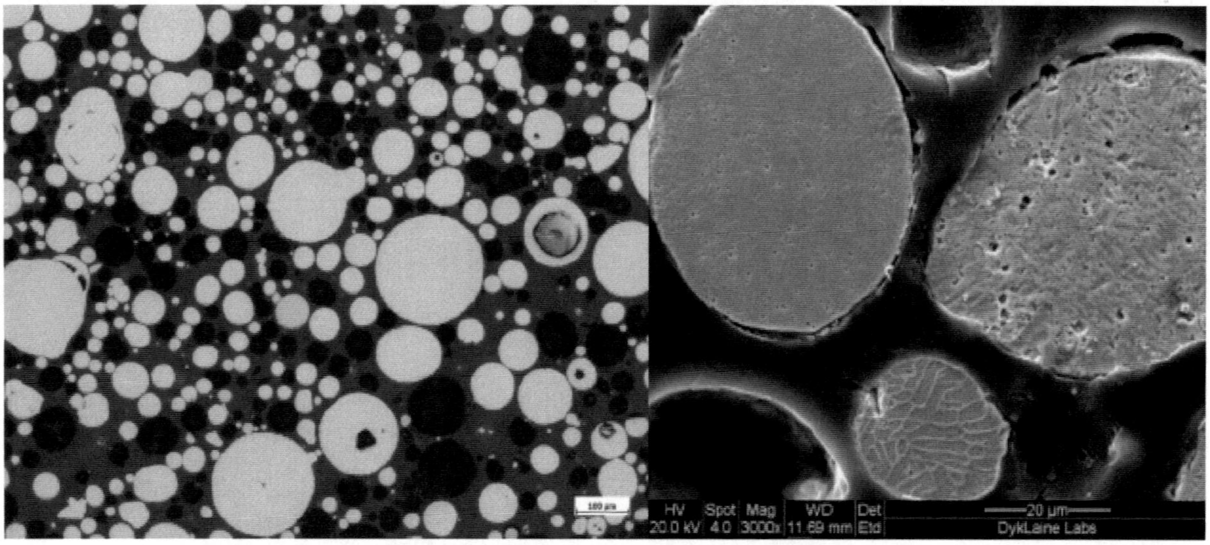

Figure 21 SEM images of the 316 stainless steel powder produced by MolyWorks in the mobile foundry (100x left, 3000x right).

The aluminum powder made entirely from aluminum battlefield scrap was also subjected to the cold gas dynamic spray (cold spray) process, and was deposited successfully as shown in Figure 22. Three grooves were machined onto the test panels. Two grooves were filled with the portable cold spray system, and one was machined back to the level of the substrate. The cold spray deposition machined nicely, and no defects were visually noted. Figure 23 shows the mechanical mixing that was noted at the interface, which is indicative of a strong adhesive bond.

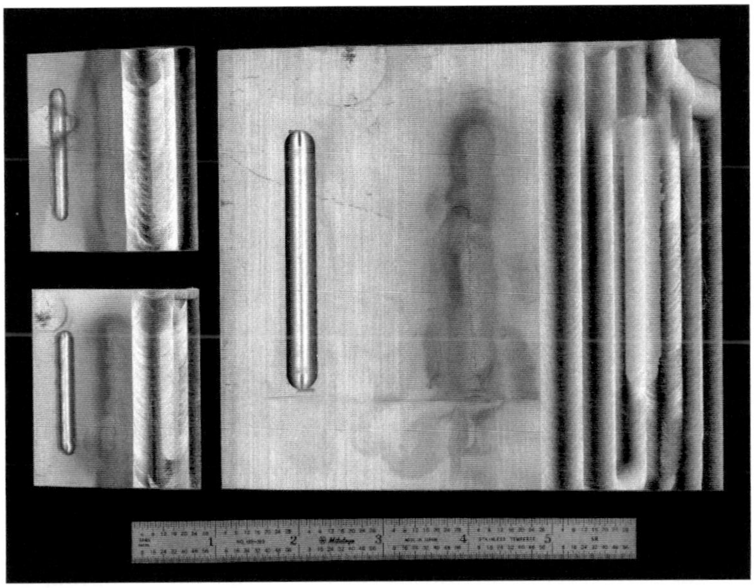

Figure 22 Cold spray deposition of aluminum powder (-325 mesh) made entirely from battlefield scrap within the MolyWorks mobile foundry onto three panels.

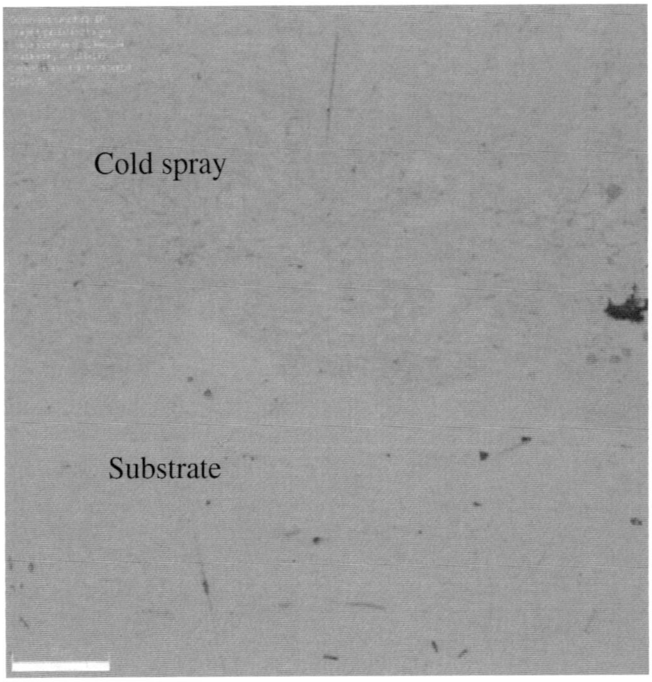

Figure 23 Section through one of the small samples shown in Figure 22 showing mechanical mixing at the interface of deposition and substrate. Micron marker = 20μm.

Conclusions

Reducing the dependence on the logistical supply chain at the point of need will not only increase operational readiness and the self-sustainability of warfighters in-theater, but also improve the safety of the warfighter by reducing threat vulnerabilities associated with our military logistics tail. The ability to fabricate or repair needed parts on demand, in-theatre in austere environments with available resources would be game-changing for the military.

Although further research is needed to optimize the process, we have shown a proof-of-concept of melting and casting recycled and reclaimed scrap metals using simple tools and equipment that can be subtractively manufactured into functional items.

We have also shown that desert sand can be 3D-printed into casting molds to produce parts via traditional foundry casting. This proof-of-concept showed that locally available sands, with little processing could be effectively utilized to produce metal casting molds of sufficient strength to cast light metal alloys. This capability may someday allow for manufacturing of parts in-theater using indigenous sand, and recycled/reclaimed battlefield scrap.

With respect to the production of AM-grade metallic powder in-theater, we have made great strides in producing powder in a mobile foundry contained within a shipping container. The batches can be made with either certified alloy or scrap metal, and the subsequent powder can be cold sprayed. The vision is to have the future capability to produce this powder in-theater for real-time repair of components (for example, with cold spray technology), or the building of spare parts (for example, combined cold spray and subtractive manufacturing).

This initial research has shown the realm of the possible with respect to the use of indigenous, recycled, and reclaimed materials at the point of need for the repair or manufacturing of parts.

Future Work

ARL is interested in any and all manufacturing processes that can be utilized in-theater with indigenous, recycled and reclaimed materials, and will pursue paths leading to the ultimate use of traditional or AM on the battlefield using these materials as feedstocks.

Acknowledgements

The authors wish to thank the following individuals for their contributions to this work: Zack Stewart, Molly Brockway, Emily Kooistra-Manning and Morgan Ashbaugh from Montana Tech; Prof. Jerry Thiel and team from UNI, Kelly Morris and team from DLA (3D-printing with desert sand for metal casting); Chris Eonta and team at MolyWorks Corp. (AM-Grade metallic powder from battlefield scrap).

Research performed by Montana Tech was sponsored by the Army Research Laboratory and was accomplished under Cooperative Agreement Number W911NF-15-2-0020 (Cooperative Agreement manager: Kyu Cho). The views and conclusions contained in this document are those of the authors and should not be interpreted as representing the official policies, either expressed or implied, of the Army Research Laboratory or the U.S. Government. The U.S. Government is authorized to reproduce and distribute reprints for Government purposes notwithstanding any copyright notation herein.

References

[1] "Demand Reduction: Setting Conditions to Enable Multi-Domain Battle – White Paper", Army TRADOC document, 2/21/18.
[2] Vitale, M., Cotteleer, M., Holdowsky, J., "An Overview of Additive Manufacturing", Defense AT&L, November-December, 2016.
[3] Langlais, Jr. R., Avdellas, N., Finfrock, C., Salley, R., Newcomb, M., "Separating Hype from Reality", Defense AT&L, November-December, 2016.
[4] Cox, M., "Army Sees Rapid Prototyping as Key to Rapid Innovation", 4/1/15.
[5] Clark, S.L., "Getting AM up to Speed Across the Army Life Cycle", Defense AT&L, November-December, 2016.
[6] Q&A with General Dennis L. Via, Readiness Provider – Responsibly Shaping the Command and Anticipating the Future Operating Environment, Military Logistics Forum 9.8.
[7] Hazeltine, J., Austin, K., Smith, G., "Additive Manufacturing Challenges and Opportunities for Military Applications", RIAC TAT Number RI-13-RMS#690/DO#290, WR13-13, November 2013.
[8] "Dr. Thomas Russell: Research Vision for 3-D Printing", Army Technology Magazine, RDECOM Public affairs, 7/5/14.
[9] Pomerleau, M., "Officials: The Time is Now for Battlefield 3D-Printing", Defense Systems, 3/2/16.
[10] Chalcraft, E., "US Military Invests in 3D Printing on the Frontline", https://www.dezeen.com/2012/11/12/us-military-invests-in-3d-printing-on-the-frontline/, 11/12/12.
[11] Scott, J., Gupta, N., Weber, C., Newsome, S., Wohlers, T., Caffrey, T., "Additive Manufacturing: Status and Opportunities", IDA Science and Technology Policy Institute, March 2012.
[12] Kurdian, A. CAPT, "Implications of AM for the Navy Supply Chain", Defense AT&L, November-December, 2016.
[13] Morris, K., "Driving Innovation to Support the Warfighter – Additive Manufacturing initiatives Within the Defense Logistics Agency", Defense AT&L, November-December, 2016.
[14] Deal, K., "Additive Manufacturing – How 3D Printing Will Transform the A&D Support Chain", Aerospace and Defense Technology, December, 2016.
[15] Bacon, L., "Here's How Marines are Using 3-D Printing to Make Their Own Parts", Marine Corps Times, 4/30/16.
[16] Gaska, M. and Clement, T., "Additive Manufacturing as a Sustainment Enabler – An Industry Perspective", Defense AT&L, November-December, 2016.
[17] Samuel, M.P., Mishra, A.K., and Mishra, R.K., "Additive Manufacturing of Ti-6Al-4V Aero Engine Parts: Qualification for Reliability", Journal of Failure Analysis and Prevention, (2018) 18:136-144.
[18] Arnold, A., "Additive Manufacturing and the Battlefield: What Next?" Defense IQ, 3/23/16.

[19] History of Metal Casting – A Brief Timeline, Metal Technologies, http://www.metal-technologies.com/docs/default-source/education/historyofmetalcasting.pdf?sfvrsn=8
[20] Strauss, J., "Feasibility Study on the Deployment of an Atomization System Contained Within Conex Shipping Containers to Operate at a Forward Operating Base", 11/25/15, special report for ARL.

A Subspace Clustering Chart Using a Reference Model for Featureless Bearing Performance Degradation Assessment

Xiaoxi Ding[a], Yimin Shao[a] and Qingbo He[b] Diego Galar[c]

[a]State Key Laboratory of Mechanical Transmission
Shazheng Street 174, Shapingba District
Chongqing, 400044, CHINA
ymshao@cqu.edu.cn

[b]Department of Precision Machinery and Precision Instrumentation
University of Science and Technology of China
Hefei, Anhui 230026, CHINA
[c]Division of Operation and Maintenance Engineering
Lulea University of Technology
Lulea 97187, SWEDEN

Abstract: The health index (HI) of machine condition must be sensitive and robust in complex working conditions. A systematic HI will assess machine performance automatically, reliably, and in a timely manner without intervention. This paper proposes a subspace clustering HI in a model using reference data on component health. Unlike the conventional HIs empirically learned from raw feature sets, a subspace clustering HI aims to automatically describe the migration and variation of the condition clustering distribution in a series of two-class subspace models derived from the raw data. First, in a featureless process, a covariance-driven Hankel matrix is directly constructed from the raw time-domain signal, and principal component analysis is used to separate the feature subspace and noise null-space. Second, in the index construction process, the reference health subspace data (from healthy data) and the monitored subspace data (from monitored data) are combined to construct a referenced model. Thus, a new spatial clustering HI with kernel operation is implemented to assess the current bearing performance and reveal discriminative features. The effectiveness of the proposed subspace clustering HI for the detection of abnormal condition is evaluated experimentally on bearing test-beds, using a mobile mapping mode. A novel subspace clustering chart, CUSUM-based spatial clustering HI, is developed to depict the real bearing performance degradation. Compared to the regular HI (e.g., root mean square), the proposed approach provides a more accurate and reliable degradation assessment profile with an early fault occurrence alarm. The experimental results show the potential of the proposed spatial clustering analysis to assess bearing degradation.

Key words: Bearing health monitoring; spatial clustering analysis; subspace clustering chart; featureless process; performance degradation assessment

Introduction: Bearings are widely used in rotating machinery but are easily damaged, creating a need to reliably assess their health in a timely manner [1-2], so that failure and loss can be prevented. Ideally, health degradation indicators will monitor and accurately assess a bearing's health condition throughout its lifecycle, giving early warning of slight degradation. But constructing a reliable and sensitive health indicator (HI) remains an elusive challenge. The dominant analyses are time-domain, frequency-domain and time-frequency analysis, with feature selection (FS) [3-4] or feature extraction (FE) [5-6] techniques often used as pre-processing methods for bearing fault diagnosis and degradation assessment.

It should be noted that, first, in conventional health monitoring schemes, these features are constructed by means of artificial experience and intervention. Second, the HIs are always directly constructed or mapped from a healthy dataset in the training process. The new obtained features are put into models for bearing fault diagnosis [7-8], performance assessment [9-11] and residual useful life (RUL) prediction [12-13]. For example, in the application of bearing health monitoring, Qiu et al. [9] trained self-organizing map (SOM) from the original feature matrix of the normal data and used the minimum quantization error (MQE) to detect the degradation. Yu [10] obtained the feature mapping from the healthy data by performing locality preserving projection (LPP) and put the mapped features of the monitored data into a Gaussian mixture model (GMM) to determine bearing performance degradation. Pan et al. [11] used the extracted features from normal data to train a support vector data description (SVDD) model and health index (HI) based on general distance. The above work mainly focuses on artificial features built from raw data and relies on mathematical models for assessment.

The following two characteristics are not well considered in the assessment process. First, according to the physical mechanism of the bearing failure, there are differences between abnormal and normal data. Second, the principle of statistical quality control indicates that the feature distribution of the real-time monitoring data will transfer from heathy to degraded status with operating time. Based on this understanding, our previous work [6,14] has shown that a significant difference between the features of healthy and unhealthy signals can improve the clustering and distinguishability of multiclass features, including identification of different degradation severities with similar feature distribution. Arguably, then, a reference model using healthy data with feature clustering distribution can be imported into performance degradation assessment (PDA).

However, artificial statistical characteristics need to be prepared in advance, and conventional vector features are distributed by noise and external interference. To detect structure damage, subspace identification (SI) [15,16] is used to mine the principal subspace, so that the desired information can be well represented and identified. Due to its simplicity and utility, this study employs SI to automatically mine the latent components of health condition. It uses a series of subspace bases to support a two-dimensional space and operating time, so that the monitoring process is a three-dimensional feature interpretation process.

The paper uses the subspace clustering distribution transition healthy data in a reference model to assess the condition of the bearing in a process called spatial clustering analysis. It proposes a subspace clustering chart using the reference model instead of the complex mathematical monitoring models with artificial features. First, a series of subspace bases are automatically reached through manifold learning from a covariance-driven Hankel matrix. Then, a reference model is built using the same reference data (healthy data) and the monitored samples at each moment. Finally, a new subspace clustering HI is implemented in Grassmann manifold space to assess the current machine performance. In this manner, a subspace clustering HI can be obtained for a featureless health condition assessment; the latent components of each health condition can be described in an automatic mode, with variations in information highlighted by the reference model. In short, spatial clustering analysis and an improved subspace clustering chart using a reference model with healthy data are proposed for featureless bearing health performance degradation identification and assessment.

The rest of this paper is organized as follows. Section II presents the framework of the subspace clustering chart using a reference model; a covariance-driven Hankel matrix is automatically constructed for spatial clustering analysis. Section III introduces the experimental setups; abnormal information is detected using a mobile mapping mode and bearing performance degradation is assessed over the full lifecycle to verify the effectiveness of the proposed model. Section IV provides conclusions.

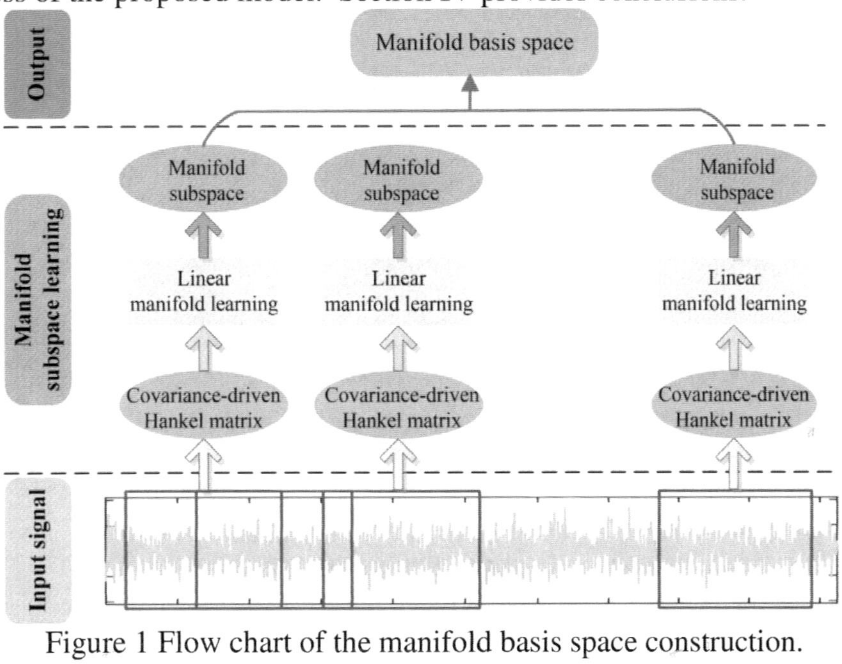

Figure 1 Flow chart of the manifold basis space construction.

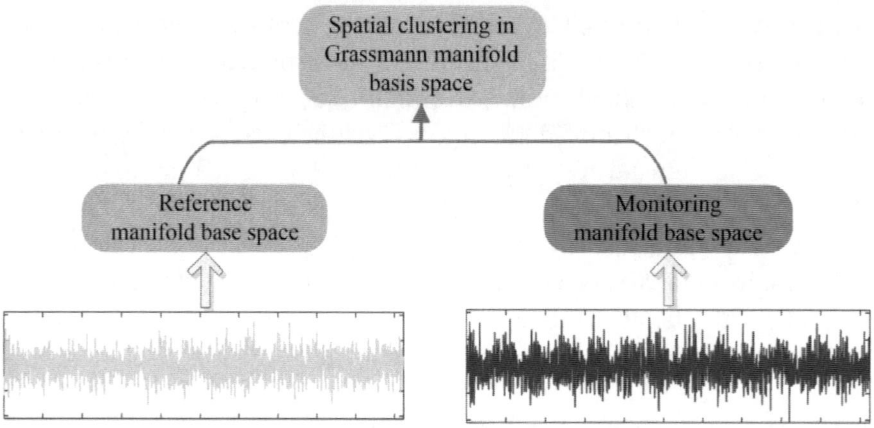

Figure 2 Flow chart of the featureless cluster analysis based on Grassmann manifold basis space.

Subspace clustering chart using reference model: The accuracy of health identification depends on the sensitivity of the extracted features and their ability to show variations from the healthy state. However, because of the complex conditions and because early degradation has a weak response, it is difficult to achieve a timely and reliable performance degradation assessment using artificial features in a conventional monitoring process. Therefore, this paper proposes a subspace clustering chart using a reference model with healthy data to monitor and assess bearing health. The process involves two steps: manifold base space construction (MBSC) and featureless clustering analysis (FCA) based on the Grassmann manifold (GM) reference model. Detailed flowcharts are shown in Fig. 1 and Fig. 2.

In the process of MBSC, as illustrated in Fig. 1, a new Hankel matrix is built from the raw time-domain signal. Then, the manifold subspace is adaptively learned from these matrixes through manifold learning. Lastly, these are employed to support the new manifold basis space, representing the part's present condition. In the process of performing the FCA of the Grassmann manifold reference model, as displayed in Fig. 2, a new reference model is built combining the healthy manifold basis space and the monitored manifold basis space. A discriminant index is later calculated to assess the spatial clustering of these two manifold bases in GM's space. A subspace clustering chart is thus obtained for online health monitoring. Unlike the classical health monitoring scheme, with the healthy data, the feature clustering distribution of the reference model can expose differences once a fault happens.

The following sections provide more detailed information on the proposed subspace clustering system, including subspace matrix construction, manifold learning, spatial clustering analysis, and CUSUM-based subspace clustering chart.

Covariance-driven Hankel matrix construction: In order to achieve an adaptive state representation from the raw signal, this paper employs subspace identification (SI) to separate and mine the desired structural subspace to construct the principal-subspace or null-subspace supported by the signal space. SI utilizes the incoherence between the feature information and noise components in the space domain. The proposed covariance-driven Hankel matrix is operated in this manner. For a given multi-channel signal, X={x_1, x_2,..., x_N} $\in R^{T \times N}$ (T and N are respectively the channel and length of the measured signal), a covariance-driven Hankel matrix can be obtained as:

$$\mathbf{H}\big|_{p\times q} = \begin{bmatrix} C_1 & C_2 & \cdots\cdots & C_q \\ C_2 & C_3 & \cdots\cdots & C_{q+1} \\ \cdots & \cdots & \cdots\cdots & \cdots \\ C_p & C_{p+1} & \cdots\cdots & C_{p+q-1} \end{bmatrix} \tag{1}$$

where p and q are the size of the matrix, p is equal to q, and C_i is the output covariance, defined as:

$$\begin{aligned} C_i &\approx \frac{1}{N-i+1}\sum_{n=1}^{N-i+1} x_n x_{n+i-1}{}^T \\ &= \frac{1}{N-i+1}[x_1 \; x_2 \; \dots \; x_i]\big|_{S\times i} \cdot [x_i \; x_{i+1} \; \dots \; x_N]^T\big|_{i\times S} \end{aligned} \tag{2}$$

Based on the matrix **H**, the corresponding principal subspace of the monitored signal can later be mined to assess the equipment condition.

Manifold learning: The latent manifold of the raw feature set can be mined in linear and nonlinear learning processes. Considering the parameters and efficiency with orthogonal bases, principal component analysis (PCA) is applied to obtain the latent condition information for the principal subspace remaining. Thus, a new manifold basis space can be built using d principal basis written as:

$$S = span[s_1, s_2, ..., s_d] \in R^{p\times d}, d \le p \tag{3}$$

Note that subspace S represents condition information in a high-domain information space; this will be further analyzed in Grassmann manifold (GM) space [17].

Spatial clustering analysis: Along with the working time, data on health status can be collected. To explore the health status and determine differences between the healthy and monitored status, a reference model with two classes is built at each moment. In other words, two kinds of basis spaces are combined after the corresponding manifold basis spaces are obtained. This means that a two-class basis space matrix at monitored moment t (t=1,2,...,T) can be created by synthesizing the set $\mathbf{X}_{\text{health}}=\{S_1, S_2, \ldots, S_{n0}\}$, with n_0 number of samples, together with the corresponding basis space set $\mathbf{Y}^t_{\text{monitor}}=\{S^t_1, S^t_2,\ldots, S^t_{nt}\}$, with n_t samples. Thus, as shown in Fig. 2, a reference model $\mathbf{Z}_t=\{\mathbf{X}_{\text{health}}, \mathbf{Y}^t_{\text{monitor}}\}$ using healthy data is constructed for the subsequent spatial clustering analysis.

Instead of using the traditional one-dimensional feature vector from a time-domain signal, this paper uses a two-dimensional basis matrix based on MBSC with Riemannian metric. A new spanning basis matrix can be regarded as a new point and a series of bases can support a subspace called Grassmann manifold (GM). The aforementioned manifold basis space can be regarded as a new point in this subspace group, $O(n) \in G^n$, and the relationship between these subspaces can be represented as:

$$d_G(\mathbf{P},\mathbf{Q}) = \|\boldsymbol{\theta}\|_2 \tag{4}$$

where is θ=[θ_1, θ_2,…, θ_m] and usually calculated as:

$$\cos(\theta_k) = \max_{\mathbf{p}\in\mathbf{P},\mathbf{q}\in\mathbf{Q}} \mathbf{p}^T\mathbf{q} = p^T q \quad s.t. \ \|\mathbf{P}\| = \|\mathbf{Q}\| = \mathbf{I}_m \tag{5}$$

From Eqs. (3)~(5), it can easily be seen that an operation is achieved by utilizing an inner product. Therefore, as in manifold learning, a kernel operation is used to retain the latent information, such as KPCA, LPP etc. In this study, we can suppose that the desired information is embedded in the GM of the reproducing kernel Hilbert space (RKHS). Three kernel operations are usually applied to the GM spaces: Binet-Cauchy kernel [18], projection kernel [19], and correction pseudo kernel [20], respectively defined as:

$$d_{bc}(\mathbf{P},\mathbf{Q}) = \det\left\|\mathbf{P}^T\mathbf{Q}\mathbf{Q}^T\mathbf{P}\right\| \tag{6}$$

$$d_p(\mathbf{P},\mathbf{Q}) = tr\left(\mathbf{P}^T\mathbf{Q}\mathbf{Q}^T\mathbf{P}\right) \tag{7}$$

$$d_{cc}(\mathbf{P},\mathbf{Q}) = \max_{p\in\mathbf{P},q\in\mathbf{Q}}\left(p^T q\right) \tag{8}$$

For a given sequence of time-domain signal $x \in R^{1\times N}$, the manifold basis space S=span{s_i, i=1,2,…,d} $\in R^{p\times m}$ can be obtained to construct the reference model. To evaluate the distribution of the reference model $\mathbf{Z}_t$, a spatial clustering factor is further proposed in the GM space with a kernel operation as mentioned above, found as:

$$G = G_b / G_w \tag{9}$$

$$G_b = \sum_{j=1}^{c} \sum_{k\neq l\in C_j} d(\mathbf{S}_k, \mathbf{S}_l) \tag{10}$$

$$G_w = \sum_{j=1}^{c} \sum_{l\in C_k, k\neq j} d(\mathbf{S}_j, \mathbf{S}_k) \tag{11}$$

where the GM-based between-class scatter G_b indicates the scattered level among different classes, while the GM-based within-scatter S_w describes the concentrated level in the same class. The GM-based discriminant factor G is a comprehensive indicator that combines the property of between-class scatter and within-class scatter. Specifically, according to Eqs. (6)~(8), three different GM-based discriminant factors can be acquired, respectively called G_B, G_P and G_C. This GM-based discriminant factor G can be used to trace and assess the distribution and transition of the manifold spaces as these changes due to performance degradation. It is foreseeable that this HI will be suitable for the detection of abnormal status. Thus, a new subspace clustering chart can be created for health monitoring.

Experimental descriptions: To verify the effectiveness of the proposed spatial clustering analysis in monitoring equipment status, two cases are tested here. One is the detection of abnormal status using a mobile measurement system to acquire the acoustic signal of a bearing; the other is the assessment of bearing performance degradation for the full lifecycle of the bearing using a test provided by PCE, IMS of University of Cincinnati. The details of two cases are given below.

Case I: First, to validate the ability of the proposed method to detect abnormal status, a mobile measurement system is built to acquire the acoustic signal of a ball bearing (Type 620 6-2RZ HRB) with an outer-race defect, as shown in Fig. 3(a). Fig. 3(b) illustrates a sound mobile measurement system, where sound signals from mobile multipaths are collected by the front-end collector in a hand-held inspection. This information is transmitted to the host computer to detect the mobile health condition. In this case study, two android mobile phones (ZTE 889D and MI 1) with specially developed acquisition software are employed to movably acquire the acoustic signal in a random way. The sampling frequency is 44.1 kHz with a sampling time of 10s.

Two different experiments (Data A and Data B) are tested: one is for mobile detection with different measuring locations, as shown in Fig. 3 and Table 1, and the other is for mobile detection with different measuring locations and tools, as in Table 2. It should be noted that two status conditions, healthy and outer-race detect, are tested. In the test for Data A, two traditional HIs, namely RMS and kurtosis, are calculated for 70 samples; their corresponding scatter plots are displayed in Fig. 4. There seems to be no valid knowledge from these two status conditions; they distribute four different scatter classes, leading to a confusing detection. There is an even worse phenomenon in Fig. 4(b). Because there were unavoidable impacts in the bearing test, the kurtosis failed to detect impact. Based on the proposed spatial clustering analysis, three HIs, G_B, G_P and G_C as introduced in Eqs. (6)~(11), are respectively obtained on the GM-based reference model. The scatter results are displayed in Fig. 5. It can be easily seen that these three HIs yield a much better clustering effect; the same types of status condition can be gathered together and normal and abnormal conditions can be clearly distinguished.

In the test for Data B with RMS and kurtosis, using the different measuring locations and tools, confusing detection results are obtained, as shown in Fig. 6. Next, using the proposed method, the GM-based subspace clustering HI is calculated, as shown in Fig. 7. According to the halving line between two status conditions, G_P and G_C, the method outperforms conventional RMS and kurtosis. By introducing the reference model using healthy data, the proposed spatial clustering HI can detect the difference between normal and abnormal conditions while ignoring the effect of the measuring distance, measuring tools, and some inference from the working conditions. Therefore, the proposed spatial clustering using the reference model can achieve mobile detection in an itinerant way and shows good potential for use in real production quality control (PQC).

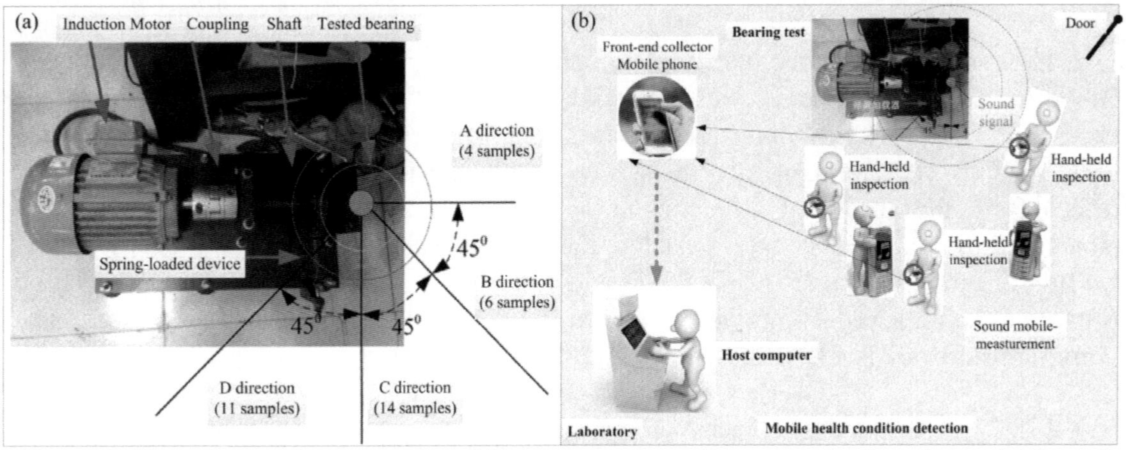

Figure 3. Experimental setup and mobile measurement for acoustic signal acquisition of ball bearing with outer-race defect: (a) bearing test, and (b) sound mobile-measurement.

Table 1: The conditions of the mobile measurement on the bearing test with the ZTE mobile phone in an android system (Data A)

Condition	Healthy	Outer-race fault
Measuring tool	An android mobile phone with a developed acquisition Software	
Measuring direction	Four direction (A_d, B_d, C_d, D_d) with 45 degrees interval	
Measuring samples	A_d-4 samples; B_d-6 samples; C_d-14 samples; D_d -11 samples	
Sampling condition	Sampling frequency 44.1 kHz with a time sampling 10s	

Table 2: The conditions of the mobile measurement on the bearing test with different measuring tools and locations (Data B)

Condition	Healthy		Outer-race fault	
Measuring tool	ZTE 889D (Person P)	MI 1 (Person Q)	ZTE 889D (Person P)	MI 1 (Person Q)
Measuring samples	25	28	56	50
Label	H-Z	H-M	F-Z	F-M

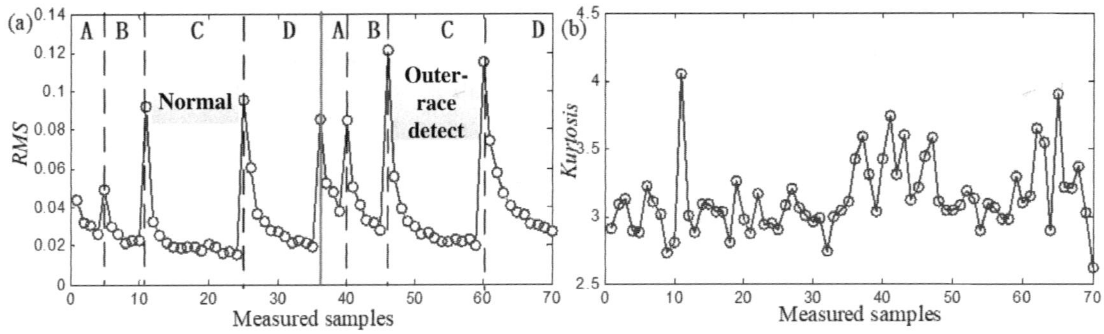

Figure 4. The health distribution of the measured Data A based on different HIs for Data A: (a) RMS and (b) Kurtosis.

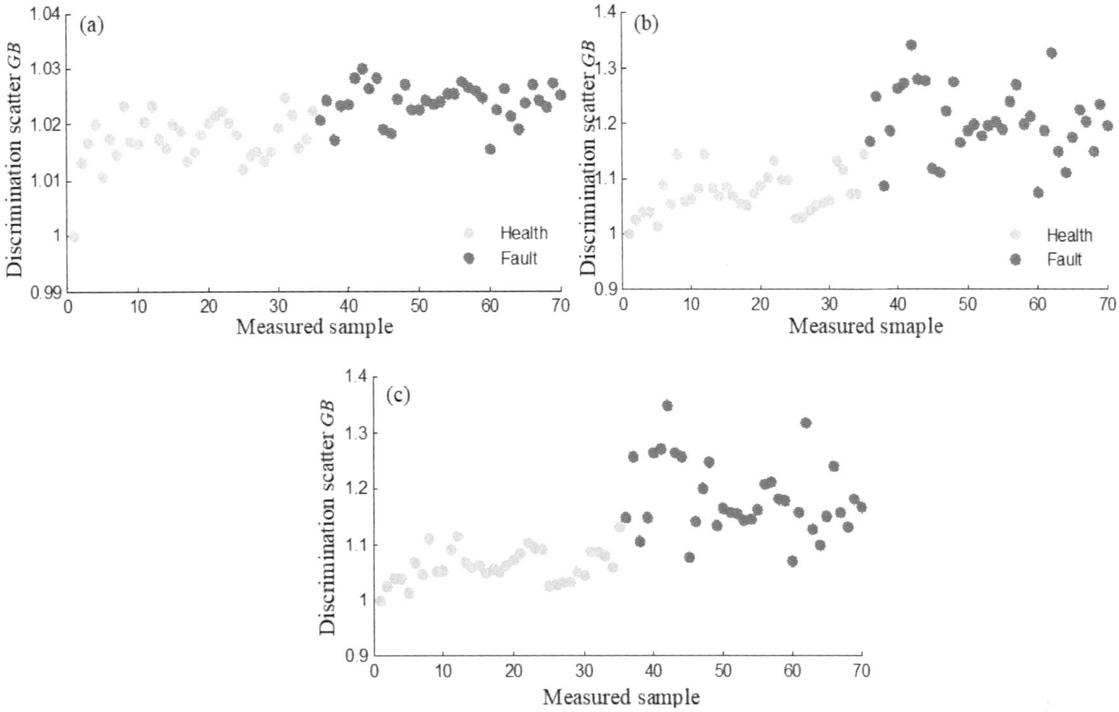

Figure 5. The health distribution of the measured Data A based on the proposed HI_{GM} with different kernel operation: (a) *GB*-Binet-Cauchy kernel, (b) *GP*-projection kernel and (b) *GC*-correction pseudo kernel.

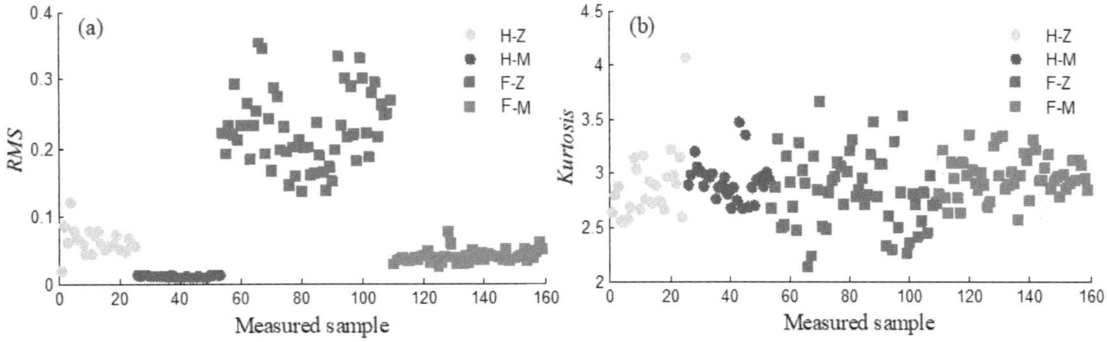

Figure 6. The health distribution of the measured Data B based on different HIs: (a) RMS and (b) Kurtosis.

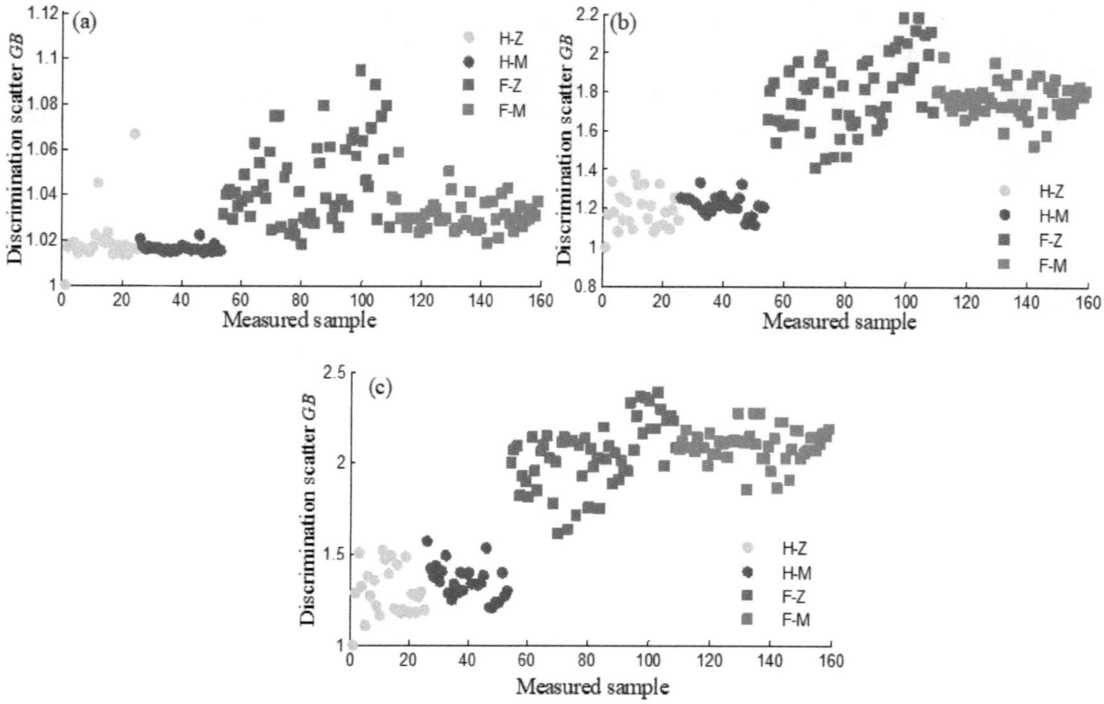

Figure 7. The health distribution of the measured Data B based on the proposed HI_{GM} with different kernel operation: (a) *GB*-Binet-Cauchy kernel, (b) *GP*-projection kernel and (b) *GC*-correction pseudo kernel.

Case II: To verify the effectiveness of the proposed method with the subspace clustering chart using a reference model, the study tests a case of a bearing running to failure. The experimental data are downloaded from Prognostics Center Excellence (PCE) through a prognostic data repository contributed by Intelligent Maintenance System (IMS), University of Cincinnati [21]. As plotted in Fig. 8, this study uses bearing 2 (2#) and bearing 4 (4#) of test 2 with outer-race failure to illustrate the performance of the proposed subspace clustering chart. The healthy data size n_0 and the monitored data size n_t are both set at 100, with a sample length of 1024. The size of the Hankel matrix is 20×20, and five principal bases are extracted to support the manifold basis space S; the reference model with the healthy subspaces and the monitored subspaces is at monitoring time t (t=1,2,…,T) is 200. It should be noted that the same healthy basis subspace is used in each monitoring process. Then, according to the spatial clustering analysis in GM space, as expressed in Eqs. (6)~(11), the GM-based discriminant factor G with three manifold kernel operations is calculated from the reference models to assess the performance degradation of the monitored bearing at each corresponding moment t.

For bearing 2#, as displayed in Fig. 9(a), the HIs based on *GB* and *GP* reveal a much clearer degradation trend; a significant and sharp change in the subspace clustering chart will occur once the monitored bearing degenerates from healthy status to slight degradation. Generally, these changing points can be used to accurately verify the occurrence of faults in a timely manner, helping to identify degradation at an early stage. An early alarm can be sounded, and effective maintenance can be arranged to restore bearing health. There is also a local fluctuation between the HI values, however, and this could have a bad effect on assessment.

In order to strengthen the sensitivity of HI to precisely assess the bearing performance degradation and detect the occurrence of degradation at an early stage, the study tests two traditional quality-control techniques: the exponentially weighted moving average (EWMA) and cumulative sum statistical control (CUSUM). These two techniques can be applied to the proposed GM-based subspace HI for a developed assessment chart. The EWMA-based HI can be defined as:

$$s_t = (1-\alpha)G_{t-1} + \alpha G_t \quad \alpha \in [0,1] \tag{12}$$

where a larger weight value of α allots more weight to current data G and less weight to older data. Unlike the consideration of the current status shown in Fig. 9(a), as displayed in Fig. 9(b), EWMA can associate the current status with the previous status, thereby achieving a much more sensitive response for PDA. It can detect a defect, and the degradation effect can be assessed in a clearer band. Furthermore, degradation is an irreversible physical process, and the status of the bearing will degrade with working time; these factors should be taken into consideration in the degraded HI. Accordingly, the CUSUM-based HI is calculated as:

$$c_i = \mathrm{CUSUM}(G_i - \mathrm{T}_{\mathrm{arget}}) \tag{13}$$

where $\mathrm{T}_{\mathrm{arget}}$ represents the average level of the health status in the past. This HI can indicate the cumulative deviation of the monitored bearing. As shown in Fig. 9(c), this improved subspace clustering HI can reveal a much more stable and clearer degradation trend with good monotonicity. It also detects different degradation rates, from slight degradation to severe degradation; this illustrates that the degradation of the full lifecycle is nonlinear.

The conclusion is shown in Fig. 10. The GM-based subspace clustering chart can accurately assess the degradation of a monotonic process, but the determination of degeneration based on the traditional HI is inaccurate, resulting in a confusing warning and incorrect diagnosis. Specifically, compared to the conventional HIs as displayed in Figs. 9 and 10, the proposed subspace clustering charts and their improvements, i.e., the HIs based on *GB* and *GP* kernel operation, can automatically reach a more accurate and sensitive PDA from the raw data by means of the GM-based spatial clustering analysis, and a significant warning will be clearly given. Therefore, spatial clustering analysis can provide a more accurate and comprehensible HI in real-word applications of bearing health monitoring.

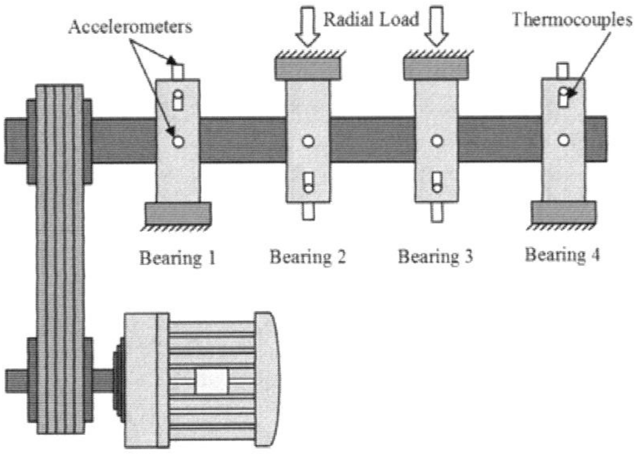

Figure 8. Bearing test rig

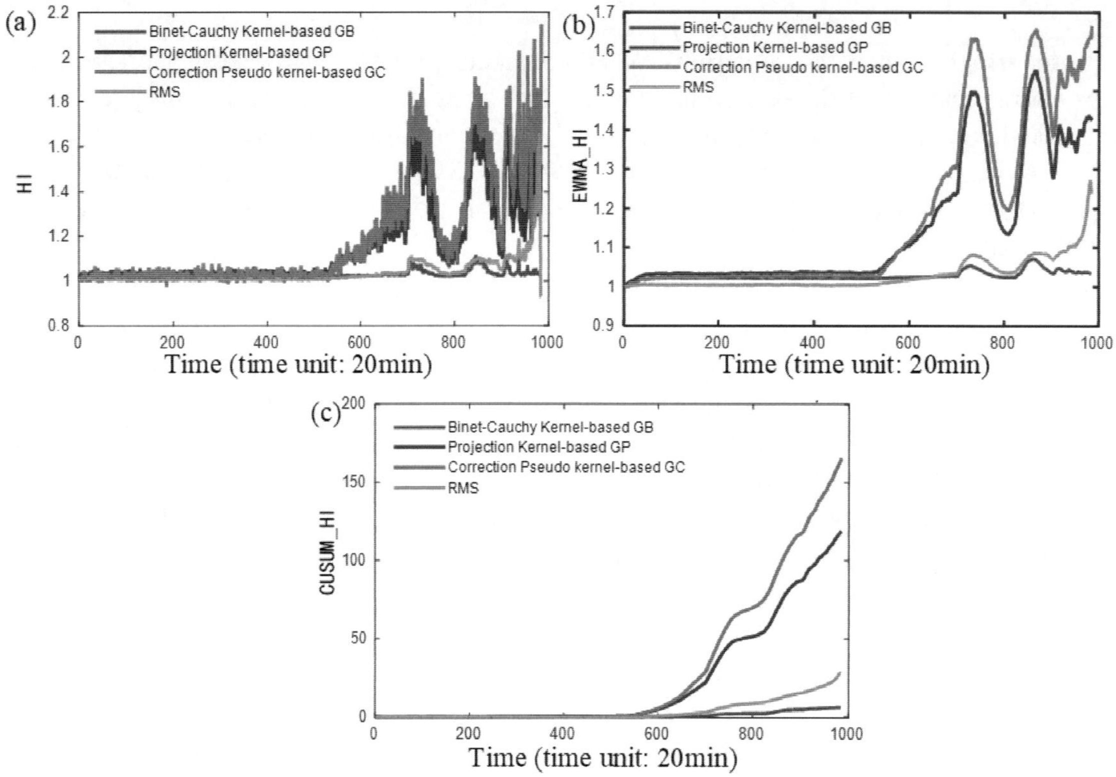

Figure 9. The PDA results for the full cycle life of the tested bearing 2#: (a) the raw HI, (b) the EWMA-based HI_w and (c) the CUSUM-based HI_c (the brown, blue, red and green lines represent different His, GB, GP, GC and RMS, respectively).

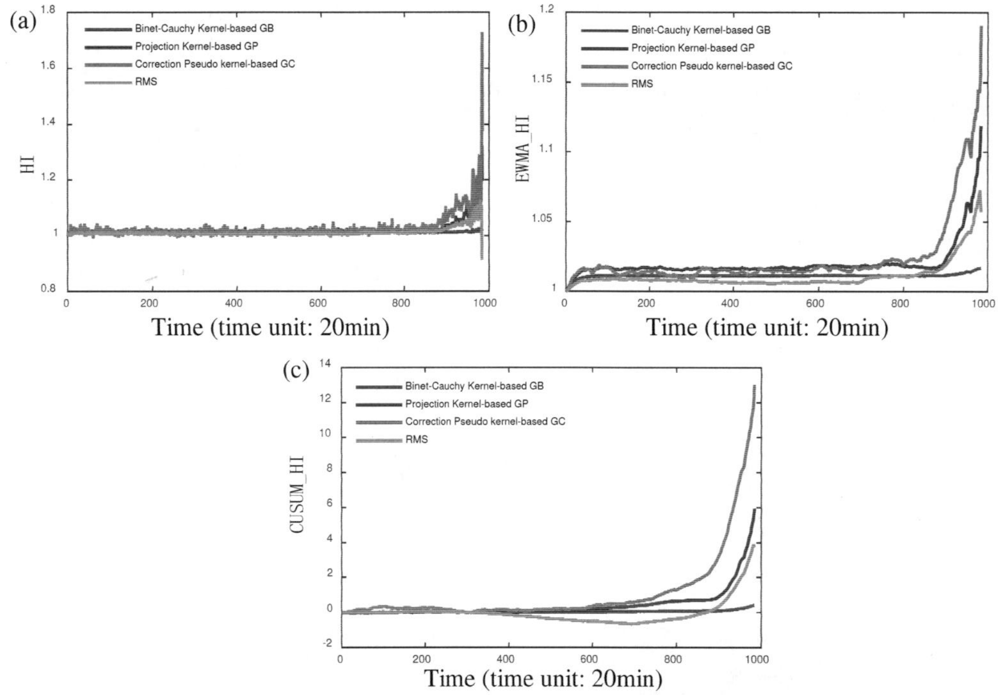

Figure 10. The PDA results for the full cycle life of the tested bearing 4#: (a) the raw HI, (b) the EWMA-based HI_w and (c) the CUSUM-based HI_c (the brown, blue, red and green lines represent different HIs, GB, GP, GC, and RMS, respectively).

Conclusion: The paper proposes the use of GM-based spatial clustering analysis to assess bearing performance degradation using the subspace clustering distribution of a two-class model to automatically depict the bearing's health condition. First, a covariance-driven Hankel matrix is constructed from the raw time-domain signal; manifold learning is then used to separate the feature subspace and noise null-space. This meaningful feature basis space can automatically represent the health conditions. Second, a reference model is built by combining the reference healthy data subspaces and the monitored subspaces. A new subspace clustering HI with kernel operation is implemented to assess the current machine performance, revealing the features that discriminate abnormality from healthy operation. Subspace clustering HI is used to experimentally detect abnormal bearing condition using the mobile mapping mode. The study finds that the spatial clustering analysis using the reference model is effective to detect abnormalities in a self-driven way. Finally, a novel subspace clustering chart, CUSUM-based HI, is developed to depict the real bearing performance degradation assessment (PDA).

Compared to the regular HI, (e.g., root mean square, kurtosis), the proposed approach can provide a more accurate and reliable degradation assessment profile with an early fault occurrence alarm. The case study comparisons show the potential of the proposed spatial clustering analysis to assess bearing health performance degradation. For an extension of the GM-based spatial clustering, this reference model can be further implemented in the PDA of other key machine components and in remaining useful life (RUL) predictions.

Acknowledgement: This work is supported by the National Natural Science Foundation of China under Contract No.51475053, Foundation of SKLMT-KFKT-201704 8: Diego Galar Lulea of technology wind farm condition monitoring and Postdoctoral Science Foundation of China under Contract No.2017M622960.

References:

[1] W. Caesarendra, A. Widodo, P.H. Thom and Y. Bo-Suk and J.D. Setiawan, "Combined probability approach and indirect data-driven method for bearing degradation prognostics," IEEE Transactions on Reliability, vol. 60, pp. 14-20, 2011.

[2] N. Trapani, M. Macchi and L. Fumagalli, "Risk driven engineering of prognostics and health management systems in manufacturing," IFAC-Papers online, vol. 48, pp. 995-1000, 2015.

[3] Y. Lei and M. Zuo. "Gear crack level identification based on weighted K nearest neighbor classification algorithm," Mechanical Systems and Signal Processing, vol. 23, pp. 1535-1547, 2009.

[4] A. Malhi and R.X. Gao, "PCA-based feature selection scheme for machine defect classification," IEEE Transactions on Instrumentation and Measurement, vol. 53, pp. 1517-1525, 2004.

[5] J. Harmouche, C. Delpha, D. Diallo, "Linear discriminant analysis for the discrimination of faults in bearing balls by using spectral features," Green Energy, 2014 International Conference on. IEEE, pp: 182-187, 2014.

[6] X. Ding, Q. He and N. Luo, "A fusion feature and its improvement based on locality preserving projections for rolling element bearing fault classification," Journal of Sound and Vibration, vol. 335, pp. 367-383, 2015.
[7] X. Jin, M. Zhao, T.W.S. Chow T W S and P. Michael, "Motor bearing fault diagnosis using trace ratio linear discriminant analysis", IEEE Transactions on Industrial Electronics, vol. 61, pp. 2441-2451, 2014.
[8] F. Li, B. Tang, R. Yang, "Rotating machine fault diagnosis using dimension reduction with linear local tangent space alignment," Measurement, vol. 46, pp. 2525-2539, 2013.
[9] H. Qiu, J. Lee, J. Lin and G. Yu, "Robust performance degradation assessment methods for enhanced rolling element bearing prognostics," Advanced Engineering Informatics, vol. 17, pp. 127–140, 2003.
[10] J. Yu, "Bearing performance degradation assessment using locality preserving projections and Gaussian mixture models," Mechanical Systems and Signal Processing, vol. 25, pp. 2573-2588, 2011.
[11] Y. Pan, J. Chen and L. Guo, "Robust bearing performance degradation assessment method based on improved wavelet packet—support vector data description," Mechanical Systems and Signal Processing, vol. 23, pp. 669–681, 2009.
[12] K. Medjaher, D.A. Tobon-Mejia and N. Zerhouni, "Remaining useful life estimation of critical components with application to bearings," IEEE Transactions on Reliability, vol. 61, pp. 292-302, 2012.
[13] T.H. Loutas, D. Roulias and G. Georgoulas, "Remaining useful life estimation in rolling bearings utilizing data-driven probabilistic e-support vectors regression," IEEE Transactions on Reliability, vol. 62, pp. 821-832, 2013.
[14] Ding X, He Q. Two-class model based on nonlinear manifold learning for bearing health monitoring[C]//Instrumentation and Measurement Technology Conference Proceedings (I2MTC), 2016 IEEE International. IEEE, 2016: 1-6.
[15] Sun C, Zhang Z, Guo T, et al. A novel manifold–manifold distance index applied to looseness state assessment of viscoelastic sandwich structures [J]. Smart Materials and Structures, 2014, 23(6): 065019.
[16] Allahdadian S, Ventura C E, Andersen P, et al. Sensitivity evaluation of subspace-based damage detection method to different types of damage[M]//Structural Health Monitoring and Damage Detection, Volume 7. Springer International Publishing, 2015: 11-18.
[17] Harandi M T, Sanderson C, Shirazi S, et al. Kernel analysis on Grassmann manifolds for action recognition[J]. Pattern Recognition Letters, 2013, 34(15): 1906-1915.
[18] Wolf L, Shashua A. Learning over sets using kernel principal angles[J]. Journal of Machine Learning Research, 2003, 4(Oct): 913-931.
[19] Hamm J, Lee D D. Grassmann discriminant analysis: a unifying view on subspace-based learning[C]//Proceedings of the 25th international conference on Machine learning. ACM, 2008: 376-383.
[20] Harandi M T, Sanderson C, Shirazi S, et al. Graph embedding discriminant analysis on Grassmannian manifolds for improved image set matching[C]//Computer Vision and Pattern Recognition (CVPR), 2011 IEEE Conference on. IEEE, 2011: 2705-2712.

[21] J. Lee, H. Qiu, G. Yu, J. Lin & Rexnord Technical Services, 2007. Bearing Data Set, IMS, University of Cincinnati. NASA Ames Prognostics Data Repository. <http://ti.arc.nasa.gov/project/prognostic-data-repository>, NASA Ames, Moffett Field, CA.

AN APPLICATION OF PATTERN ANOMALY DETECTION METHODS TO FLEET-WIDE ASSET LEVEL DIAGNOSTICS

Chance M. Kleineke, Nilimb Misal and Michael T. Santucci
Engineering Consultants Group, Inc.
3394 W. Market Street
Akron, OH 44333
330.869.9949
kleinekec@ecg-inc.com

Abstract: Centralized monitoring techniques have become more widely used as business demands and budgetary cuts for companies require streamlined operation and maintenance of a company's assets. These assets may be located at a single site where the monitoring is taking place, or they may be located all over a state, country or the world. Local data collection with consolidated servers allows a central maintenance center to pool big data for fleet-wide monitoring purposes. Advanced pattern recognition (APR) software solutions have been on the forefront of managing big data for dealing with a multitude of assets. APR techniques can provide evidence that a machine is not operating as expected, but the condition detected could indicate many possible underlying faults. The root cause may still be unknown.

Causal network analysis has been widely used in providing differential diagnosis in the medical field when a set of symptoms are known. This method is based on Bayesian probability which can handle uncertainty in the data, both input and output, and has a good theoretical foundation. This paper discusses methods to utilize pattern anomalies as symptoms for a causal network to diagnose asset conditions and to mitigate failures for predictive maintenance programs.

Key Words: Advanced Pattern Recognition; causal networks; asset condition diagnostics; health management; condition-based maintenance; predictive maintenance; centralized monitoring; fleet-wide monitoring; diagnostics; prognostics

Introduction: With ever increasing business demands in the energy utilities industry, many organizations look for methods to maximize departmental efficiency. Rather than depend solely on crews at a single location, power generation companies now look to consolidate maintenance and performance optimization roles to a single entity monitoring a fleet of plants or units. Breaking this down further, the monitoring center must be able to concentrate their efforts on optimizing performance and understanding the condition of each individual asset across the fleet.

At this level, a monitoring center may be inundated with tens of thousands of individual asset variables. A method is required to manage big data to best accomplish the monitoring objectives. There can be terabytes of new data generated daily that these centers must use to determine the health and performance of the organization's many assets. A single gas turbine, for example, may have ~2,000 variables being stored, with about 200 specific to the performance of the turbine, like gas temperatures and exhaust pressures. While there can be another 200 variables more specifically for mechanical integrity, i.e. vibration readings or bearing and oil temps. A single coal-fired generating unit may have 40-50 different assets to monitor, each with anywhere from a few recorded variables to hundreds.

Advanced Pattern Recognition: A simple approach to performance monitoring is setting hard-limit thresholds to alarm when data values extend outside of a predetermined "optimal" range. A drawback to this, however, is that the optimal range must be one that includes all operating loads and external input to avoid false alarms. For example, a bearing temp high limit would need to be set high enough to include the highest operating temperature on a hot day when ambient temps are affecting the bearing and the unit is operating at full capacity. But at different operating conditions, such as a cooler day when the ambient isn't lending to higher bearing temps and load is low, the bearing temp could be higher than it historically was, while remaining under the high threshold. This method would not alarm on faults occurring within the defined normal operating range.

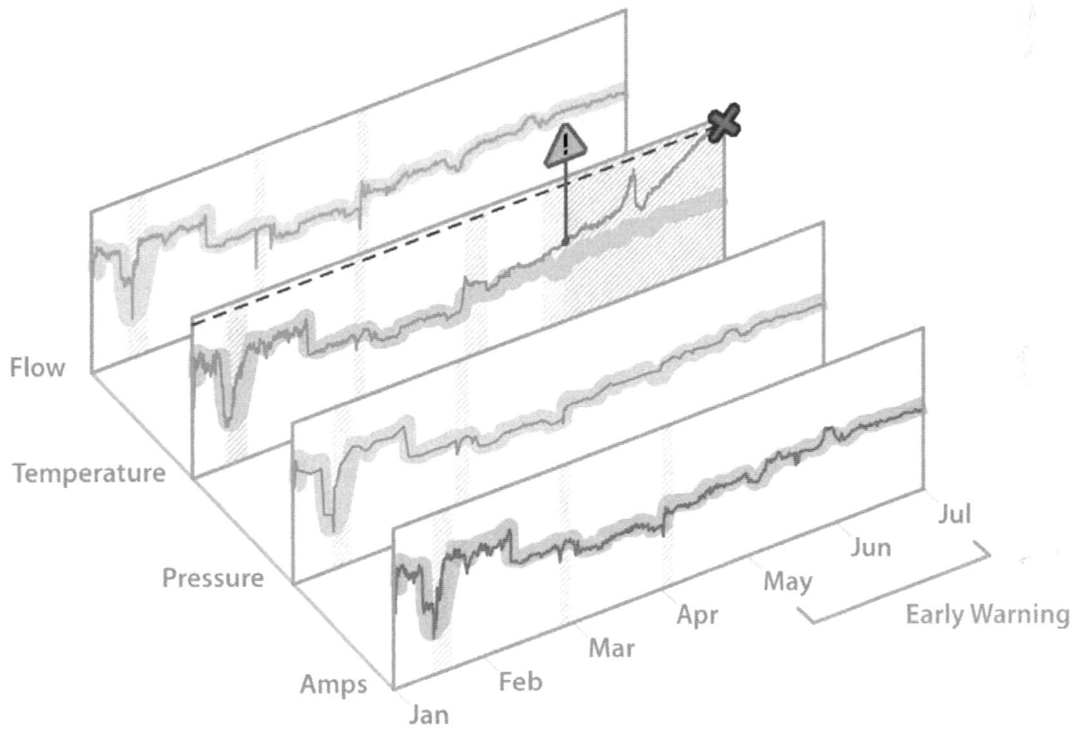

Figure 1: Early fault detection compared to hard-limit methods.

APR methods rely on the historical data of multiple variables of an asset to predict what the value of each variable should be at any given moment. The predictions are based on the activity of the other variables and relates to the performance of the variables in the past. APR methods are able to find faults much sooner than threshold monitoring techniques widely used in maintenance programs and can predict values dependent on the other variables, see Figure 1. The methods work best when there is at least some correlation among the variables being monitored, and so independent assets are modeled separately and the calculations utilize only variables relevant to that particular asset.

APR has become a common application for organizations that manage large amounts of performance and maintenance data on their assets. An engineer or data analyst building an APR model for the first time must determine boundary conditions for each asset, and subsections of those assets if necessary. Data variable selection can be completed using knowledge of the sensors fitted to each piece of equipment. Asset management and variable mapping can be completed easily by templating assets in a spreadsheet, Table 1. The models can then be imported into an APR software suite, such as Predict-It™ by ECG, Inc. [1].

Table 1: Building asset model templates.

Process Variable	1A Fan	1B Fan	2A Fan	2B Fan
Inlet Pressure	1APT1060	1BPT1060	2APT1060	2BPT1060
Outlet Pressure	1APT1065	1BPT1065	2APT1065	
Amps	1ACT2556	1BCT2556	2ACT2556	2BCT2556
Flow	1AFT1559	1BFT1559	2AFT1559	2BFT1559
Air Temperature	1ATT1205	1BTT1205	2ATT1205	2BTT1205
In Bearing Temp	1ATT1215	1BTT1215	2ATT1215	2BTT1215
Out Bearing Temp	1ATT1216	1BTT1216		2BTT1216
Damper Position	1AZT3324	1BZT3324	2AZT3324	2BZT3324
Vibration 1	1AVT0056	1BVT0056	2AVT0056	2BVT0056
Vibration 2	1AVT0057	1BVT0057	2AVT0057	2BVT0057
Vibration 3	1AVT0058		2AVT0058	2BVT0058

Once the boundary conditions are set and asset variables have been selected, the models that will be monitoring the assets can be created and trained on historical data. Relevant data should be selected that best represents optimal or normal operating conditions, Figure 2. Next, viewing the correlations among all the variables, the engineer can remove outliers using standard X-Y scatter plots, or other statistical charts that aid in narowing down the best data to be used for training, Figures 3 and 4.

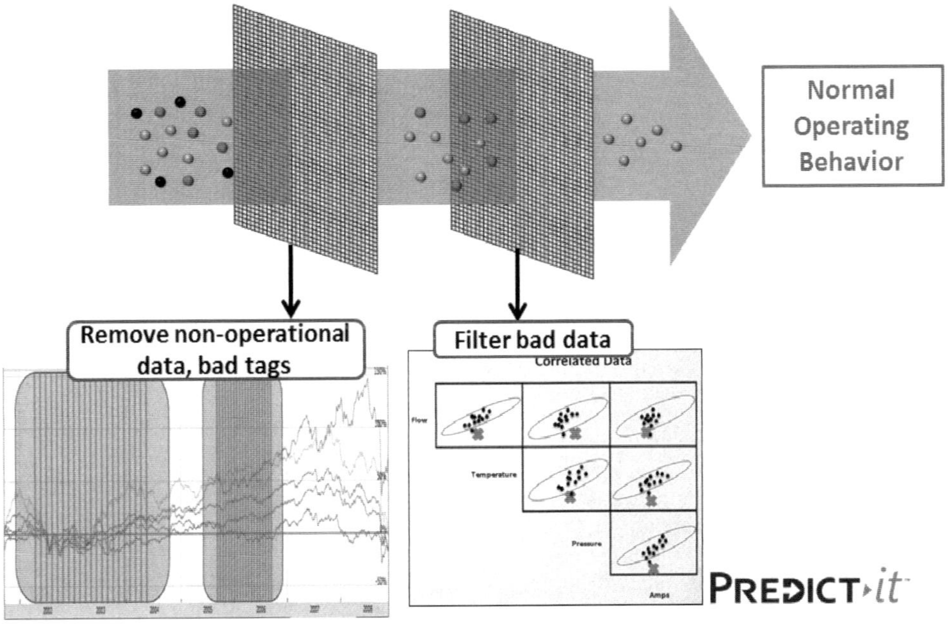

Figure 2: Selecting training time frames that best represent optimal conditions.

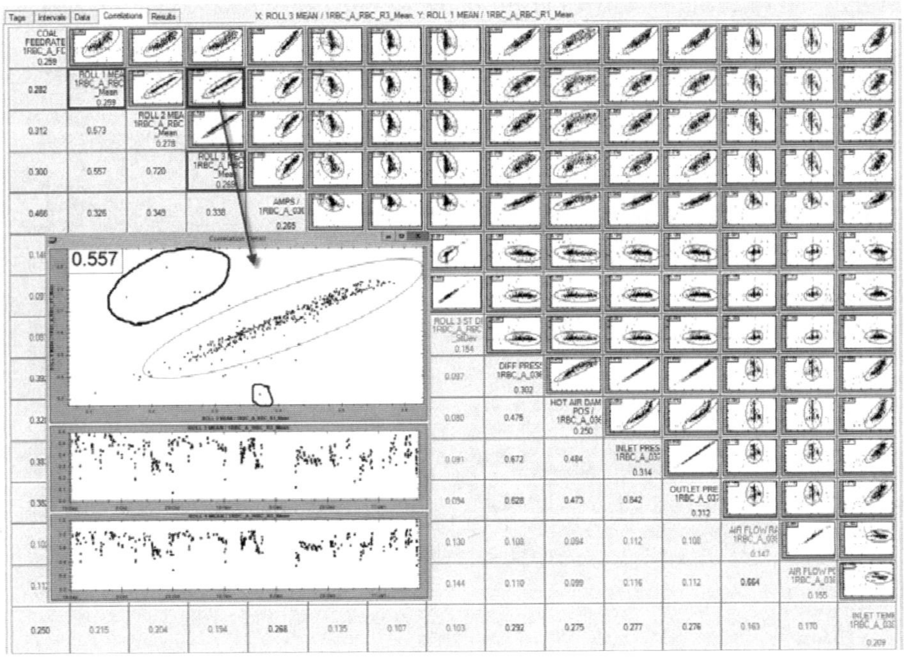

Figure 3: Correlation charts of all asset variables for the model. The model builder can manually remove outliers from individual plots (black circles).

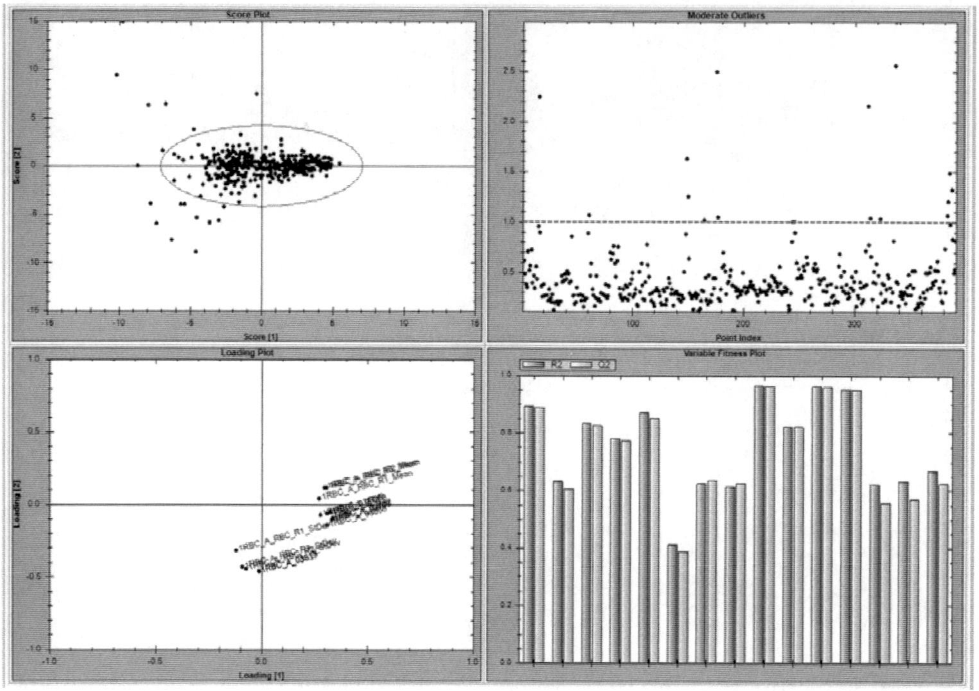

Figure 4: Alternate statistical plots for narrowing down training data.

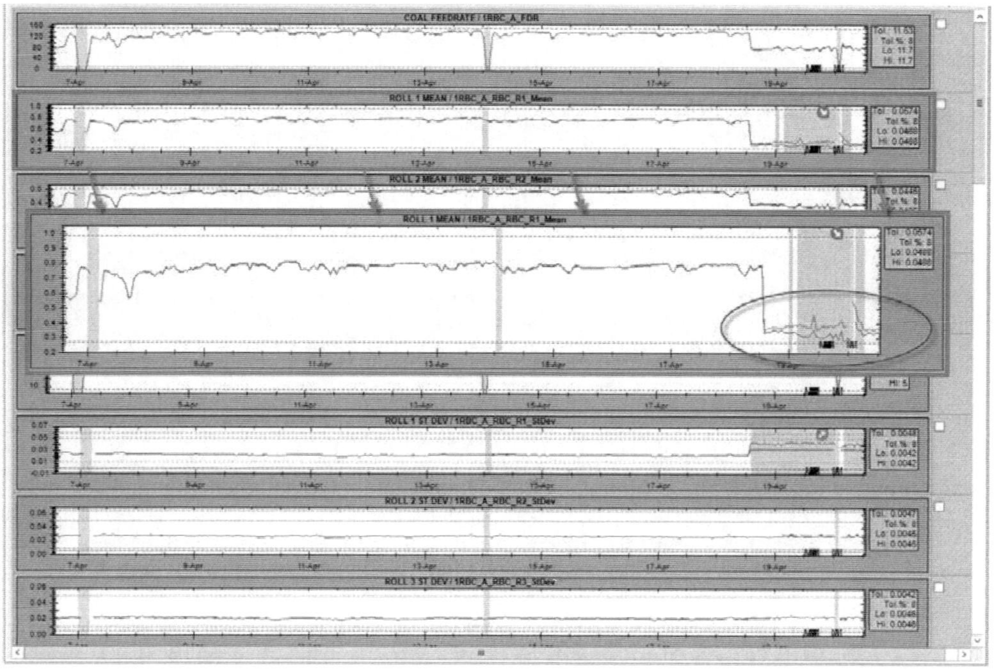

Figure 5: Asset variable in alarm; actual value is lower than expected.

Once the data is cropped to an acceptable extent, the model can be programmatically trained. It can then be run against existing data as well as run against incoming current data. Figure 5 shows an example of the results of running a model against 2 weeks of past data. One of the variables is lower than expected and in alarm. Any APR solution typically produces a result similar to this, revealing an early warning when a particular variable, or variables, are not performing as expected. But the engineer or data analyst may be left wondering what this means. For the purpose of this paper, these variable deviations will be used as inputs to a Diagnostic Advisor that can be used as decision support for the end user. The supporting diagnostics are based on the theory of Bayesian networks.

Bayesian Networks: Bayesian networks are directed acyclic graphs (DAG) that map variables to each other with their associated dependencies. A DAG is a graph with directed arrows but which do not contain directed cycles. These dependencies are represented by conditional probability tables (CPT). The CPTs are a reformulation of the joint probability distribution between two variables. Joint probability is the probability of two or more events or variables occurring together. Conditional probability is the probability of a variable/event given that we know another state of another variable/event.

There is no distinction between independent and dependent variables and inference can be done in any direction. The method is non-parametric, nonlinear and can handle numerical and categorical variables. In statistics these models are also called directed graphical models.

The term "Bayesian networks" was coined by Judea Pearl in 1985 to emphasize three aspects [2, 3]:

1. The often subjective nature of the input information.
2. The reliance on Bayes' conditioning as the basis for updating information.
3. The distinction between causal and evidential modes of reasoning

Bayesian networks consist of nodes connected by arrows representing real causal relations. Bayesian networks have a big advantage in that they are direct representation of the world. Unlike neural networks, deep learning and other "black box" methods, Bayesian networks are transparent, intuitive to understand and are capable of providing clear explanations.

Bayesian network use the Bayes theorem to update the network. Mathematically Bayes' theorem is stated as the following equation [4]:

$$P(A \mid B) = \frac{P(B \mid A)\, P(A)}{P(B)},$$

Where A and B are events and $P(B) \neq 0$:

- P(A|B) is conditional probability of event A occurring given B is true.
- P(B|A) is the conditional probability of event B occurring given that A is true
- P(A) and P(B) are the probabilities of observing A and B independent of each other.

Humans naturally can think of problems by going from cause to effect. For example, it is known that the flu causes a cough. But a cough could be caused by the flu or the common cold. Bayes' theorem allows a reverse reasoning from effect to cause using sound probability theory. Since an engineer has to usually start with effects and then try to figure out the cause, the Bayes approach is very useful in this scenario. Starting from prior probabilities of faults, as soon as new evidence is available, Bayes theroem allows for these probabilities to be updated.

Bayesian networks can be converted to causal network when constrained in the sense that the parents of each node are its direct causes. Historically structural equations are commonly used to causally represent the different mechanisms active in a system. Bayesian representation has been shown to be equivalent to structural equations [5, 6].

Asset Management: Organization is crucial when dealing with such large amounts of data. Predict-It™ utilizes an asset hierarchy system that allows users the ability to order assets by unit or site. In addition, particularly large assets may be subdivided into smaller segments. Or one can monitor different modes of operations, such as mechanical soundness and process performance. Figure 6 represents an asset hierarchy of 5 coal mills on a single steam unit with the parent folder labeled as Mill Data. There would be other parent folders for other assets on this unit. Other units or plants could be similarly set up.

Defining an asset involves programming any possible symptoms the user believes could result from the pattern recognition calculations. The end user must then outline any possible faults for that particular asset. The faults must be defined with the corresponding directional symptoms, i.e. deviation high, deviation low, absolute high, absolute low, etc. User questions may also be set-up that allows input regarding data that may not be electronically recorded or tracked as an analog variable. Examples of this might be an asset's service time or observational items like noise level of the asset.

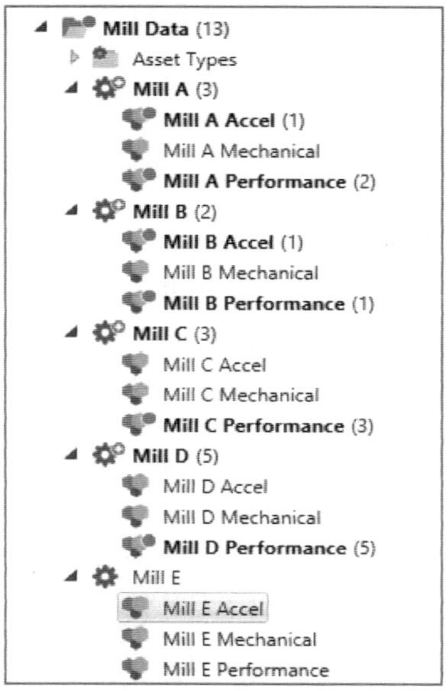

Figure 6: Asset hierarchy.

Actual instances that have been documented can be entered into the diagnostic calculations as well. When a fault occurs, the user should make note of any symptoms present to record this as a case. When these cases are entered, the probability of each fault is adjusted accordingly. Figure 7 shows what a fully configured coal mill might look like.

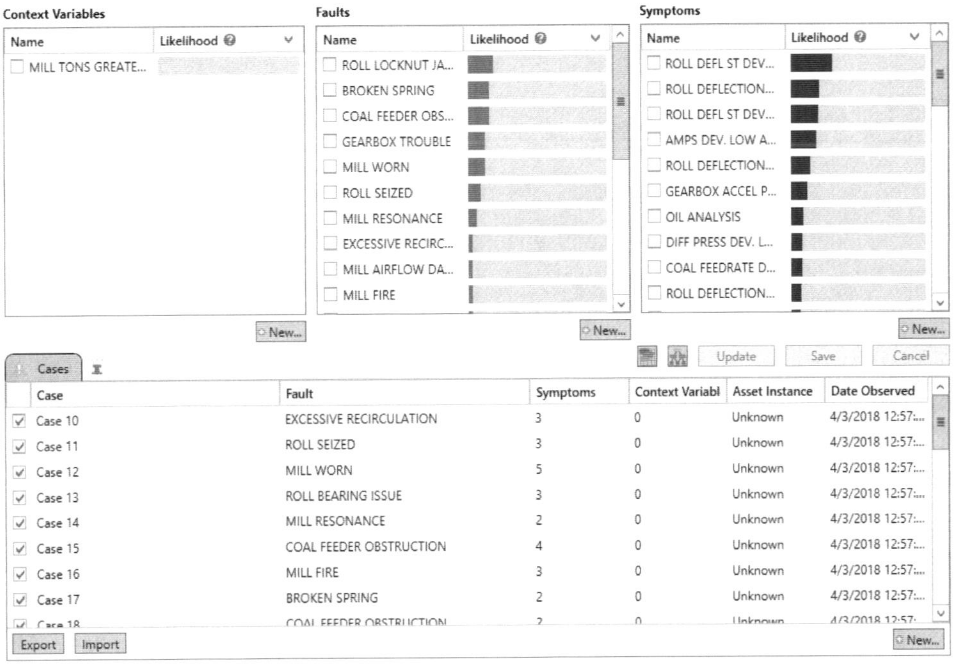

Figure 7: Defining an asset with context variables, possible faults and symptoms. The end user can submit actual cases to further solidify the diagnostic calculations.

Defining assets types allows the Diagnostic Advisor to create the Causal Asset Network (CAN). The CAN is represented as a network of Faults and Symptoms, with arrows linking associations from each column. These networks can become quite complex as the many variations of symptom combinations are set up. Figure 8 illustrates a simple CAN for an arbitrary asset. The red boxes are faults, blue boxes are symptoms and the green boxes represent context variables.

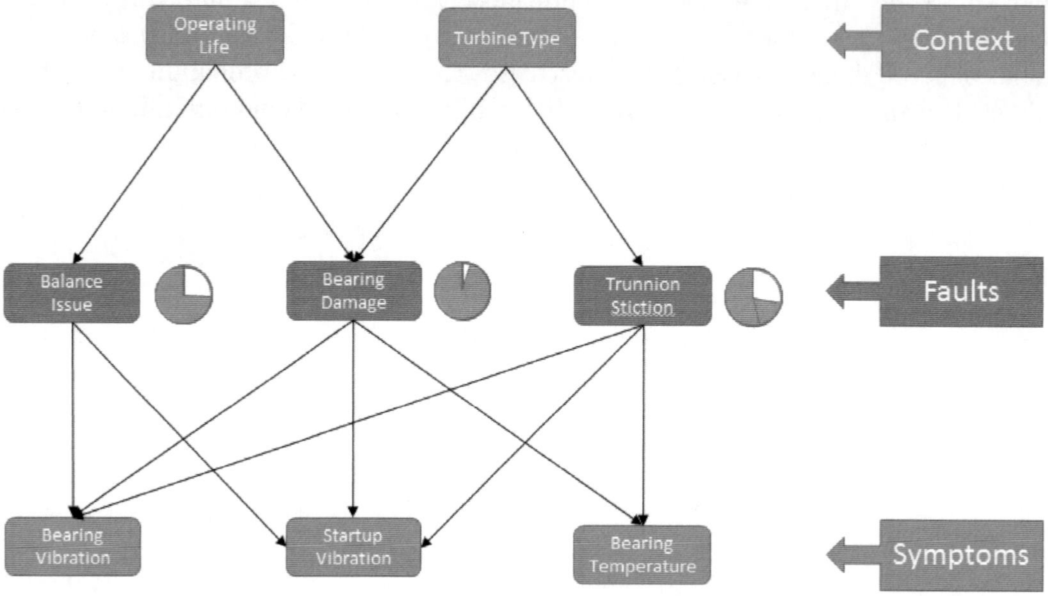

Figure 8: An example of a Causal Asset Network (CAN) created with an asset's faults and symptoms.

Diagnostic Reasoning: The Diagnostic Advisor will combine the functionality of the Bayesian networks with the results of pattern recognition fault detection. It will be used as a tool to visualize the diagnostics steps, experiment with diagnostic scenarios and answer questions related to asset conditions. Figure 9 shows an example list of pattern anomaly alarms that have been triggered for Mill A used in examples above. It shows that Roll 1 mean deflection is lower than expected and the standard deviation for that roll is higher than expected. It also shows that the accelerometer on the worm bearing is in alarm as an absolute high (hard-limit). These deviations from the expected are read into the diagnostic engine as symptoms.

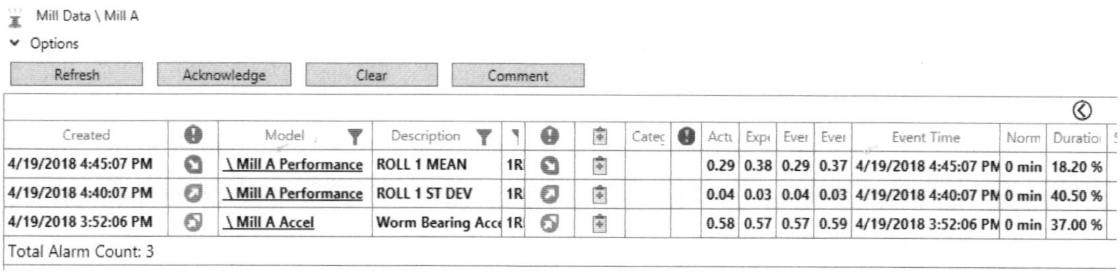

Created		Model	Description				Categ		Actu	Expe	Ever	Ever	Event Time	Norm	Duratio
4/19/2018 4:45:07 PM		\ Mill A Performance	ROLL 1 MEAN	1R					0.29	0.38	0.29	0.37	4/19/2018 4:45:07 PM	0 min	18.20 %
4/19/2018 4:40:07 PM		\ Mill A Performance	ROLL 1 ST DEV	1R					0.04	0.03	0.04	0.03	4/19/2018 4:40:07 PM	0 min	40.50 %
4/19/2018 3:52:06 PM		\ Mill A Accel	Worm Bearing Acce	1R					0.58	0.57	0.57	0.59	4/19/2018 3:52:06 PM	0 min	37.00 %

Total Alarm Count: 3

Figure 9: Current faults from the APR deviations.

The Diagnostic Advisor is divided into four main viewing blocks, see Figure 10. These blocks include Ranked Faults, Evidence, User Questions and Ranked observations. The Ranked Faults box will include any possible faults that have been added to the asset. If the asset has evidence from pattern recognition or user questions, the faults will be ranked by likelihood of presence. Any symptoms from pattern recognition or context variables that are currently present will be shown in the Evidence box, in order of influence to the faults.

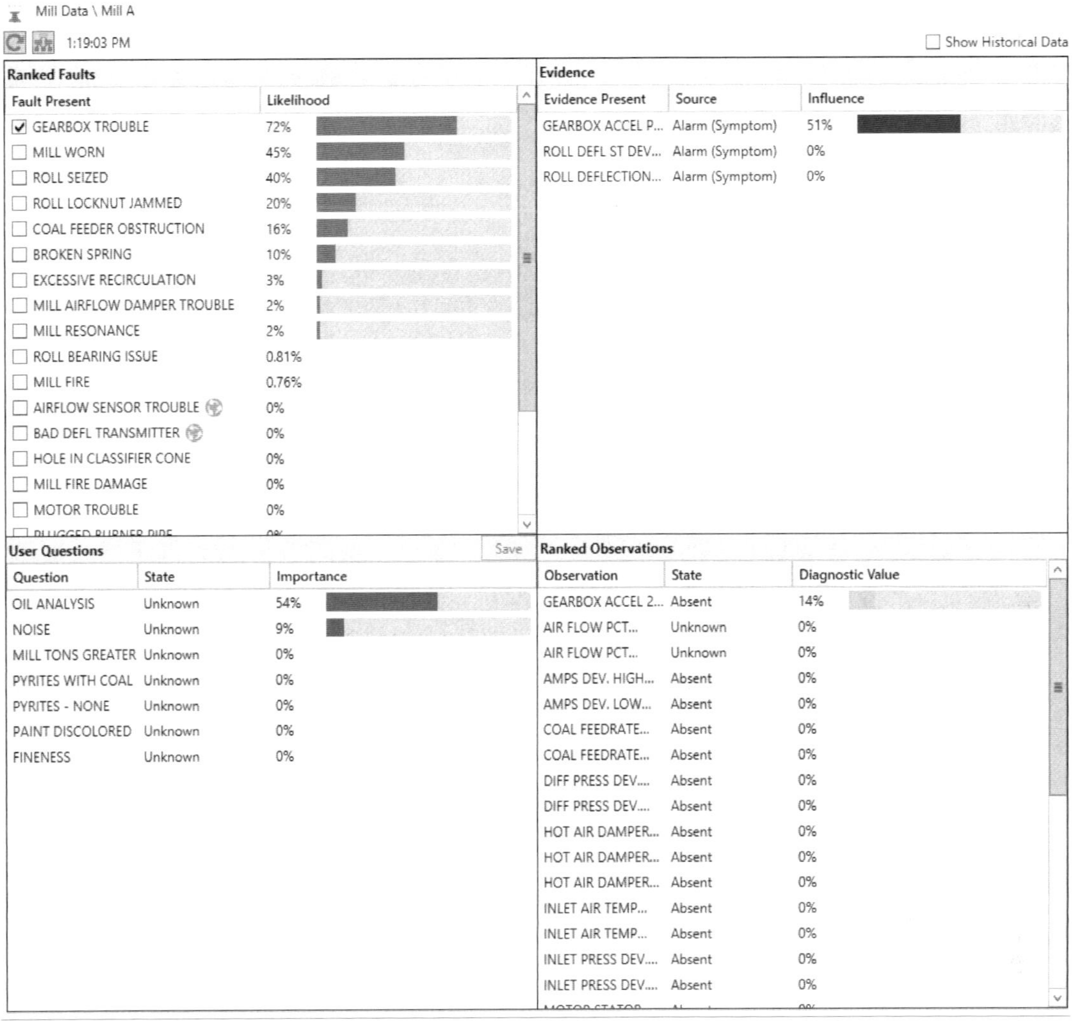

Figure 10: The four-block view of the Diagnostic Reasoner.

The User Questions block will contain any questions that are not present or are unknown; otherwise they will be included in the Evidence block. The importance of the remaining questions will also be shown in this box. The Ranked Observations contains any other possible symptoms that are absent or unknown.

The Ranked Faults box can guide decision making for the end user based on the likelihood of possible faults. When first opened, it will contain the current state of the diagnostic for that asset based on available information. The tool can be used for experimentation by tweaking the other inputs and answering user questions in the other boxes. The likelihood of the faults will adjust accordingly, without disrupting the live system.

Application to Fleet-Wide Monitoring: APR technologies have been employed by many organizations to aid in organizing and managing multiple assets with large amounts of sampled data. Utilizing the results of APR as observations in a causal network will allow these organizations not only the early detection of faults, but aid in diagnosing the root cause. This can assist in outage planning and preparation, as well as help mitigate and avoid catastrophic failures, preventing the possibility of hundreds of thousands, or even millions, of dollars of lost revenue.

References:

[1] Misal, N., Kleineke, C. and Santucci, M., 2010, Adapting Advanced Pattern Recognition Software to Predict and Optimize Pulverizer Performance, *MFPT Proceedings 2010*

[2] Pearl, J., 1985, Bayesian Networks: A Model of Self-Activated Memory for Evidential Reasoning (UCLA Technical Report CSD-850017). *Proceedings of the 7th Conference of the Cognitive Science Society,* University of California, Irvine, CA., pp. 329–334.

[3] Pearl, J., 2002, *Causality: Models, Reasoning and Inference*, Cambridge University Press, New York, NY, ISBN 0-521-77362-8, 384 p.

[4] Wikipedia contributors. "Bayes' theorem." *Wikipedia, The Free Encyclopedia.* Wikipedia, The Free Encyclopedia, 18 Apr. 2018. Web. 18 Apr. 2018.

[5] Druzdzel, M. J. and Simon, H. A., 2013, Causality in Bayesian Belief Networks, *Proceedings of the Ninth Annual Conference on Uncertainty in Artificial Intelligence (UAI-93)*, Morgan Kaufmann Publishers, Inc., San Francisco, CA., pp. 3-11.

[6] Stuart, A. and Ord, K., 1994, *Kendall's Advanced Theory of Statistics, Volume 1, Distribution Theory*, Wiley, Hoboken, NJ, 676 p.

COMPARATIVE ANALYSIS OF BEARING HEALTH MONITORING METHODS FOR MACHINE TOOL LINEAR AXES

N. Jordan Jameson and Gregory W. Vogl
Engineering Laboratory
National Institute of Standards and Technology (NIST)
100 Bureau Drive
Gaithersburg, MD 20899
Telephone: (301) 975-8966
noel.jameson@nist.gov

Abstract: The study of rotating machinery ball bearing diagnostics and prognostics is quite mature and an abundance of methods/algorithms are available to perform these functions. However, extending these algorithms to other ball bearing applications is challenging and may not yield usable results. This work used a linear axis testbed to study the ability of an inertial measurement unit to measure changes in geometric error motions. Faults were introduced on the recirculating ball bearings of one carriage truck with increasing severity. The inertial measurement unit data was analyzed using a variety of methods proposed and used in the rotating machinery community, including auto-regressive filtering, minimum entropy deconvolution, and spectral kurtosis. The results reveal an ineffectiveness of the methods for the induced faults, for this one experiment, which have low signal-to-noise ratio and/or weaker periodicity than faults in rotating machinery.

Keywords: Linear axis, ball bearing, degradation, diagnostics, smart manufacturing

Introduction: Linear translational axes are important systems in manufacturing, often installed as a subsystem of a computer numerical control (CNC) machine. They are used to move cutting tools and workpieces to their desired positions for part production [1]. In a typical machine, there can be multiple such axes, each of which can degrade, producing inaccuracies that affect the quality of the parts produced. A single linear axis is shown in Figure 1(a), with each of the main components labeled. Four trucks (also called "linear motion guides") are used to guide the carriage along the axis rails using recirculating ball bearings.

As a linear axis operates over time, abrasion and adhesion between parts causes material fatigue, pitting, cracking, and wear. This can result in faults developing in a variety of components such as the axis rails, rolling element bearings, and/or ball screw [2, 3]. If not properly mitigated, these faults will develop to such an extent as to affect the quality of the parts produced and/or precipitate a failure of the machine [4]. As demands for versatility and batch volume increase for manufacturing processes, machines are experiencing higher loads, and as a result the potential for faults and failures are becoming more common.

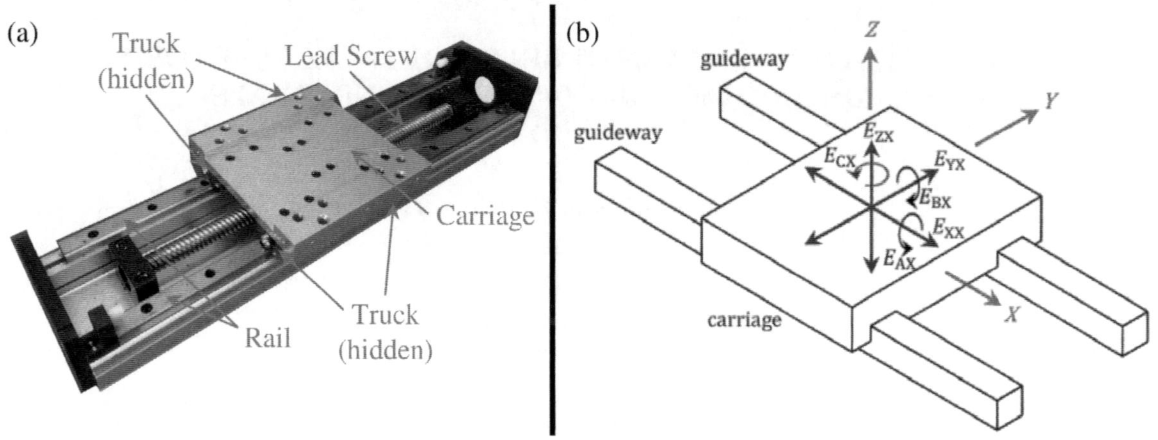

Figure 1: (a) Linear axis. The carriage travels along the axis of the lead screw and four trucks guide the carriage along the two rails (or guideways). (b) Angular and linear error motions of a carriage commanded to move along a (nominal) straight-line trajectory parallel to the X-axis.

The health assessment process consists of measuring the error motions of the linear axis machine and comparing those with specifications for allowable errors. In a linear axis, there are six degrees of freedom corresponding to six error directions, as shown in Figure 1(b), where the second variable in the subscript denotes the direction of travel (in this case, the carriage is traveling in the X-direction). In the schematic, E_{XX} denotes the linear positioning error motion of the X-axis, E_{YX} denotes the straightness error motion in the Y-direction, E_{ZX} denotes the straightness error motion in the Z-direction, E_{AX} denotes the angular error motion around the X-axis (roll), E_{BX} denotes the angular error motion around the Y-axis (pitch), and E_{CX} denotes the angular error motion around the Z-axis (yaw).

Ideally, after a linear axis is installed on a machine tool, all errors are zero, but in practice nonzero errors exist resulting in the errors on the workpiece. Hence, the allowable errors are dependent upon the tolerance needed for the workpiece. As the machine ages or crashes occur as a result of programming mistakes, these errors can increase leading to the workpieces becoming more and more out of tolerance. Instructive machine tool monitoring coupled with proper maintenance regimes can be used to mitigate this accumulation of degradation and minimize the costs imposed by imperfect production and scrapped parts.

Mature methods for the fault detection and diagnostics of linear axes are manual, time consuming, and often cost prohibitive. The state-of-the-art instruments in linear axis error measurement (the basis for diagnostics) are explained in ISO 230-1 [5]: straightedge and linear displacement sensor, microscope and taut wire, alignment telescope, alignment laser, and laser straightness interferometer. These time-consuming instruments require the shutdown of the machine and thus cannot provide in-situ diagnostics [6].

To ease the burden of implementing the traditional monitoring technologies, manufacturers need automated and efficient methods for diagnosing machine tool linear axes without halting production. In 2010, Teti *et al.* [7] identified that new sensors and sensor systems, advanced sensor signal data processing, and intelligent sensor monitoring need to be further developed to help achieve decreased machine downtime, higher productivity, higher product quality, and enhanced knowledge of manufacturing processes.

Previous work has shown that a possible advance in sensor technology lies in the use of an inertial measurement unit (IMU) consisting of a 3 degree-of-freedom (DOF) accelerometer and a 3 DOF rate gyroscope [8-10] as shown in Figure 2. The bandwidths and noise properties of the sensors used in the IMU are shown in Table 1. Data from the IMU can be used to detect changes in the positioning, straightness, and angular error motions. IMU measurements can be made quickly and with little intrusion into the operation of the machine, resulting in data that provides insight into the condition of the linear axis. It has been shown to be effective at detecting rail degradation to similar levels of accuracy delivered by a laser interferometer [8]. However, a challenge remains in verifying the IMU's capabilities for detecting degradation in the truck bearings of a linear axis.

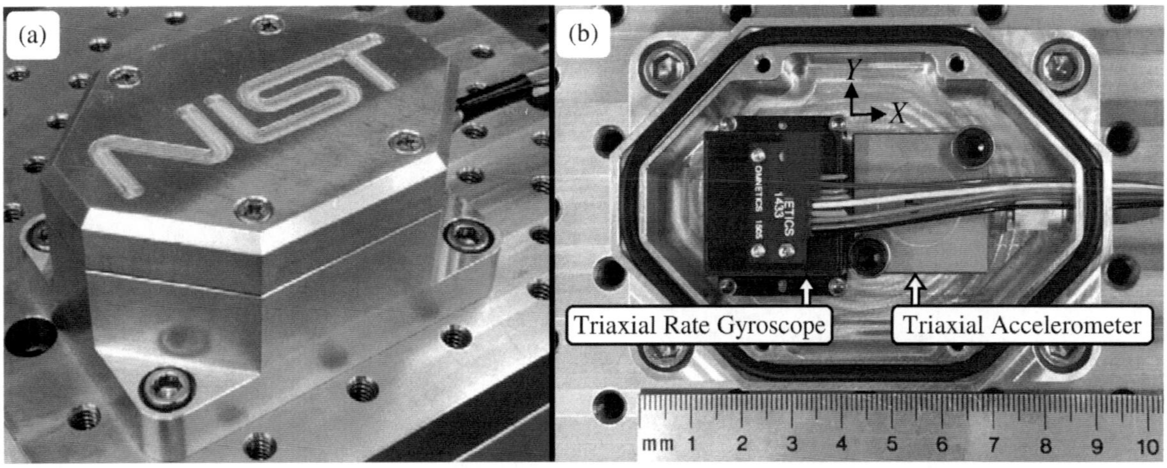

Figure 2: (a) Isometric view of industrial IMU and (b) top view of industrial IMU without its cover.

Table 1: Specified properties of sensors used in the IMU

Sensor	Bandwidth*	Noise
Accelerometer	0 to 400 Hz	$69\ (\mu m/s^2)/\sqrt{Hz}$
Rate Gyroscope	0 to 200 Hz	$35\ (\mu rad/s^2)/\sqrt{Hz}$

*frequencies correspond to half-power points, also known as 3 dB points

One complication that presents itself in this task is the multiplicity in the rolling element bearings within the bearing system, which causes a low signal-to-noise ratio (that can be less than 1) in the health monitoring data. In the linear axis shown in Figure 1(a), there are four trucks, each with two loops of ball bearings, and each loop has 32 ball bearings. The ball bearings rotate in and out of contact with the rail, with about 13 ball bearings per loop contacting the rail at any given instance. The two loops interact with different raceways (or grooves) in the rails, one inner and one outer. Hence, at any given instance, there are about 104 bearings (13 bearings per loop × 2 loops per truck × 4 trucks) in contact with the rails. The convolution of so many bearing ball signatures into the sensed effects on the carriage make it difficult to isolate small indications of damage from a single bearing ball. This interaction between the bearings and rails is illustrated in Figure 3(a) and Figure 3(b). It was with this in mind that an experiment was designed to examine the sensitivity of the IMU-based error motions to artificially-induced damage on the bearings.

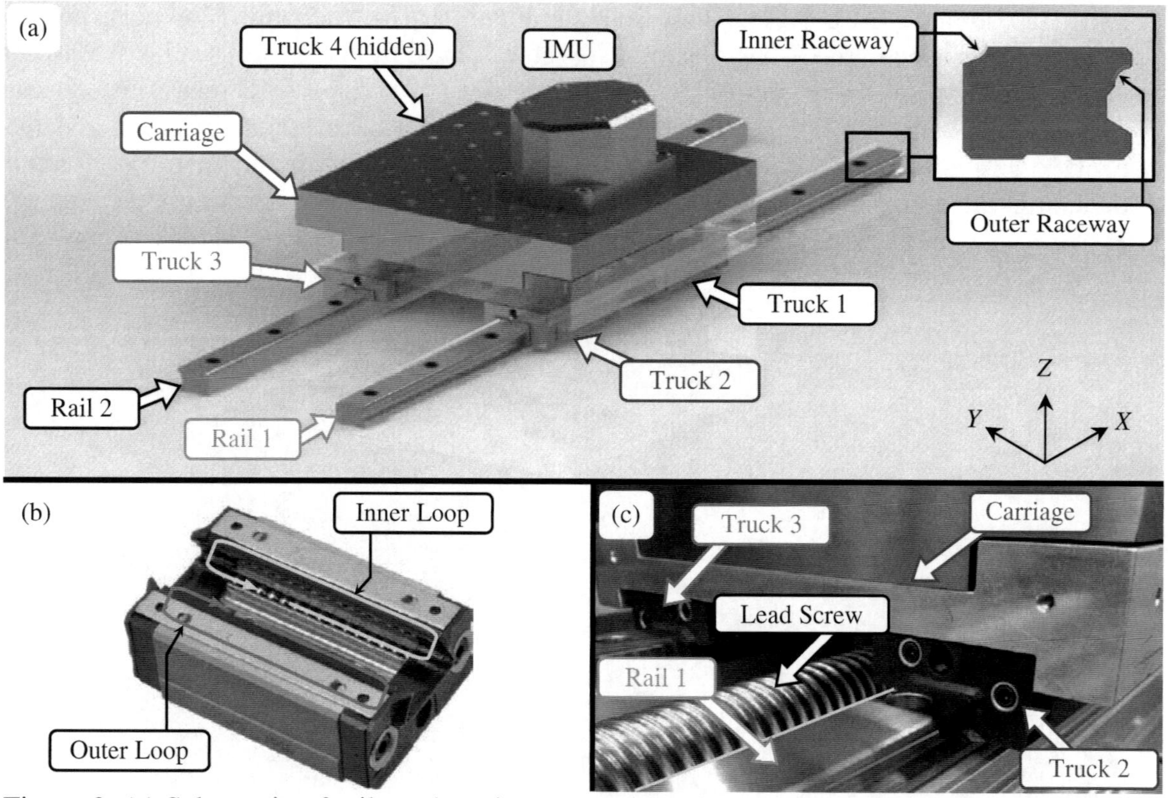

Figure 3: (a) Schematic of rails and carriage assembly of one-axis testbed; (b) underside view of one truck with loops of ball bearings exposed, where the "inner loop" contacts the inner raceway and the "outer loop" contacts the outer raceway; and (c) view of rail, carriage, lead screw, and trucks assembly showing the attachment of the trucks to the rails.

In this study, ball bearings from a single truck were removed sequentially, one-at-a-time, intentionally degraded, then replaced in their original position within the truck. The balls were degraded one at a time to progressively alter the error motions of the carriage. IMU data was collected after each ball bearing was damaged and the data was analyzed to develop diagnostic features. The data analysis procedure progressed from less to more complex culminating in a series of algorithms recommended by Randall and Antoni in their 2011 bearing diagnostics tutorial [11].

Experimental Setup: Truck 2 was modified to allow the removal of one ball at a time. This permitted the opportunity to inflict damage on to a single ball and then replace it into the truck, while leaving the carriage/truck/rail/lead screw system effectively unchanged. This was necessary because if the trucks were removed and then reassembled at each stage of degradation, then the load placed upon each of the trucks would have changed from one assembly procedure to the next, which would have changed the error motions as well. Hence, to eliminate the need for disassembly and reassembly of the trucks during the experiment and thus ensure that changes in error motions are due only to bearing ball damage in truck 2, the truck was modified with an access hole and slot, as shown in Figure 4(a). The set screw covered the access hole from which balls were removed to introduce damage. Bearing balls were removed one-at-a-time and a "flat" was introduced by dragging the ball across sandpaper. The flat damage for one bearing is shown in Figure 4(b). After damaging the bearing, it was replaced in Truck 2 at its original position within the outer ring.

Figure 4: (a) Truck 2 as modified for this experiment with a slot for viewing the bearing movement and an access hole covered by a set screw, which allowed for the removal of single bearing ball; and (b) diagram showing the damage introduced to the balls in Truck 2.

Over the course of the experiment, 26 of the 32 bearing balls in the outer loop of Truck 2 were damaged. Figure 5 shows the pattern of twenty-six metal balls with six (black) silicon nitride balls. The pattern was chosen so that at any time, visual inspection of the balls through the slot, as seen in Figure 4(a), yields a unique pattern of black and silver (metallic) balls for identification. The twenty-six metal balls were made of chromium steel with a nominal diameter of 3.962 mm ± 0.001 mm, while the six ceramic balls had a nominal diameter of 3.969 mm ± 0.0008 mm. The average depth of material removed for the chromium steel balls was 30 μm with a standard deviation of 2 μm. The maximum depth was 35 μm and the minimum depth was 26 μm. The Truck 2 outer loop balls were degraded in a specific order, as shown in Figure 5, that attempts to increase general frequency content of defect-induced impulses.

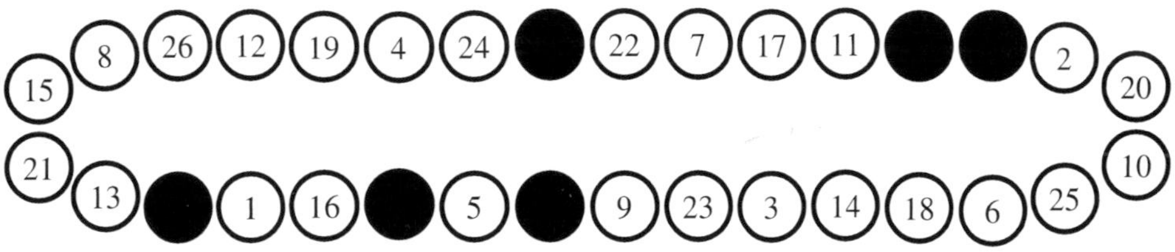

Figure 5: Truck 2 outer loop bearing degradation arrangement. The filled in circles represent ceramic bearing balls that were not degraded, while the numbers in the open circles denote the order in which the remaining metal balls were damaged.

After a single bearing was degraded and placed back into Truck 2, 50 runs of IMU data was gathered and processed to yield the estimated error motions for the carriage. Each run consists of moving the carriage over the linear axis, forward and backward, at three different speeds: 0.02 m/s, 0.1 m/s, and 0.5 m/s. These three speeds are meant to capture different ranges in the spatial frequency spectrum, with the slowest speed capturing the highest spatial frequencies and the highest speed capturing the lowest spatial frequencies. With the rate gyroscope and accelerometer used, the available spatial frequency range is between 0 and 2 cycles/mm (this can also be written as 2000 m^{-1}). (For more information on how the rate gyroscope and accelerometer data was processed to achieve the error motion data, see [8, 9].) Hence, for each of 27 degradation "stages", there are fifty (50) data runs to be analyzed. Stage 1 is collected prior to any balls being degraded, and data collection ended at stage 27 with 26 of the balls having been damaged. The first sample for stages 1, 9, 18, and 27 are shown in Figure 6 for all six degrees of freedom.

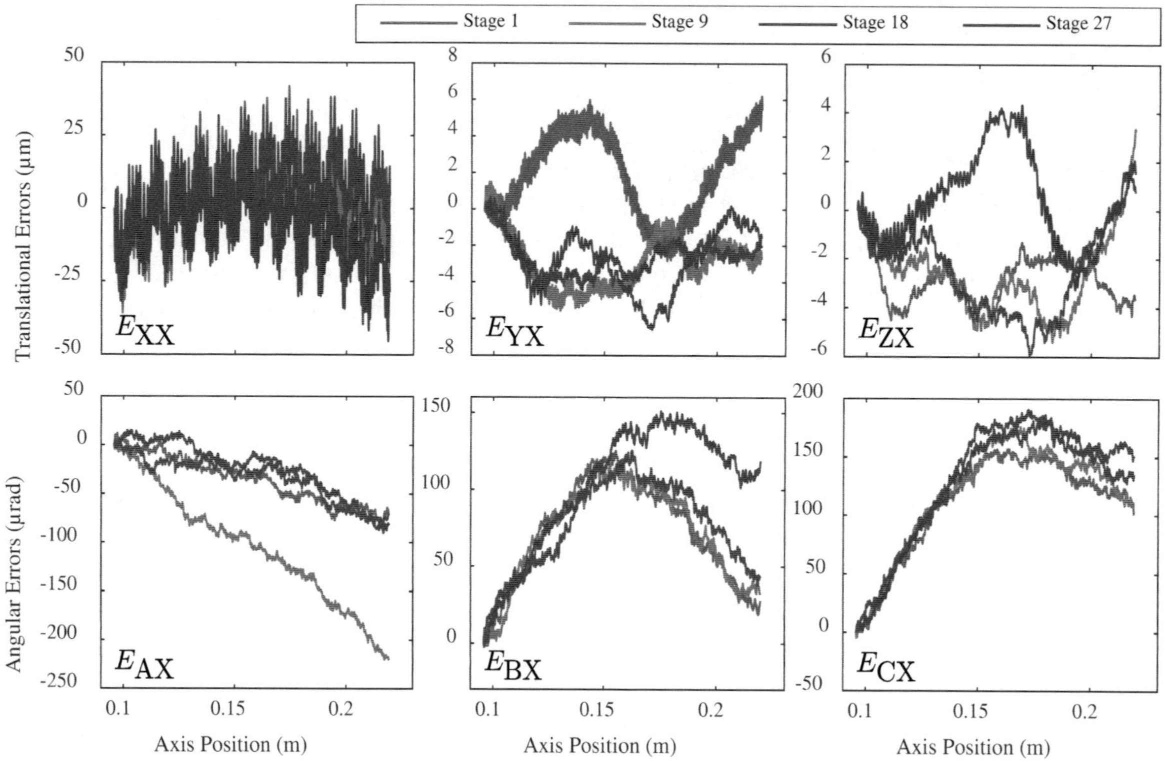

Figure 6: Unfiltered error motions of the carriage at four stages of bearing degradation. The error components are noted in the bottom left corner of each sub-figure.

Analysis Methods: The simplest analysis methods use only statistical and frequency domain features to analyze each instance of data. The statistical time-domain features used in the analysis of this experiment are given in Table 2, where $\bar{y}$ denotes the signal average and N denotes the number of data points in the signal. For the current implementation, each 'signal' of Table 2 is an error motion or some transformation of the sampled error motion, e.g., via filtering. Each of the values provided were computed for every instance of error motion data generated at the incrementally increasing stages of degradation.

Table 2: Description of statistical time-domain features used in bearing diagnostic analysis [12-15]

Feature name	**Brief description**	**Formula**
Peak value	The peak value measures the amplitude of the signal, y. As bearings faults develop, there are more impacts producing higher average amplitude, hence the peak value is expected to increase.	$PV(y) = \frac{1}{2}\left(\max(y) - \min(y)\right)$
RMS	The root mean square (RMS) of a signal is measure of the energy of the signal.	$rms(y) = \sqrt{\frac{1}{N}\sum_{i=1}^{N} y_i^2}$
Standard deviation	The standard deviation measures the dispersion of the signal around the mean value, which is another measure of the amplitude and energy in the signal.	$\sigma(y) = \sqrt{\frac{1}{N-1}\sum_{i=1}^{N}(y_i - \bar{y})^2}$
Skewness	Skewness quantifies the asymmetry of the signal with respect to its probability density function.	$\gamma_1(y) = \frac{\frac{1}{N}\sum_{i=1}^{N}(y_i - \bar{y})^3}{\sigma^3}$
Kurtosis	Kurtosis quantifies the "tailedness" of a signal. As a bearing degrades, more impacts are expected. This leads to a signal having more frequent and extreme deviations than a signal from a healthy bearing.	$\gamma_2(y) = \frac{\frac{1}{N}\sum_{i=1}^{N}(y_i - \bar{y})^4}{\sigma^4}$
Crest factor	Crest factor is a ratio of the peak value to the RMS. If there are impacts from a degraded bearing, then this value will be high, due to the discrepancy between the high peak produced from the impacts and the nominal amplitude of the waveform.	$CF(y) = \frac{PV(y)}{rms(y)}$
Shape factor	Shape factor is a ratio between the RMS and the average absolute value. Since RMS measures the square of the signal, then it will be more sensitive to outliers, such as impacts. So as bearings get more degraded, the shape factor is expected to increase.	$SF(y) = \frac{rms(y)}{\frac{1}{N}\sum_{i=1}^{N}\lvert y_i\rvert}$

Table 2: Description of statistical time-domain features used in bearing diagnostic analysis [12-15]

Feature name	Brief description	Formula
Impulse factor	As bearings degrade, they are expected to produce more impacts, which will greatly affect the peak value, but will have a smaller effect on the average absolute value. Hence, as bearings degrade, the impulse factor is expected to increase.	$IF(y) = \frac{PV(y)}{\frac{1}{N}\sum_{i=1}^{N}\lvert y_i\rvert}$
Clearance factor	The clearance factor is also a measure that relates the peak value to a measure of the average amplitude. As the bearings degrades and more impulses are produced, the numerator should rise while the denominator remains almost constant.	$CLF(y) = \frac{PV(y)}{\left(\frac{1}{N}\sum_{i=1}^{N}\sqrt{\lvert y_i\rvert}\right)^2}$

Beyond computing these statistical features from the error motions, many signal processing procedures have been proposed in the bearing community. Most of the proposed techniques aim to increase the kurtosis of the signal, in order to isolate changes due to impulses from degradation. These techniques include autoregressive (AR) filtering, self-adaptive noise cancellation (SANC), minimum entropy deconvolution (MED), and spectral kurtosis (SK). The AR filtering and SANC are both methods of separating the deterministic portion of the signal from the random portion of the signal [16]. In both cases, a prediction is computed, then the residual between the predicted signal and the real signal is considered as the bearing signal. In this study, only AR filtering was used.

The AR filter predicts the deterministic pattern of the signal, but is not capable of adapting to the sudden impulses caused by bearing faults. Consequently, the fault signal is expected to remain in the residual of the AR filter, which is the difference between the measured signal and the estimated signal (i.e., the output of the AR filter). An $AR(L)$ (i.e., an Lth order AR process) process is one in which the signal is assumed to be given as a linear combination of L past values and some input, e_n, as shown in (1):

$$x_n = -\sum_{l=1}^{L} a_l x_{n-l} + e_n \tag{1}$$

By assuming the input is unknown, the signal can be approximately predicted from the linear summation of past samples. This is computed as shown in (2).

$$\hat{x}_n = -\sum_{l=1}^{L} a_l x_{n-l} \tag{2}$$

The residual is then computed as $e_n = x_n - \hat{x}_n$. The model coefficients can be computed through a variety of methods, though perhaps the most used is the Yule-Walker equations, which relate the model coefficients to the autocovariance of the signal [16]. The order of the AR model (i.e., the number of model coefficients) can be chosen using a variety of criteria, such as Akaike information criteria (AIC), Bayesian information criteria (BIC), or cross validation. However, in the present study, the AR model order was chosen as the number of coefficients that maximized the kurtosis of the residual, in an attempt to isolate the defect-caused terms from the error motion data.

After separating out the bearing signal via AR (or SANC) filtering, the impulsive fault signal can be enhanced using MED, which was originally proposed by Wiggins [17] to extract information from seismic recordings. MED operates under the assumption that the original excitation was impulsive (with high kurtosis), then designs a filter that attempts to reconstruct the original signal. This is a means to overcome the effect of the transmission path through which the impulses reach the sensor(s). Since impulsive signals are highly structured (low entropy), the desired filter should minimize the signal entropy. Sawalhi *et al.* [18] applied MED to bearing faults, Endo and Randall [19] used it with gear faults, and Randall and Antoni [11] recommended its use as part of a "semi-automated bearing diagnostic procedure". Lee and Nandi [20] and McDonald *et al.* [21] have provided comprehensive methods for computing MED. The procedure used herein is shown below.

Step 1: Assume a centered impulse filter, $\boldsymbol{f} = [0 \quad \cdots \quad 0 \quad 1 \quad 0 \quad \cdots \quad 0]^T$, of length L

Step 2: Compute the $L \times L$ autocorrelation matrix $\boldsymbol{R}_x$; a Toeplitz matrix with the first row as the first L terms of the autocorrelation of $\boldsymbol{x}$

Step 3: Compute filter output, $\boldsymbol{y} = \boldsymbol{X}_0^T \boldsymbol{f}$

Step 4: Compute new filter: $\boldsymbol{f} = \frac{\sum_{i=1}^{N} y_i^2}{\sum_{i=1}^{N} y_i^4} \boldsymbol{R}_x^{-1} \boldsymbol{X}_0 \boldsymbol{y}^3$

Step 5: Repeat steps 3-4 until stopping criterion are satisfied

Here, $\boldsymbol{y}^3$ is an element-wise cubing of the filtered signal, and $\boldsymbol{X}_0$ is an $L \times N$ matrix of the delayed input signal:

$$\boldsymbol{X}_0 = \begin{bmatrix} x_1 & x_2 & x_3 & \cdots & x_N \\ 0 & x_1 & x_2 & \cdots & x_{N-1} \\ 0 & 0 & x_1 & \cdots & x_{N-2} \\ \vdots & \vdots & \vdots & \ddots & \vdots \\ 0 & 0 & 0 & \cdots & x_{N-L+1} \end{bmatrix}$$

The stopping criterion can be based upon the change in the filtered signal kurtosis (i.e., once the change in filtered signal kurtosis reaches a sufficiently small value) as proposed by McDonald *et al.* [21]. Or the stopping criterion can be based upon changes in the filter as proposed by Lee and Nandi [20], stating that when $E[\boldsymbol{\epsilon}]$ gets below a certain threshold, the routine is complete.

$$\epsilon = \left(\boldsymbol{f}^{(i)} - \mu \boldsymbol{f}^{(i-1)}\right) / \mu \boldsymbol{f}^{(i-1)} \tag{3}$$

$$\mu = \sqrt{\frac{E[(\boldsymbol{f}^{(i-1)})^2]}{E[(\boldsymbol{f}^{(i)})^2]}}$$

In (3), the expressions are shown for the ith iteration, $\left(\boldsymbol{f}^{(i)}\right)^2$ is an element-wise squaring of the vector $\boldsymbol{f}^{(i)}$, and $E[\cdot]$ denotes the expected value. In this study, the criterion proposed by Lee and Nandi was used with the threshold for $E[\boldsymbol{\epsilon}]$ set to $5(10^{-4})$.

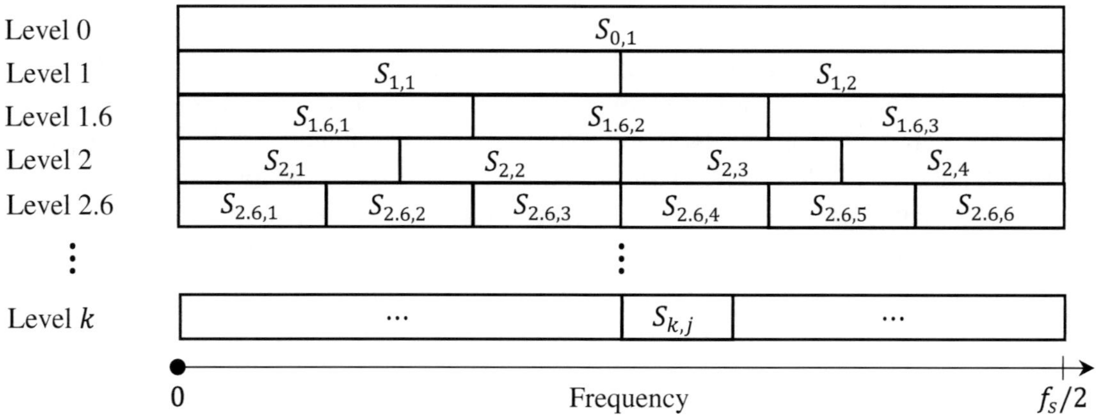

Figure 7: Frequency and decomposition levels for construction of kurtogram.

The last step in the process is to use spectral kurtosis and the kurtogram to locate the frequency band and frequency resolution that maximizes the signal kurtosis. The kurtogram is a method that uses a finite impulse response (FIR) filter bank to decompose the signal into a set of subsignals with different frequency bands [22]. This decomposition consists of taking the signal, then low-pass- and high-pass-filtering to the frequency bands $[0, 1/4]$ and $[1/4, 1/2]$, respectively (in normalized frequency). This is considered as Level 1 decomposition. The original signal is also decomposed into the frequency bands $[0, 1/6]$, $[1/6, 1/3]$, and $[1/3, 1/2]$. This is considered as Level 1.6 decomposition. Then the resulting signals are sent through further replicate filter banks, producing an arborous structure as shown in Figure 7, where $S_{k,j}$ represents the signal decomposed at level k in the jth frequency band, and f_s is the sampling frequency. After the signals are decomposed, the kurtosis of each $S_{k,j}$ can be computed, and the level and frequency band that maximizes kurtosis chosen.

Experimental Analysis Results: The error motions were first high-pass filtered to frequencies above 500 samples/m, which was done to eliminate any data that simply captured the shape of the rails. Bearing fault signatures are assumed to occur with wavelengths less than about 12.4 mm (circumference of a single ball), therefore sampling between 0.5 samples/mm and 2 samples/mm is sufficient to capture any impacts from bearing faults. Most of the statistical metrics yielded no indications of the increasing number of faults being introduced to the balls of truck 2. As an example, the kurtosis of each sample is shown in Figure 8. There are 27 total stages, and at each stage, there are 50 runs, translating to 50 values of kurtosis. At each stage, the gray box represents the middle 50% of these 50 values, and the whiskers extend to the largest value that falls below $q_3 + 1.5(q_3 - q_2)$, or the smallest value that falls above $q_2 - 1.5(q_3 - q_2)$, where q_2 and q_3 are the 25th and 75th percentile respectively. Any value that falls outside this range is classified as an outlier, and the outliers are plotted as red circles. The black circle in the center of each box is the median value for all 50 runs in a given degradation stage.

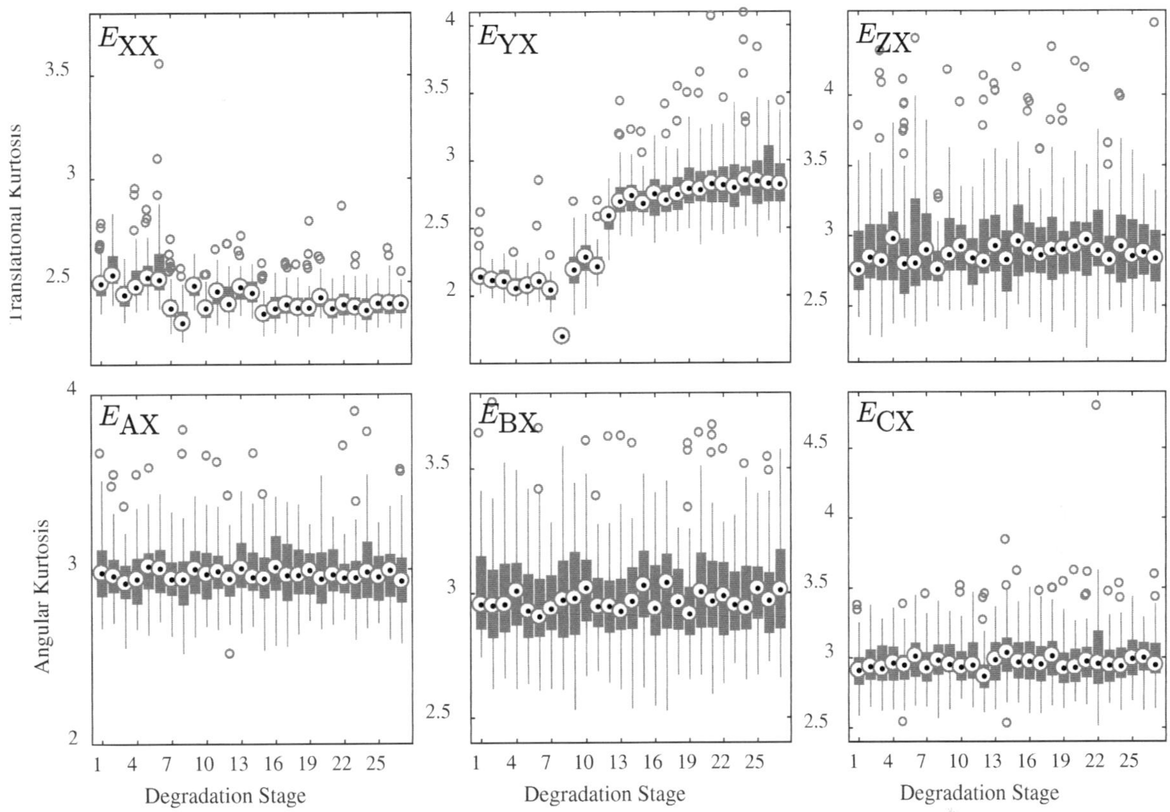

Figure 8: Boxplots showing how the filtered signal kurtosis changed as additional degradations were introduced into Truck 2 outer ring.

The statistical features can be evaluated using the Pearson correlation coefficient between the median of each distribution of 50 samples and the degradation stages (1-27). Because Pearson correlation measures the strength of a linear relationship between two variables, it is expected that as a feature is more responsive to bearing degradation, it will register a higher Pearson correlation coefficient magnitude. The sign of the Pearson correlation coefficient indicates the direction of the line, where $+1$ indicates the line is increasing and -1 indicates the line is decreasing. The Pearson correlation coefficient for each feature and DOF is shown in Figure 9.

From Figure 8 and Figure 9, it is apparent that E_{YX} trends the most as the bearings of Truck 2 outer loop are degraded, since most of its statistical features have Pearson correlation coefficients with magnitudes significantly above 0.5. Furthermore, E_{XX} has a moderate trend in kurtosis and E_{ZX} has a moderate trend in skewness, each with Pearson correlation coefficient magnitudes around 0.5. Moreover, in the case of E_{YX}, the kurtosis starts as lower than that of the normal distribution (which has a kurtosis of 3) and ends with most runs still below a kurtosis of 3, but having a higher kurtosis than at Stage 1. Generally, when bearings are faulty, the impacts that the bearings have with the runways produce kurtosis values much higher than that of normal distribution (see e.g., [18, 23]). This physical trend should be further examined for further understanding. Also, it must be noted that the error motion data under examination are high-pass filtered and not unprocessed.

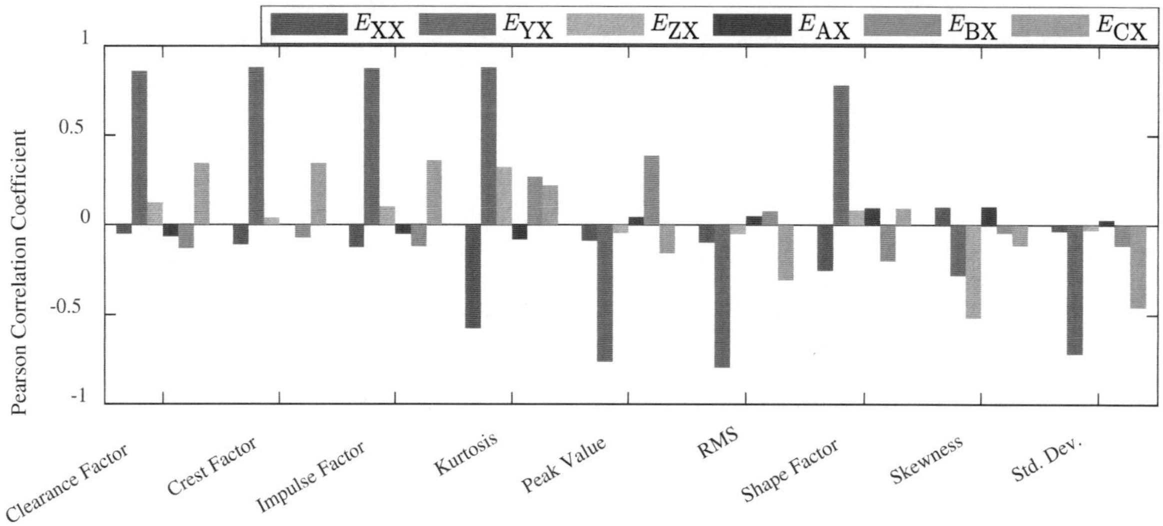

Figure 9: Pearson correlation coefficient for each statistical feature computed.

After performing the procedure recommended by Randall and Antoni, the results for one response are shown in Figure 10. The data was first high-pass-filtered to above 500 m^{-1}, then an $AR(26)$ filter was used to eliminate the deterministic portion of the signal. The length of the AR filter was found by exhaustively searching all error directions. Each signal was filtered using filter lengths between 2 and 400, and the kurtosis was measured for each filtered signal. For all error directions, the optimal length that maximized filtered signal kurtosis was between 25 and 28, with an average of 26. After AR filtering, the MED filter of length 10 was used. The length of the MED filter was also chosen according to the length that maximized signal kurtosis. Finally, the kurtogram was used on the MED filtered signal.

After processing all six DOF error motion data in the manner shown in Figure 10, the kurtosis values are shown using boxplots in Figure 11, à la Figure 8. In most cases, the kurtosis achieved using the MED filter was very close to the kurtosis achieved using the kurtogram. More importantly, like the simple statistical analysis results shown in Figure 8, the straightness errors in the Y-direction exhibit the most movement in Figure 11 as the ball bearings were degraded. Also, as shown in comparison of Figure 11 to Figure 8, this signal kurtosis was increased by decomposing the MED filtered signal to Level 1 with center frequency of $f_s/8 = 2500$ m^{-1}. However, it appears that the kurtogram may have decreased the robustness for monitoring changes in degradation, because the kurtogram increased the variance of the kurtosis of E_{YX}, as evidenced by the significantly longer gray boxes for E_{YX} in Figure 11 compared with those for E_{YX} in Figure 8.

These results show that certain metrics lack no significant trends with increasing degradation, which could result from several possibilities. First, the faulty bearings occurred in only one of two bearing loops in one of four trucks on the linear axis, so it is possible that the degradation in one bearing ball loop does not significantly affect certain error motions due to the influence of seven healthy bearing ball loops and the geometry of the raceways. Second, the signal processing techniques used were designed for temporal measurements (i.e., vibration as a function of time) that typically have many cycles for each dataset due to the rotary nature of the applications (e.g., spindles and motors). On the other hand, each run collected by the IMU has on the order of only one cycle of ball motion, in which the balls move in and out of contact between the truck and rail. Furthermore, because of the approximately one cycle of ball motion and recirculation of balls, the spatial locations of the defect-caused impulses along the rail are constantly moving, sometimes to outside of the spatial measurement range. Third, the ceramic balls were nominally 7 μm larger in diameter than the metal balls, which means that as more metal balls were degraded, the ceramic balls were bearing more of the load, which could account for apparent approaching of relative equilibria in Figure 8.

Finally, the results raise many questions for future work:

1. What is the lowest amount of damage that can be detected with the IMU via the methods explored or any other?
2. Why is the Y-direction most sensitive to changes in the bearings?
3. What signal processing techniques could be used to increase the signal to noise ratio for earlier detection of degraded linear axis bearings?
4. How can signal processing techniques be adapted to dealing with spatial measurements, rather than temporal measurements?

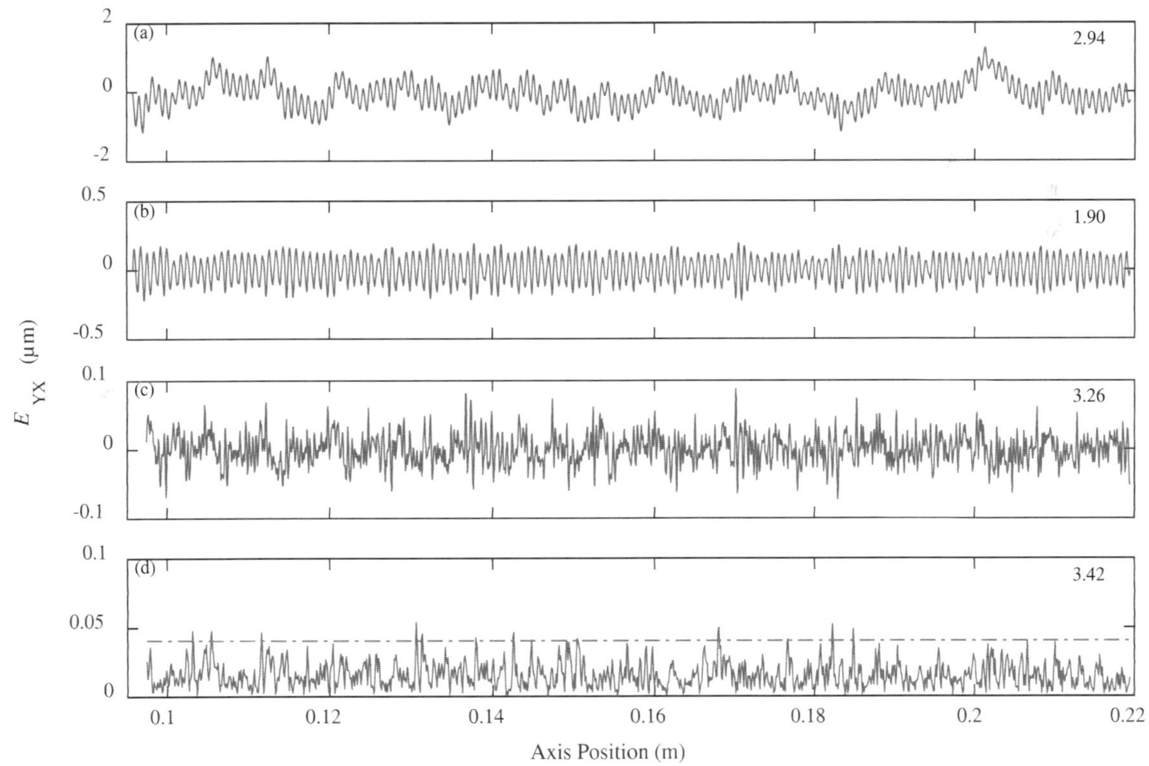

Figure 10: Semi-automated bearing diagnostic procedure for degradation stage 16, run 6, in y-axis translation, with the respective kurtosis values noted in the upper right corner of each

plot. (a) Error response filtered to spatial frequencies above 500 m^{-1}. (b) Residual of signal after $AR(26)$ filtering of the signal from (a). (c) Signal after MED (filter length of 10) of signal from (b). (d) Magnitude of complex envelope of the signal from (c) which maximized the kurtogram (level 1 decomposition) and its 0.1% significance threshold (shown in red).

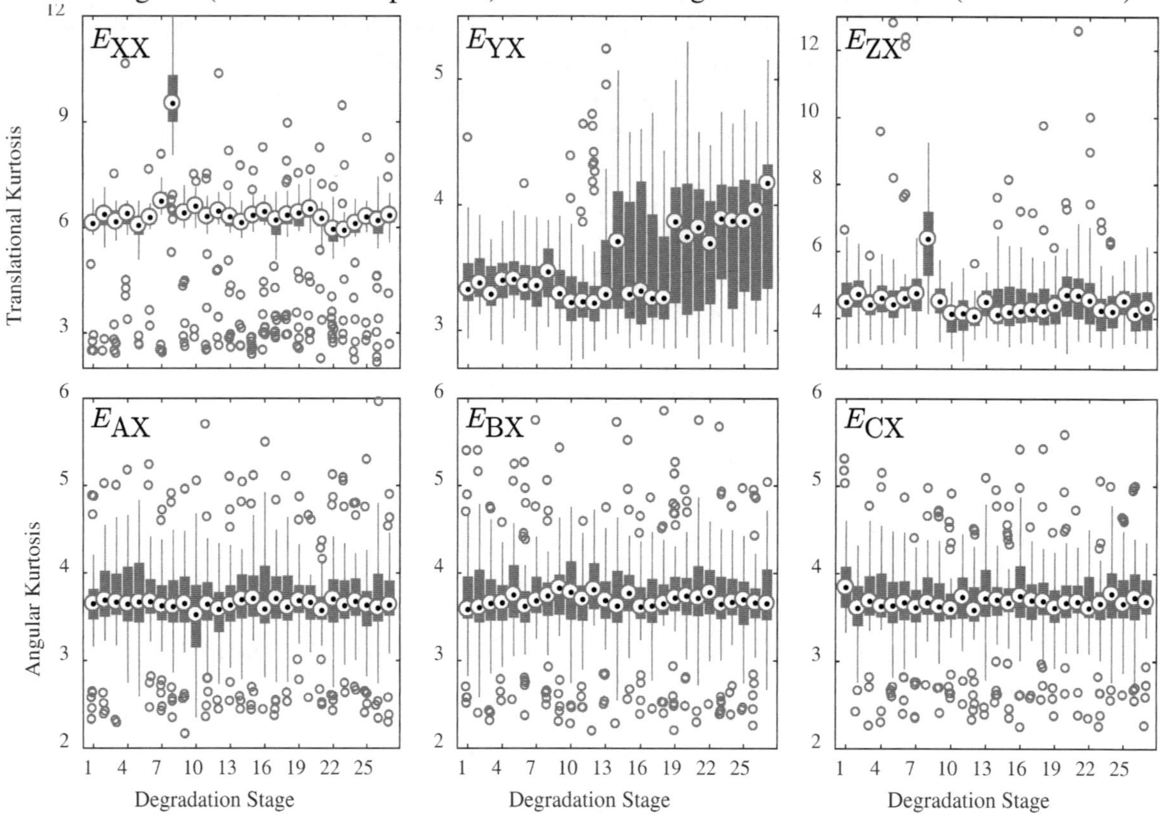

Figure 11: Kurtosis values resulting from the bearing diagnostics procedure.

Conclusion: The linear axis is used in a variety of machining processes and their bearings are subject to wear. This paper presented an experiment wherein an inertial measurement unit (IMU) was employed to detect degraded bearings in a linear axis. The six DOF error motions for the carriage were first analyzed for statistical features and then signal processing techniques were used to amplify the signal of degraded bearings. The results showed the error motions, determined from processing IMU data, experienced some evolution over time, particularly in the Y-direction (error motion described as E_{YX}). Pearson correlation coefficients showed additional trends detected by kurtosis for the translation in the X-direction (E_{XX}) and skewness for translation in the Z-direction (E_{ZX}). However, certain metrics lacked no significant trends with increasing degradation, which could have resulted from the experimental setup (lack of cycles per run), physical phenomena (effects of degradation, recirculation of balls), and the analytical techniques themselves. The results motivate further research into fault detection and diagnostics for linear axes, with an emphasis on identifying the optimal techniques suited to the unique problem of linear translation.

NIST Disclaimer: Certain commercial equipment, instruments, or materials are identified in this paper to foster understanding. Such identification does not imply recommendation or endorsement by the National Institute of Standards and Technology, nor does it imply that the materials or equipment identified are necessarily the best available for the purpose.

References:

[1] Y. Altintas, A. Verl, C. Brecher, L. Uriarte, and G. Pritschow, "Machine tool feed drives," *CIRP Annals - Manufacturing Technology,* vol. 60, no. 2, pp. 779-796, // 2011.

[2] Y. Li, X. Wang, J. Lin, and S. Shi, "A Wavelet Bicoherence-Based Quadratic Nonlinearity Feature for Translational Axis Condition Monitoring," *Sensors,* vol. 14, no. 2, pp. 2071-2088, 2014.

[3] Y. Zhou, X. Mei, Y. Zhang, G. Jiang, and N. Sun, "Current-based feed axis condition monitoring and fault diagnosis," in *4th IEEE Conference on Industrial Electronics and Applications, ICIEA 2009*, Xi'an, China, 2009, pp. 1191-1195: IEEE Computer Society.

[4] E. Uhlmann, C. Geisert, and E. Hohwieler, "Monitoring of slowly progressing deterioration of computer numerical control machine axes," *Proceedings of the Institution of Mechanical Engineers, Part B: Journal of Engineering Manufacture,* vol. 222, no. 10, pp. 1213-1219, 2008.

[5] *ISO 230-1 - Test code for machine tools – Part 1: Geometric accuracy of machines operating under no-load or quasi-static conditions*, 2012.

[6] A. W. Khan and W. Chen, "Calibration of CNC milling machine by direct method," in *2008 International Conference on Optical Instruments and Technology: Optoelectronic Measurement Technology and Applications*, Beijing, China, 2009, vol. 7160, p. 716010: SPIE.

[7] R. Teti, K. Jemielniak, G. O'Donnell, and D. Dornfeld, "Advanced monitoring of machining operations," *CIRP Annals - Manufacturing Technology,* vol. 59, no. 2, pp. 717-739, 2010.

[8] G. W. Vogl, M. A. Donmez, and A. Archenti, "Diagnostics for geometric performance of machine tool linear axes," *CIRP Annals - Manufacturing Technology,* vol. 65, no. 1, pp. 377-380, 2016.

[9] G. W. Vogl, M. A. Donmez, A. Archenti, and B. A. Weiss, "Inertial Measurement Unit for On-Machine Diagnostics of Machine Tool Linear Axes," presented at the Annual Conference of the Prognostics and Health Management Society 2016, Denver, CO, October 3 - 6, 2016, 2016.

[10] G. W. Vogl and M. E. Sharp, "Diagnostics of machine tool linear axes via separation of geometric error sources," presented at the Annual Conference of the Prognostics and Health Management Society, St. Petersburg, FL, October 2-5, 2017, 2017.

[11] R. B. Randall and J. Antoni, "Rolling element bearing diagnostics—A tutorial," *Mechanical systems and signal processing,* vol. 25, no. 2, pp. 485-520, 2011.

[12] W. Caesarendra and T. Tjahjowidodo, "A review of feature extraction methods in vibration-based condition monitoring and its application for degradation trend estimation of low-speed slew bearing," *Machines,* vol. 5, no. 4, p. 21, 2017.

[13] I. Howard, "A Review of Rolling Element Bearing Vibration'Detection, Diagnosis and Prognosis'," Defence Science and Technology Organization Canberra (Australia)1994.

[14] B. Sreejith, A. Verma, and A. Srividya, "Fault diagnosis of rolling element bearing using time-domain features and neural networks," in *Industrial and Information Systems, 2008. ICIIS 2008. IEEE Region 10 and the Third international Conference on*, 2008, pp. 1-6: IEEE.
[15] C. T. Yiakopoulos, K. C. Gryllias, and I. A. Antoniadis, "Rolling element bearing fault detection in industrial environments based on a K-means clustering approach," *Expert Systems with Applications,* vol. 38, no. 3, pp. 2888-2911, 2011/03/01/ 2011.
[16] R. Randall, N. Sawalhi, and M. Coats, "A comparison of methods for separation of deterministic and random signals," *International Journal of Condition Monitoring,* vol. 1, no. 1, pp. 11-19, 2011.
[17] R. A. Wiggins, "Minimum entropy deconvolution," *Geoexploration,* vol. 16, no. 1, pp. 21-35, 1978/04/01/ 1978.
[18] N. Sawalhi, R. B. Randall, and H. Endo, "The enhancement of fault detection and diagnosis in rolling element bearings using minimum entropy deconvolution combined with spectral kurtosis," *Mechanical Systems and Signal Processing,* vol. 21, no. 6, pp. 2616-2633, 2007.
[19] H. Endo and R. B. Randall, "Enhancement of autoregressive model based gear tooth fault detection technique by the use of minimum entropy deconvolution filter," *Mechanical Systems and Signal Processing,* vol. 21, no. 2, pp. 906-919, 2007.
[20] J.-Y. Lee and A. Nandi, "Extraction of impacting signals using blind deconvolution," *Journal of Sound and Vibration,* vol. 232, no. 5, pp. 945-962, 2000.
[21] G. L. McDonald, Q. Zhao, and M. J. Zuo, "Maximum correlated Kurtosis deconvolution and application on gear tooth chip fault detection," *Mechanical Systems and Signal Processing,* vol. 33, pp. 237-255, 2012.
[22] J. Antoni, "Fast computation of the kurtogram for the detection of transient faults," *Mechanical Systems and Signal Processing,* vol. 21, no. 1, pp. 108-124, 2007/01/01/ 2007.
[23] J. Antoni and R. Randall, "The spectral kurtosis: application to the vibratory surveillance and diagnostics of rotating machines," *Mechanical Systems and Signal Processing,* vol. 20, no. 2, pp. 308-331, 2006.

COMPREHENSIVE CONDITION MONITORING ANALYSIS FOR POWER PLANT BOILER CIRCULATOR PUMPS

Eric Bechhoefer
GPMS Inc
40 Ridge Rd
Cornwall, VT 05753
1 (802) 377-0160
eric@gpms-vt.com

Ed Spence
The Machine Instrumentation Group LLC
4605 Myra Glen Place
Durham, NC 27707
1 (781) 439-1277
ed.spence@machineinstrumentation.com

Abstract: The boiler circulator pump (BCP) is an integral part of the power plant operations in both conventional (coal/gas fired steam plant) or nuclear plants. The BCP moves super heated water under pressure to the steam generator. These pumps are glandless, megawatt induction power pumps. If the pump fails, the power plant must be removed from service to replace the pump. In many applications, these pumps serve power plant providing the base load power requirements of a community. The failure of a pump requires the operator to buy power from other generating plants at much higher rates.

Glandless BCP pose a difficult problem for automated analysis, as there is no way to introduce a tachometer to measure pump speed. Further, while a relatively simple machine, the failure modes include: out of balance/bearing wear that is best measured by vibration, and rotor bar/opens/shorts/eccentricity that is best measured by current. Further, these machines are asynchronous and typically very well balanced, so that is difficult to determine shaft RPM from vibration.

This paper discusses the analysis of BCP using both vibration and current analysis methods and the processing needed to automate fault detection, diagnostics, and prognostics.

Key words: CBM; Diagnostics; Induction Motor; Impact Detection: Rotor Bar;

Introduction: Boiler circulator pumps (BCP) are ubiquitous in steam power generating plants for forced circulation of the working fluid to the steam generator, in both conventional and nuclear facilities. A glandless BCP improves safety and reliability in application where there is high pressure and temperatures for which the uses of a shaft seal is technically, or economically, infeasible.

BCPs designs typically have vertical pumps section suspended from the pipe line without external support. This allows for thermal expansion without distortion or strain on the pump casing. The pump casing is flanged, based on the customer's specification and application. The pump shaft is held in place radially with two water lubricated journal bearings. A tilting pad thrust bearing supports the axial load of the pump hydraulically.

The driving motor and the pump are an integral unit, with the stator and rotor surrounded by water. The water temperature is controlled, typically to 80C, with an auxiliary impeller through a heat exchanger outside of the pump. A thermal heat bearing is provided between the pump and motor section to prevent the heat from the hot pump section to the cooler motor section, thus protecting the motor insulation of the windings. A typical pump is powered by three phase, 4 to 11 KV power of 100 amps and are relatively compact for a MW pump. The capacity of many of these pumps is (depending on head), up to 26,000 gpm, with operating pressures up to 4000+ psi and temperatures of 788F.

The requirement for high availability of power plants coupled with the change in the way these older plants are being operated, give rise to the need for condition monitoring. In the past, coal fired plants were used for base load, with gas fired plants used for peak load. With the low cost of natural gas extraction, gas fired plants are used for base loads, with coal fired plants are being dispatched to supply power for peak loads. A coal fired, base load plant might run for a year without shutting down. Being run to support peak loads, coal fired plants have many more start/stop cycles than they were designed for. As such, BCPs are failing at higher rates and thus have lower reliability. Condition monitoring of BCPs can facilitate opportunistic maintenance and allow for greater flexibility in the operation of a coal-fired plant.

The BCP has a number of wear/damage modes. Electrically, the induction motor can have rotor bar failures, stator winding faults, and dynamic eccentricity. Mechanically, erosion/corrosion over time can cause imbalance. Most concerning is the wear of the journal bearing which will cause whip/whirl or the load of the axial trust bearing due to repeated starts/stops.

A comprehensive condition monitoring system was developed under contract by the pump manufacturer to address the wear/damage modes of the BCP. For the induction motor, both motor current analysis and motor vibration analysis are needed. Current allows for a direct measure of power consumed, rotor bar faults and static/dynamic eccentricity. Vibration allows a measure of the imbalance (shaft order 1), whip/whorl, and a direct measure of slip. To measure a jarring or impact event, continuous monitoring of acceleration is needed. Finally, some capability to measure the direction of rotation could be used. This would afford the ability to check if the power terminal connections were correct.

This paper addresses the analysis the use of a current sensor for motor current analysis, a MEMS smart sensor for vibration analysis, and a MEMS triaxial sensor for impact detection on a BCP.

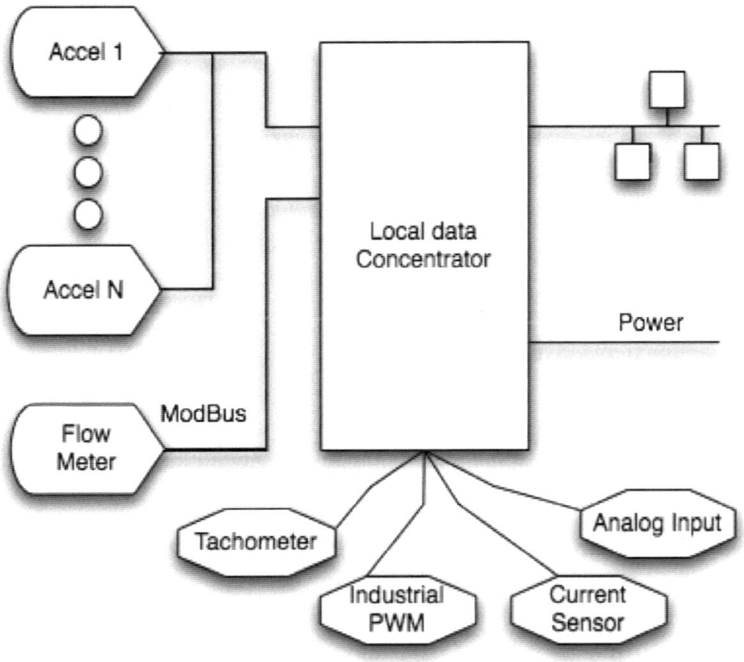

Figure 1 All testing was performed in collaboration with the manufacturer *in situ*, and use the GPMS *Foresight FX* Condition Monitoring System to extract the data, and generate CI's.

Impact Detection: This function requires a duty cycle of 100% as it is never known when an air slug will enter the line, when the pump will be turned off, and if off, when the pump will be energized. Further, the environmental noise is not known, which is needed for threshold setting/impact detection. Finally, direction of the impact force is not know – for example, for a motor start, the force may be tangential or radial, but for a shut down, it might be axial.

The lack of knowledge of the orientation of the sensor, lack of a level of process noise requires a dynamic, self learning system to evaluate what is a valid impact. As such, a tri-axial accelerometer was designed into the onboard control unit (OBCU) which is mounted on the BCP. There are a number of considerations in the design of the system for this implementation.

The design is based on a three channel, tri-axial accelerometer, that uses an initial bias correction at startup. An Alpha tracker then calculates the evolving bias and process noise of the each axis of the tri-axial sensor. The Alpha tracker is related to the Kalman filter, but with the assumption that the plant and process noise do not change. Typically, Kalman filters converge to a steady state relatively quickly. Using this assumption (e.g. that the plant and process noise approach a limit as time approaches infinity), it is possible to define the filter gains, α.

Given the plant noise, σ_w^2 (e.g. now fast the system can change nominally over time) and process noise, σ_v^2 (e.g. the measurement noise of the sensor), the fixed gain for α is calculated as:

$$\lambda = \frac{\sigma_w dt^2}{\sigma_v}, \quad \text{(eq 1)}$$

$$\alpha = {-\lambda^2 + \sqrt{\lambda^4 + 16\lambda^2}}/{8} \quad \text{(eq 2)}$$

During initialization, the bias and process noise σ_v^2, is calculated as:

- For Each Axis (x, y, z)
  - n = n +1
  - delta = x – bias;
  - bias = bias + delta/n;
  - V2 = V2 + delta *(x - bias);
- σ_w^2 = V2/(n-1);

The plant noise represents the "forgetfulness" of the system. Consider that bandwidth (that is, how fast it is anticipated that system could change), may be on the order of 5 minutes, then $\sigma_w^2 = 1/(60 * 5)^2$.

The filtered, biased corrected signal, for each axis, is then:

- For each axis, *i*:
  - residual(*i*) = bias(*i*) - measurement(*i*);
  - bias(*i*) = bias(*i*) + α * residual (*i*);

In general the residual(*i*) should have a Gaussian distribution.

Threshold Detection for Shock: There are two potential strategies for detection. If a G limit is know, then a threshold can be set such that the norm vibration level is directly compared to a value, such as 1.0 G. In this case, the norm vibration is:

$$norm = \sqrt{residual_1^2 + residual_2^2 + residual_3^2} \quad \text{(eq 3)}$$

In most cases, this limit is not known. Taking advantage of the assumption that the residual of each measurement from the bias signal is Gaussian, then the sum of the normalized square residuals is a Chi Square statistic with three degrees of freedom. The Chi Square statistic is then:

$$\chi^2 = \left({residual_1}/{\sigma_w}\right)^2 + \left({residual_2}/{\sigma_w}\right)^2 + \left({residual_3}/{\sigma_w}\right)^2 \quad \text{(eq 4)}$$

The threshold is set by taking inverse cumulative distribution function of Chi Square. For a PFA of 10e-6, the threshold is 25.9

The sample rate of each channel was set to 50 Hz, giving a Nyquist of 25 Hz. This is below the 30 Hz shaft rate of the rotor. If the sample rate were higher, the Alpha tracker would need to model the 30 Hz sine wave due to any imbalance on the system.

Motor Analysis: This analysis focuses on the detection of fault relating the drive motor of the pump. While there is a long history of motor analysis (1, 2 and 3), there are few automated systems to report the condition of a motor. Because Motor Current Signature Analysis (MCSA) is ineffective in the detection of journal/fluid bearings (as used in BCP), this system was designed both with a Rogowski Coil (a current transformer used in the MCSA) and two accelerometers (for the upper and lower journal bearings). As such, both MCSA and vibration based motor analysis were implemented. While the analysis is similar, it is felt that using both analysis are synergistic: vibration can supply information that is not available from current analysis (such as slip), and current can supply valuable information not available from vibration alone (e.g. motor amperage).

Vibration can be used to detect motor problems due to the effect of unbalanced magnetic forces. In general, the current in the motor stator produces a rotating magnetic field. Current is produced in the rotor conductors, which is proportional to the difference between the speed of the rotor and the stator magnetic fields. The rotor does not rotate at a synchronous speed, instead "slipping" backwards through the stators' rotating field.

Because the stator synchronous speed is a function of line frequency and the number of pole pairs in the machine, slip is fined per unit as:

$$s = {(RPM_{rotor} - RPM_{stator})}/{RPM_{stator}} \quad \text{(eq 5)}$$

Typical slip values of 0.01 to 0.02 are not unusual.

Torque is produced from balanced forces on the rotor. If the forces of attraction are not balanced, vibration occurs. This vibration is then the result of current or air gap variances in the motor. The cause of these magnetic vibration are related to:

- The static and/or dynamic air-gap irregularities
- Broken rotor bars or cracked rotor end rings
- Stator faults (open or short of one coil or more of a stator winding)
- Poor/abnormal connections to the stator windings.

Air Gap Eccentricity: When the distance between the rotor and the stator is not uniform, there will be a change in the magnetic flux over time. This will affect both the rate of change of current in the stator, and also exert a periodic force that can be measured with an accelerometer. Formally, this can be the result of either static or dynamic forces. Static eccentricity occurs then the radial air gap is fixed. This could be due to manufacturing error (e.g. the bearing is not centered on the stator), or in the case of a BCP, due to the force of the pump discharge.

For BCP, which use fluid bearings, the force exerted on the shaft from the bearing increases with the reduction in distance to the bearing face. Under normal operations, the reaction to the pump discharge will asymmetrically load the bearing, causing the machine to run eccentrically. As the bearing material wears, the eccentricity will increase.

Dynamic eccentricity occurs when the minimal air gap follows the rotor. This may be due to imbalance phase power.

The change in current caused by the changing magnetic fields results in side band components in both vibration and MCSA, as in (eq. 6):

$$f_e = fR\left((R \pm 1)\left(1 - {}^{s}/_{p}\right) \pm n\right) \quad \text{(eq. 6)}$$

where:

- f_e is the eccentric frequency
- f is the grid frequency
- R is the number of rotor bars
- p is the number of pole pairs and
- n is the odd harmonics values: 1, 3, 5, 7…

For example, with s of 0.011, four pole-pair and 34 rotor bars, 60 Hz, (eq 6) would frequency centered at 564 Hz, with 120 Hz separations for static eccentricity (204, 324, 444, 564, 685, 802 Hz). For dynamic eccentricity, these frequencies are modulated by the shaft rate, approximately 14.85 Hz.

Rotor Bar Faults: A broken rotor bar is typically the result of a broken weld, due to the period mechanical/vibrational loads induced by the changing magnetic flux. The relative motion of the stator magnetic flied and the rotor generates current in the rotor bars. This current produces the motor torque. When a rotor bar cracks/fails, both current in the stator windings and acceleration can be detected at the frequencies in (7):

$$f_{rb} = f\left(k\left(\left(1 - {}^{s}/_{p}\right) \pm s\right)\right) \quad \text{(eq 7)}$$

where:

- f_{rb} is the rotor bar frequency
- f is the grid frequency
- p is the number of pole pairs and
- k is the harmonics values: 1, 2, 3, …

Typically, these sidebands are formed 2x the slip frequency (e.g.. s x f). Figure 1 shows a spectrum of the current, comparing the rotor bar fault frequencies for a nominal motor, with one broken rotor bar and with two broken rotor bars.

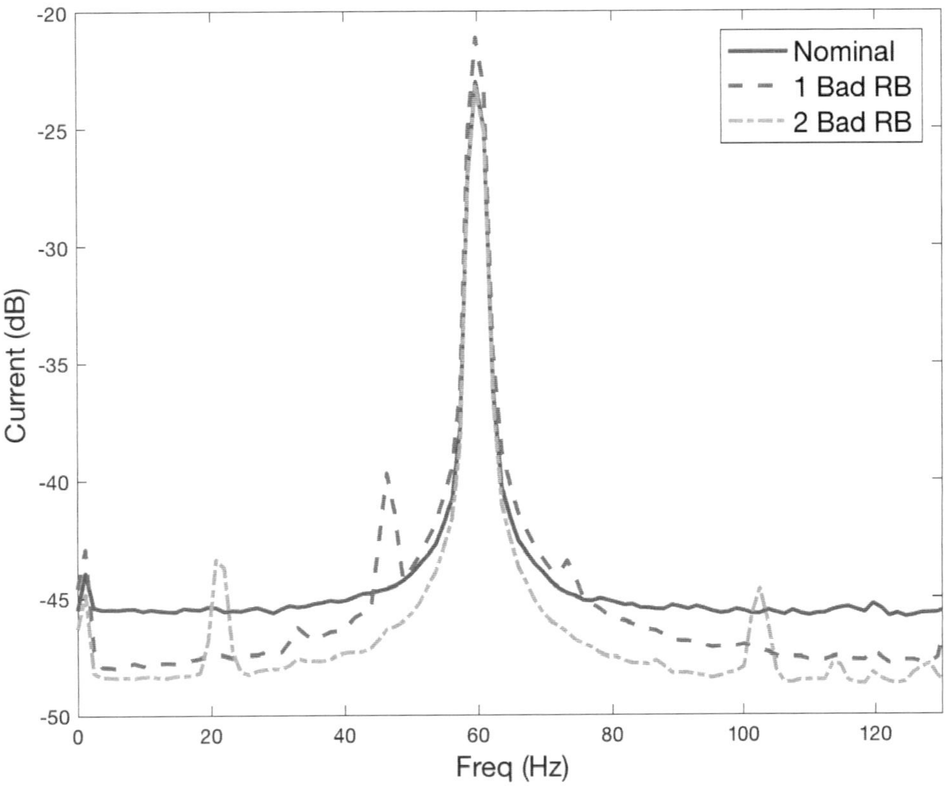

Figure 1. Example of Current Spectrum Comparing Rotor Bar Damage

Note that current clearly shows the fault, but there is little if any indication of what the slip is, especially in the nominal case. The use of vibration allows for a determination of slip simply by measuring the 1/Rev vibration. Figure 2 shows a single phase motor with two broken rotor bars. It is a 4 pole pair design fan, turning nominally at 15 Hz.

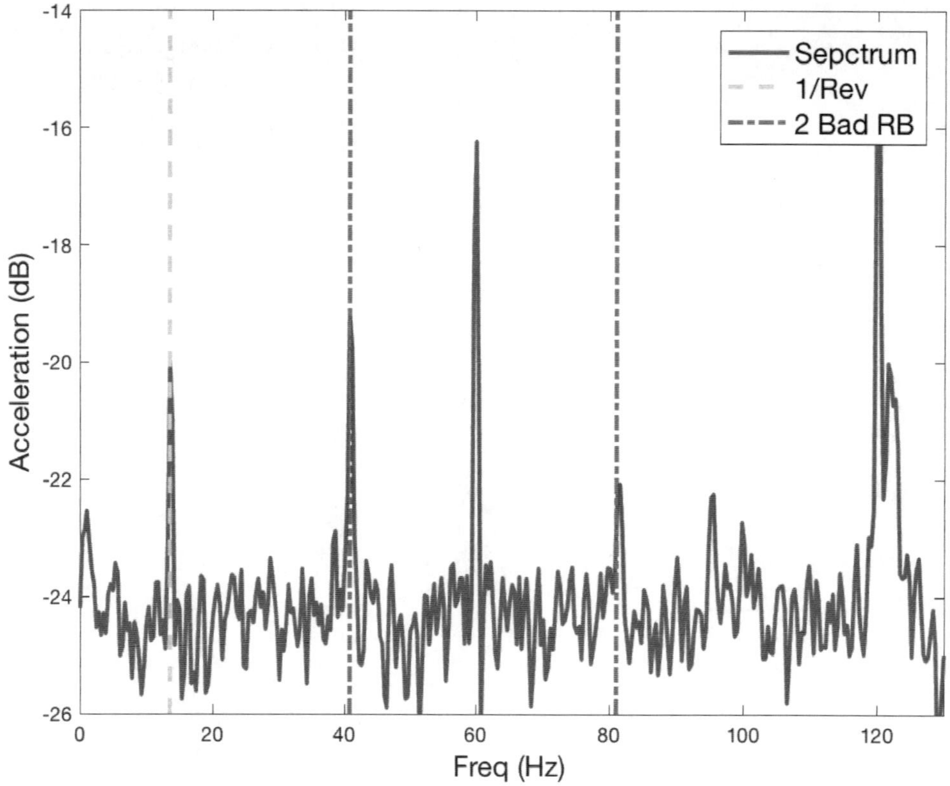

Figure 1. Example of Vibration Spectrum of Fan with Rotor Bar Damage

The measured 1/Rev is 13.56 Hz, giving a large slip of 0.11 – clearly damaged. From (eq 7), sides bands would be expected at [40.22, 52.96, 67.02 and 79.78]. This example shows the side bands at 40.22 and 79.78.

Stator Shorts: Short from turn to turn results in excessive heat and current imbalance. This type of fault typically is not be observed in a BCP, as the working fluid would short the winding to ground rather than a to stator winding. That said, the frequencies associated with MCSA, are in (eq 8).

$$f_s = f\left({}^{n}/_{p}\left((1-s) \pm k\right)\right) \qquad \text{(eq 8)}$$

where:

- f_s is the short frequency
- f is the grid frequency
- p is the number of pole pairs and
- k is the harmonic values: 1, 2, 3, …
- n is the harmonic values: 1, 2, 3, …

Note that for vibration analysis, a short, stator looseness, or unbalance phase, etc., is typically seen at 2 x the line frequency.

Mechanical Problems: Many mechanical problems are not well analyzed with current, but are clearly detectable with vibration. Common problems such as imbalance are seen as the 1/Rev (e.g. 1 x RPM) radial vibration. Bent Shaft or Misalignment can be seen in both the radial and axial direction as 1x, and 2x RPM. Bearing faults for rolling element bearings are well documented. The BCP typically uses a fluid bearing, where the damage and wear is associated with whip/whirl (0.43 to 0.48 x RPM).

Rotor Direction: Surprisingly, it rather easy to connect a three-phase motor and have it run in reverse. It is also surprising difficult to determine the direction of rotation of the pump once it is running. As such, there are three indications of a pump running backwards: reduced impeller signature, cavitation and reduced power consumption.

As the pump impeller is designed to run in only one direction, running in reverse causes, as noted, three issues. First, the impeller blades impart a period impulse on the flow of the water which can be measured. Running in reverse, this is much reduced. Second, the pump impeller will generate cavitation when run in reverse, due to a lower pressure on the backside of the impeller. This can be measured as a generalized broadband noise. This noise can be estimated envelope spectrum (e.g. say from 500 to 1000 Hz) as the kurtosis of the spectrum. Increased broadband noise will result in a larger than expected kurtosis at high frequency. The envelope spectrum is used 500 Hz to 1000 Hz, as it is above any of the frequencies that would be expected with the nominal operation of the pump.

As the pump is doing less work (e.g. it is not as effective at moving water) when running in reverse, another indicator is lower power. Measuring motor current is a direct measure of this condition.

Automated Fault Detection: There is a need for automation for fault detection, as most operators do not have the expertise to implement their own condition monitoring program. This requires the automated method to quantify the analysis for specific fault modes on the BCP. This process is exemplified by the generation of condition indicators (CIs), which are then used for component health determination.

CI algorithms implement analysis on the spectrum taken on data. The CI analysis is based on the components being monitoring.

Table 1: Vibration Based Shaft Analysis

Shaft Analysis	Vibration	Comment
SO1 Magnitude	1 x RPM	Unbalance Rotor Shaft
SO2 Magnitude	2 x RPM	Bent Shaft or Angular Misalignment
SO3 Magnitude	3 x RPM	Bent Shaft or Coupling Failure
Impeller	Blades x RPM	Indication of wear/damage to pump/rotation direction
Bearing Analysis	0.43 to 0.48 RPM	Whip/Whirl of fluid bearings
Cavitation	500 to 1000 Hz	Kurtosis of Envelope Spectrum

RPM	Calculated RPM	Pump Status, Used for Slip calculation

Table 2: Vibration Based Motor Analysis

Motor Analysis	Vibration	Comment
Slot Magnitude	# Slots x RPM	Energy Associated the MMF
Short Indicator	2 x Grid Freq	Indication of winding/wind short
Rotor Bar Fault	Eq 7	Sum of Frequency Energies
Rotor Bar Ratio	Eq 7 Normalized	RBF / line frequency energy
Dynamic Eccentricity	Eq 6	Sum of Frequency Energies
Eccentricity Ratio	Eq 6 Normalized	DE / line frequency energy
Slip	Eq 5	Easy to calculate based on vibration

Table 3: Current Based Motor Analysis

Motor Analysis	Current	Comment
Line Amperage	FFT of Line Freq	Magnitude; Pump Condition
Line Frequency	FFT of Line Freq	Calculated Freq.
Slip	Eq 5	Hard to calculate without RBF
Rotor Bar Fault	Eq 7	Sum of Frequency Energies
Rotor Bar Ratio	Eq 7 Normalized	RBF / line frequency energy
Number of Bad RB	Eq 7 Normalized	Log of RBR normalized by line energy
Dynamic Eccentricity	Eq 6	Sum of Frequency Energies
Eccentricity Ratio	Eq 6 Normalized	DE / line frequency energy

Component Health Calculation: While the CI is a used as descriptive statistic of the component under analysis, to be useful, it requires threshold setting. Byington [4] teachers that for single CI, a probability density function (PDF) for the Rician/Rice statistical distribution can be used to set a threshold based on an probability of false alarm (PFA). This was further explored by Dempsey [5], who validated the relationship between CI threshold and PFA to describe the receiver operating characteristics (ROC) of the CI for a given fault. Additionally, Dempsey used the ROC to evaluate the performance of the CI for a fault type. As no single CI has been identified that works with all fault modes, the approached used by Bechhoefer [6] was implemented, which fuses *n* number of CI into a gear health indicator (HI).

Computationally, the use of an HI is attractive. The HI provides a decision making tool for the end user on the status of system health. Health indicators consist of the integration of several condition indicators into one value that provides the health status of the component to the end user [6].

To simplify presentation and knowledge creation for a user, a uniform meaning across all components in the monitored machine was developed. The measured CI statistics (e.g. PDFs) will be unique for each component type (due to different rates, materials, loads, etc). This suggests that the critical values (thresholds) will be different for each monitored component. By using the HI paradigm the HI value is independent of the component. In this implementation, the HI has two alert levels: warning and alarm, allowing a common nomenclature for the HI:

- The HI ranges from 0 to 1, where the probability of exceeding an HI of 0.5 is the PFA,
- A warning alert is generated when the HI is greater than or equal to 0.75. Maintenance should be planned by estimating the RUL until the HI is 1.0.
- An alarm alert is generated when the HI is greater than or equal to 1.0. Continued operations could cause collateral damage.

Note that this nomenclature does not define a probability of failure of the component, or that the component fails when the HI is 1.0. Rather, it represents a reduction in the design reliability of the system and a change in operator behavior to a proactive maintenance policy: perform maintenance to restore the design reliability of the system.

The HI Algorithm: The HI is based on a generalized whitening solution using the Cholesky decomposition (see [7]). The Cholesky decomposition of a Hermitian, positive definite matrix results in:

$$\mathbf{A} = \mathbf{L}\mathbf{L}^{*} \qquad \text{(eq 9)}$$

where **L** is a lower triangular, and $\mathbf{L}^{*}$ is its conjugate transpose. By definition, the inverse covariance is positive definite Hermitian. As such:

$$\mathbf{L}\mathbf{L}^{*} = \mathbf{\Sigma}^{-1}\text{, and: } \mathbf{Y} = \mathbf{L} \text{ x } \mathbf{CI}^{\mathbf{T}}. \qquad \text{(eq 10)}$$

The vector **CI** is the correlated CIs used for the HI calculation, and **Y** is 1 to *n* independent CI with unit variance (one CI representing the trivial case). The Cholesky decomposition, in effect, creates the square root of the inverse covariance. This in turn is analogous to dividing the CI by its standard deviation (the trivial case of one CI). In turn, $\mathbf{Y} = \mathbf{L} \text{ x } \mathbf{CI}^{\mathbf{T}}$ creates the necessary independent and identical distributions required to calculate the critical values for a function of distributions.

HI Based on Rayleigh PDFs: The CIs used for this example have Rayleigh like PDFs (e.g. heavily tailed). Consequently, the HI function was designed using the Rayleigh distribution. The PDF for the Rayleigh distribution uses a single parameter, β, resulting in the mean ($\mu = \beta*(\pi/2)^{0.5}$) and variance ($\sigma^2 = (2 - \pi/2) * \beta^2$). The PDF of the Rayleigh is: $x/\beta^2\exp(x/2\beta^2)$. Note that when applying these equations to the whitening process, the value for β for each CI will then be: $\sigma^2 = 1$, and $\beta = \sigma^2 / (2 - \pi/2)^{0.5} = 1.5264$ (ref [8]).

The HI is then the norm of *n* whitened CIs. As can be show, this function defines a Nakagami PDF [8]. The statistics for the Nakagami are: $\eta = n$ (the number of CIs), and $\omega = 1/(2-\pi/2)*2*n$. The scale value is then calculated from the inverse CDF of the Nakagami for a PFA of 10e-6 at a HI value of 0.5, with η and ω, a

The HI algorithm it then:

$$\boldsymbol{Y} = \boldsymbol{L} \times (\boldsymbol{CI} - \boldsymbol{offset}) \qquad \text{(eq. 10)}$$

$$HI = \sqrt{\boldsymbol{Y}^T\boldsymbol{Y}\left({}^{0.5}/_{scale}\right)^2} \qquad \text{(eq. 11)}$$

The covariance and offset values for the n CI were calculated by sampling healthy data from four BCP pumps located at an operators site. This was done by randomly selecting 50 data points from each pump, the calculating the covariance and offsets over the resulting 200 data points. The selected CI's PDF were not Gaussian, but exhibited a high degree of skewness. Because of this, the PDFs were "left shifted" by subtracting an offset such that the PDFs exhibited Rayleigh like distributions.

Deployment: The system has been deployed on four BCP at a coal fired plant. In general, no faults were observed. The Alarm that was observed on a BCP that was undergoing maintenance.

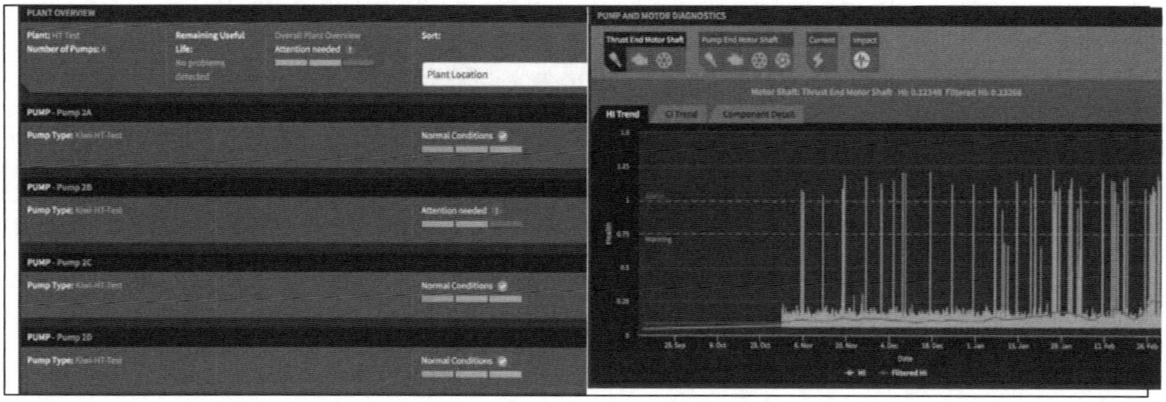

Conclusion: Using an integrated approach to condition monitoring, an attempt was made to build a comprehensive system for monitoring of BCP. Using both accelerometer and current data, analysis was developed to capture all of the fault modes that have been historically observed on BCPs. This integrated approach allows the manufacture of the pump to provided a heighten level of service and reliability to their customer. The pump manufacture is now procuring 22 additional systems for test.

References:

[1] Bates, G. H,. "Vibration Diagnostics for Industrial Electric Motor Drives", Bruel & Kjaer Applications Note, 1980.

[2] Parveenkumar, T., Saimurugan, M., Ramachandran, K. I., "Comparision of Vibraiton, Sound and Motor Current Signature Analysis for Detection of Gear Bos Faults", International Journal of Prognostics and Health Management, 2017

[3] Miljkovic, D., "Brief Review of Motor Current Signature Analysis", Hrčak Portal, Ministry of Science, Education and Sports. 2017.

[4] Byington, C., Safa-Bakhsh, R., Watson., M., Kalgren, P. (2003). *Metrics Evaluation and Tool Development for Health and Usage Monitoring System Technology.* HUMS 2003 Conference, DSTO-GD-0348

[5] Dempsy, P., Keller, J. (2008). Signal Detection Theory Applied to Helicopter Transmissions Diagnostics Thresholds. *NASA Technical Memorandum* 2008-215262

[6] Bechhoefer, E., He, D., (2011). A Process for Data Driven Prognostics, MFPT Publications, Dayton OH

[7] Bechhoefer, E., He, D., Dempsey, P. (2011). *Gear Threshold Setting Based On a Probability of False Alarm.* Annual Conference of the Prognostics and Health Management Society.

[8] Bechhoefer, E., Bernhard, A. (2007). *A Generalized Process for Optimal Threshold Setting in HUMS*. IEEE Aerospace Conference, Big Sky.

CRACK PROPAGATION IN A COMPACT TENSION SPECIMEN SUBJECTED TO GAUSSIAN RANDOM VIBRATIONS WITH OCCASIONAL OVERLOADS

Julian Raphael
J R Technical Services, LLC
Fairway Drive
Abingdon, VA 24211
Telephone: (212) 804-8172
drjulianraphael@hotmail.com

Peter Liaw
University of Tennessee
406 Ferris Hall
Knoxville, TN 38163
Telephone: (423) 974-6356
pliaw@utk.edu

Abstract

It is well known that cracks in structural components subjected to overloads manifest delayed growth for some period of time, slowly reverting to the initial rate on a crack-growth rate versus stress-intensity-factor-range curve thereafter. Frequently, constant amplitude fatigue tests with occasional programmed overloads are used to demonstrate the phenomenon. However, in response to random loading in which the sequence of loads is uncertain, different growth rates result from various spectra, even if the distributions from which the loads are taken are identical.

This paper examines variations in the number of fatigue cycles necessary for a crack to pass through the plastic zone after a single overload. A Monte-Carlo simulation generates a single initial overload on a compact-tension specimen, creating a large plastic zone, or enclave, ahead of the crack. Following the overload, both minimum and maximum random loads are generated from Gaussian distributions. The Forman equation is used to calculate linear crack growth, i.e., without any retardation, and the Generalized Willenborg retardation model is used to calculate the reduction in the crack-driving potential, K_R. The Forman equation is then modified to use the effective stress intensity factor range, ΔK_{eff}, and the effective load ratio, R_{eff}, to calculate the reduced growth rate, which is compared to the linear result. This procedure is repeated for each load cycle until the crack passes through the overload zone, and the number of cycles is recorded. The entire process is then repeated a total of 1,000 times.

Using standard hypothesis testing, the resulting crack-growth data are analyzed to determine which distributions cannot be excluded from consideration at the 5% significance level. 95% two-sided confidence intervals are determined for the distribution parameters. The best candidate distributions are overlaid on a histogram to provide graphical results. Finally, the first four statistical moments calculated from the simulation data are compared to those derived using the distribution's parameters. It was found that lognormal and Birnbaum-Saunders distributions are good fits.

The underlying basis of the Willenborg model is that large compressive residual stresses, which reduce the effect of applied tensile loads, exist in the vicinity of the crack tip after an overload. Finite Element Analysis was performed to demonstrate the magnitude and extent of that compressive stress distribution. Normal stress and effective plastic strain results are discussed.

Key words: Finite element analysis; fracture; Gaussian white noise; Monte Carlo simulation; probabilistic crack growth; statistical analysis; variable amplitude fatigue; Willenborg retardation

Introduction:

In this paper, the probabilistic crack-growth response to Gaussian white noise loading preceded by a single random overload is studied in an ASTM E 399 compact tension specimen. It is well-known that crack-growth retardation occurs after a single tensile overload (1-3). Willenborg (1) attributed the delay to residual compressive stresses in the vicinity of the crack tip, resulting in a reduction of the crack-driving potential. The Willenborg overload zone model is shown in Figure 1.

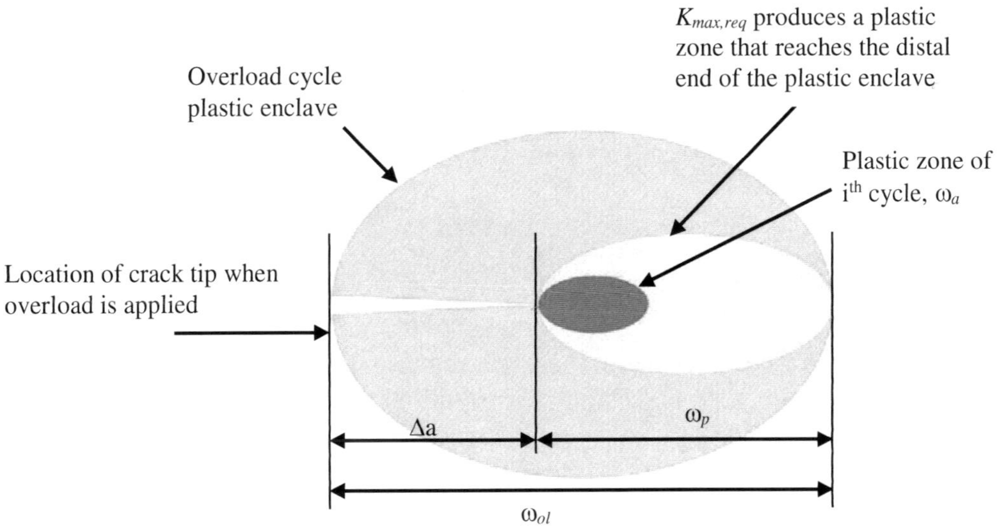

Figure 1: Willenborg Overload Zone

ω_{ol} is the plastic enclave created when the overload is applied. Δa is the total crack growth into the plastic enclave due to the application of the overload. $K_{max,req}$ is known as the virtual stress intensity factor, and represents the stress intensity factor necessary to produce a plastic zone that reaches from the current crack tip to the distal end of the plastic enclave, defined as ω_p. The quantity, ω_a, is the length of the plastic zone produced by the current load.

Mathematically, the Willenborg model is given by Equations (1) - (7):

$$\Delta K_{max,req} = K_{max}^{ol}\sqrt{1-\frac{\Delta a}{\omega_{ol}}} \tag{1}$$

$$\omega_p = \frac{\pi}{8}\left(\frac{K_{max,req}}{\alpha_p \sigma_{ys}}\right)^2 \tag{2}$$

$$K_{R_i} = K_{max,req} - K_{max_i} \tag{3}$$

$$K_{max,eff_i} = K_{max_i} - K_{R_i} \tag{4}$$

$$K_{min,eff_i} = K_{min_i} - K_{R_i} \tag{5}$$

$$R_{eff_i} = \frac{K_{min,eff_i}}{K_{max,eff_i}} \tag{6}$$

$$\begin{aligned} \Delta K_{eff_i} &= \Delta K_i && \text{if} \quad \Delta K_{max,eff_i} > 0 \quad \text{and} \ \Delta K_{min,eff_i} > 0 \\ \Delta K_{eff_i} &= K_{max,eff_i} && \text{if} \quad \Delta K_{min,eff_i} \le 0 \\ \Delta K_{eff_i} &= 0 && \text{if} \quad \Delta K_{max,eff_i} \le 0 \end{aligned} \tag{7}$$

K_{max}^{ol} represents the maximum stress-intensity factor during the overload. α_p is the plastic constraint factor, which is approximately 1.15 for plane stress and 2.55 for plane strain (4). K_R represents the residual stress-intensity factor, i.e., the reduction in the stress-intensity factor caused by the crack tip residual compressive stress distribution. $K_{max,eff}$ and $K_{min,eff}$ are the cyclic effective maximum and minimum stress intensity factors. ΔK_{eff} represents the cyclic-operating effective stress range. R_{eff} is the effective R-ratio, defined as K_{min}/K_{max}, and σ_{ys} is the material's yield strength. K_{max} and K_{min} are the maximum and minimum cyclic stress intensity factors, which are both reduced by K_R to obtain $K_{max,eff}$ and $K_{min,eff}$ as shown in equations (4) and (5).

Under normal circumstances, the Willenborg model derives its retardation effect only from the change in R-ratio. Therefore, the crack-growth model should include the R-ratio as a variable. The Forman model, modified for use with Willenborg retardation, is given by

$$\frac{da}{dN} = \frac{C\left(\Delta K_{eff}\right)^m}{\left(1 - R_{eff}\right)K_c - \Delta K_{eff}} \tag{8}$$

in which C and m are the Forman coefficient and exponent, respectively, and K_c is the critical stress intensity factor appropriate for the thickness under consideration.

The Willenborg model predicts zero crack growth when the shutoff ratio reaches a value of 2.0. In 1974 Gallagher (5) proposed a modification to K_R, shown in Equations (9) and (10), which obviates the shutoff-ratio limitation. Thus, $(K_R)^G$ replaces K_R in Equations (4) and (5).

$$\left(K_R\right)^G = \phi K_R \tag{9}$$

$$\phi = \frac{1 - \frac{\Delta K_{th,0}}{\Delta \mathrm{K}}}{R_{so} - 1} \tag{10}$$

$\Delta K_{th,0}$ is the threshold ΔK at an R-ratio of zero, and R_{so} is the actual shutoff ratio.

Statistical approach:

Monte-Carlo simulation of the loading history produced a single record of crack growth, terminating in the number of cycles for the crack to pass through the overload zone, ω_{ol}. A solid model of the compact tension specimen is shown in Figure 2, and its dimensions are shown in Figure 3.

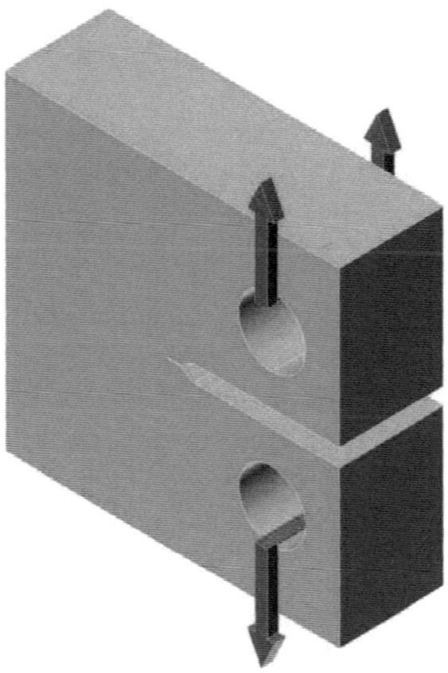

Figure 2: Solid Model of Compact-Tension Specimen

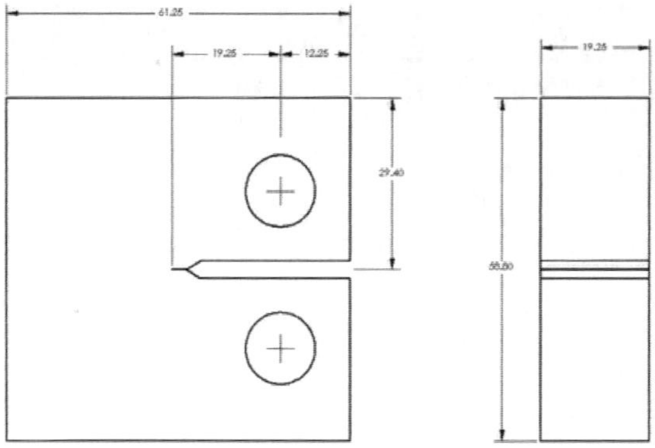

Figure 3: Dimensions of Compact-Tension Specimen

The overload, maximum and minimum cyclic loads were taken from the Gaussian distributions shown in Figure 4.

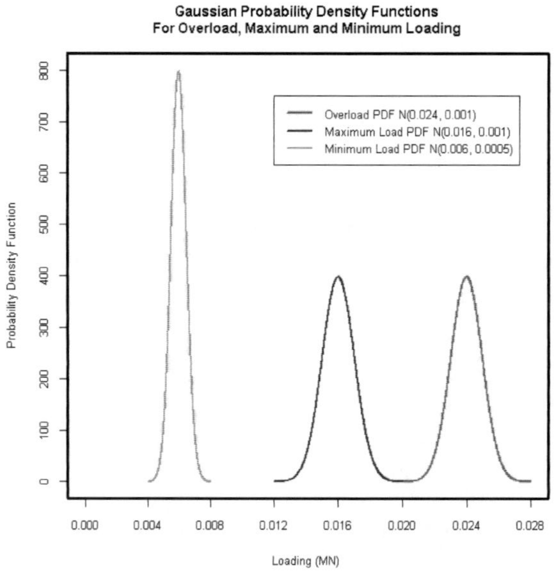

Figure 4: Gaussian distributions from which overload, maximum and minimum cyclic loads were taken

Overloads were randomly extracted from N(0.024,0.0010), while maximum cyclic loads were taken from N(0.016,0.0010), and minimum cyclic loads came from N(0.006,0.0005). Units for the mean and standard deviation are MN.

A sample history of the maximum cyclic load that required nearly 7,000 cycles for the crack to pass through the overload zone is shown in Figure 5.

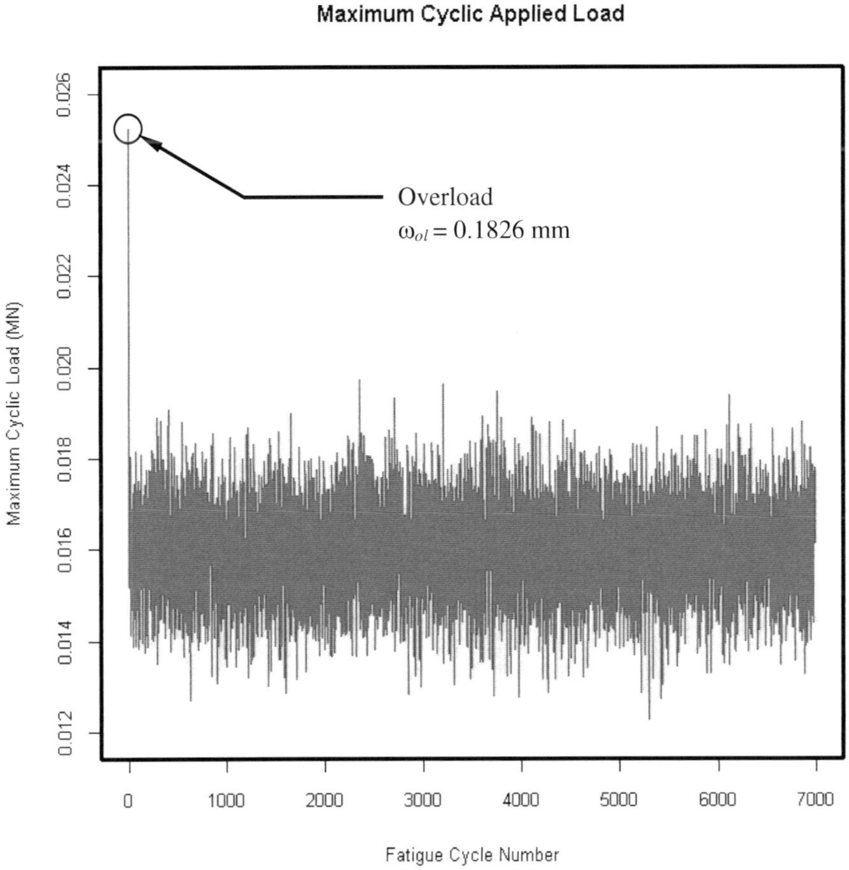

Figure 5: Sample cyclic load history of maximum load

After the list of fatigue cycle times was completed, statistical analysis, consisting of the following steps, was conducted:

1. Hypothesis tests and associated p-values to determine admissible distributions at the 5% significance level
2. Calculation of distribution parameters
3. Determination of 95% 2-sided confidence intervals for the distribution parameters

Four distributions were examined for the sample size of 1,000 using the Anderson-Darling test: Birnbaum-Saunders, Lognormal, Weibull, and Normal. Figure 6 shows the histogram of results with overlays of the four distributions. All statistical work was performed in *R* (6), which is statistical software.

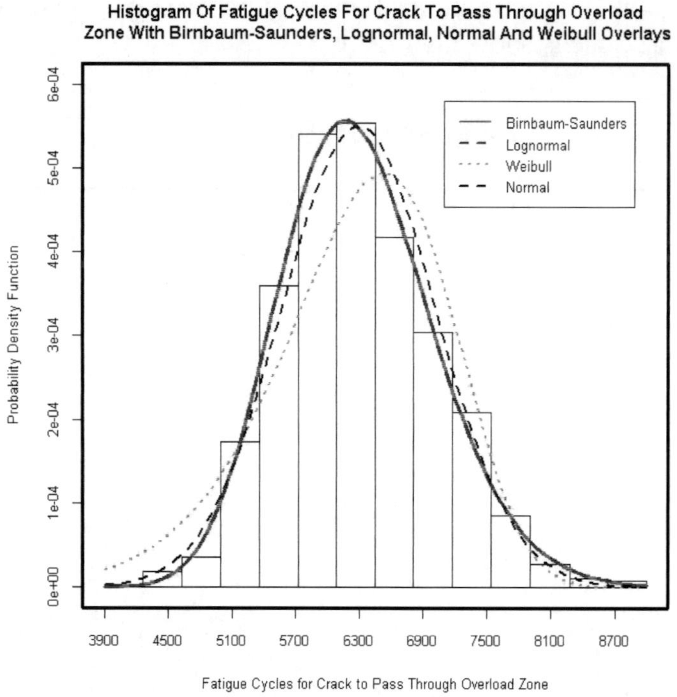

Figure 6: Histogram of fatigue cycles for the crack to pass through overload zone

Table 1 compares the Anderson-Darling p-values for each of the distributions under consideration.

Table 1: Comparison of Anderson-Darling p-values for various distributions

Distribution	p-value
Birnbaum-Saunders	0.881
Lognormal	0.871
Normal	0.060
Weibull	0.000

As can be seen from Table 1, neither Birnbaum-Saunders nor Lognormal can be excluded as an admissible distribution at the 5% level. The Birnbaum-Saunders distribution is to be preferred over the lognormal due to the slightly-greater p-value. Although the Normal distribution cannot be excluded at the 5% level, Birnbaum-Saunders is still to be preferred. Table 2 is a comparison between the simulation data and the Birnbaum-Saunders model. Table 3 gives the 95% 2-sided confidence limits for the Birnbaum-Saunders parameters, α and β, whose CDF, $F(x)$, is given by

$$F(x) = \Phi\left[\frac{1}{\alpha}\left(\sqrt{\frac{x}{\beta}} - \sqrt{\frac{\beta}{x}}\right)\right] \quad (11)$$

$$\Phi(\bullet) = N(0,1) \quad (12)$$

$F(x)$ represents the probability that the crack will pass through the overload zone in fewer than x cycles.

Table 2: Comparison between the simulation data and Birnbaum-Saunders model

	Simulation Data	Birnbaum-Saunders Model
Mean Value	6,294	6,294
Median Value	6,246	6,252
Variance	526,881	530,060
Standard Deviation	725	728
Skewness	0.2883	0.0195
Kurtosis	0.0141	0.2044

Table 3: 95% 2-sided confidence intervals for Birnbaum-Saunders parameters

	Parameter Value	Confidence Interval
Shape Parameter, α	0.1155	(0.1106, 0.1208)
Scale Parameter, β	6252.7	(6208.4, 6297.7)

Finite-element results:

The existence of a residual compressive stress distribution at the crack tip forms the basis of the Willenborg retardation model. Figure 7 shows the mesh and Figure 8 presents the residual stress distribution (σ_y) in a Ramberg-Osgood material. Kinematic hardening was used to simulate the Bauschinger effect. The midplane hydrostatic crack-tip stress was -642 MPa, and the maximum residual effective strain was 0.049. A vertical load of 1,976 N was applied to the model at the hole. This FEA model is different from the model used for statistical analysis. Its function is to show the magnitude and extent of the residual compressive stress distribution near the crack tip.

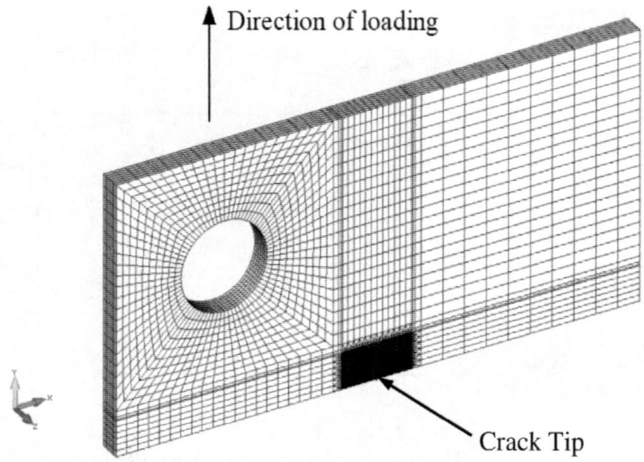

Figure 7: View of elements used in the finite-element analysis. Element spacing is 0.065 mm near the crack tip. 1/4 model.

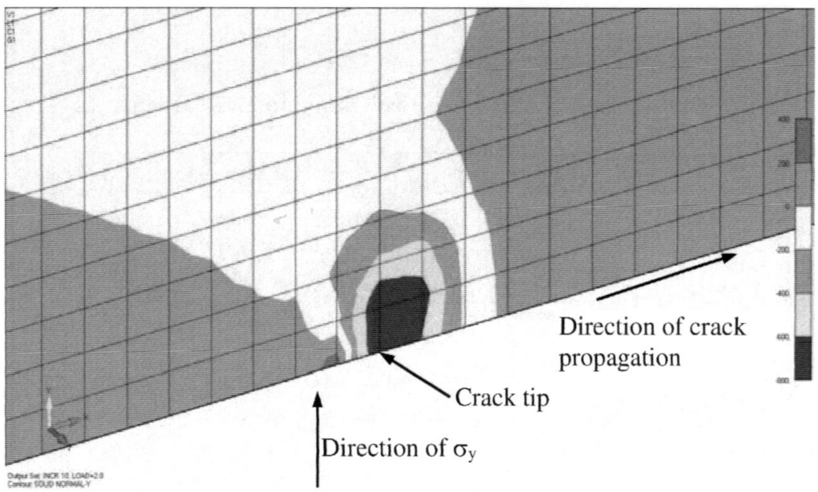

Figure 8: Residual-stress distribution of σ_y in the vicinity of the crack tip showing large compressive residual stresses

A comparison of the linear crack growth, in which retardation is not considered, to crack growth using the Willenborg model is shown in Figure 9, from which it can be seen that the Willenborg model predicts about 20% more cycles are required for the crack to traverse the overload zone of 0.183 mm than does the linear model. In the context of this study, no retardation occurs when K_R is less than or equal to zero. That is why the two curves are parallel after about 4,000 cycles.

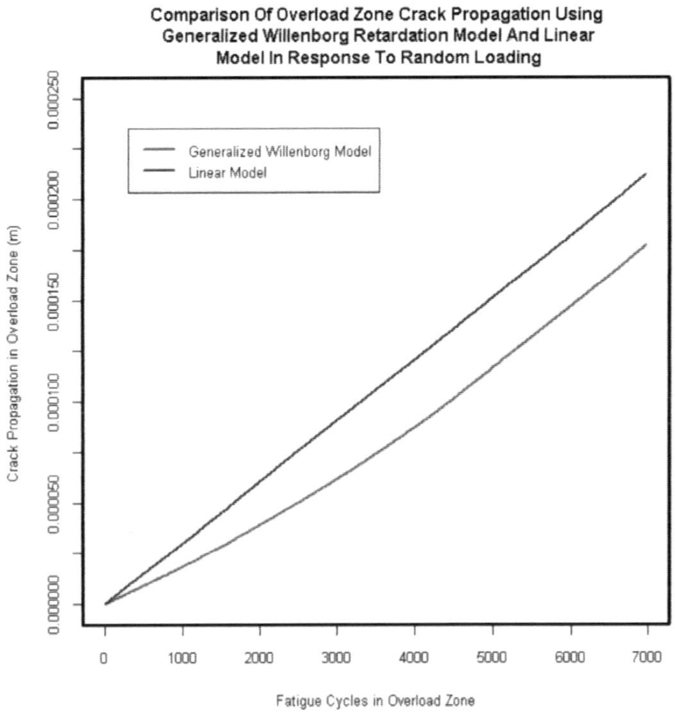

Figure 9: Comparison of Generalized Willenborg crack growth to that of linear model

Summary:

The Generalized Willenborg model accounts reasonably well for the observed retardation effects. Variations in the number of fatigue cycles needed for a crack to pass through the overload zone after a single overload is well-modeled by both Birnbaum-Saunders and lognormal distributions. Finite element results verify the underlying assumption of the presence of large compressive residual stresses at the crack tip after a single overload in a Ramberg-Osgood material.

References

1. Willenborg, J, Engle, RM, Wood, HA, (1971), *A Crack Growth Retardation Model Using an Effective Stress Concept*, Air Force Flight Dynamics Laboratory, Wright Patterson Air Force Base

2. Ribeiro, AS, Jesus, AP, Costa, JM, Borrego, LP, Maeiro, JC, *Variable Amplitude Fatigue Crack Growth Modeling*, (2011), Mechanica Experimental, Vol 19, 33-44

3. Suresh, S, (1998), *Fatigue of Materials* 2nd edition, Cambridge University Press

4. Sander, M, *Comparison of Fatigue Crack Growth Concepts With Respect to Interaction Effects*, University of Paderborn, Institute of Applied Mechanics

5. Gallagher, JP,(1974), *A Generalized Development of Yield Zone Models*, Wright-Patterson Air Force Laboratory

6. R Core Team (2012), *R: A language and environment for statistical computing.* R Foundation for Statistical Computing, Vienna, Austria. ISBN 3-900051-07-0, URL http://www.R-project.org/.

7. Carlson, RL, Kardomateas, GA, Bates, PR, (1991), *The Effects of Overloads in Fatigue Crack Growth*, (1991), International Journal of Fatigue

8. Nahlik, L, Pokorny, P, Hutar, P, (2014), *Influence of Crack Retardation on Fatigue Crack Propagation in Steels for Railway Axles*, doi:10.4028/www.scientific.net/AMR.891-892.351

9. ASTM E399-90 (1997), *Standard Test Method For Plane-Strain Fracture Toughness of Metallic Materials*, ASTM International, West Conshohocken

Degradation Modeling from Condition-based Data to Functional Failure Signature Data

James P. Hofmeister
Ridgetop Group, Inc.
3580 West Ina Road
Tucson, AZ 86741
520-742-3300x107
hoffy@ridgetopgroup.com

Douglas L. Goodman
Ridgetop Group, Inc.
3580 West Ina Road
Tucson, AZ 86741
520-742-3300
doug@ridgetopgroup.com

Ferenc Szidarovszky
Ridgetop Group, Inc.
3580 West Ina Road
Tucson, AZ 86741
520-742-3300
szidarka@gmail.com

***Abstract*—This paper presents approaches to degradation modeling starting with condition-based data (CBD) and progressing to functional-failure signature (FFS) data: FFS data forms a transfer curve that is very amenable to processing by prediction algorithms in support of a Prognostic Health Monitoring (PHM) system. The approach uses degradation signal models that are previously developed, validated, and presented: for example, an MFPT 2018 paper "Degradation Signal Modeling." Degradation signal modeling transforms curvilinear, CBD-based signature data into signature data that is much more linearized, which increases the accuracy of prediction information such as remaining useful life (RUL) and state of health (SoH). The focus of this paper is transforming CBD signatures into fault-to-failure progression (FFP) signatures, degradation-progression signatures (DPS), and then into FFS [1]-[3].**

1. INTRODUCTION

A Prognostic Health Monitoring (PHM) system uses sensors to acquire data and processes that data for the purpose of detecting condition indicators (precursors to failure) and using that data to prognose the health of a system by producing prognostic information to include, for example, the state-of-health (SoH) and remaining useful life (RUL) of the system. Important goals include maximizing the time between when a less than 100% healthy state is first detected and the time when functional failure occurs. Functional failure is defined to occur at the level of degradation at which a device, component, or assembly is no longer operating within specifications. Condition-based data (CBD comprises features and noise: features being anything that can be used for diagnostic and/or prognostic purposes and noise is everything else. The left side of Figure 1 is an oscilloscope capture of the output of a switch-mode power supply: feature data (FD) includes ripple voltage, voltage droop (see the upper trace), and what is often termed to be spikes and glitches, such as that circled in red. Typically, though, those spikes and glitches are what is called a damped-ringing response (right side of Figure 1).

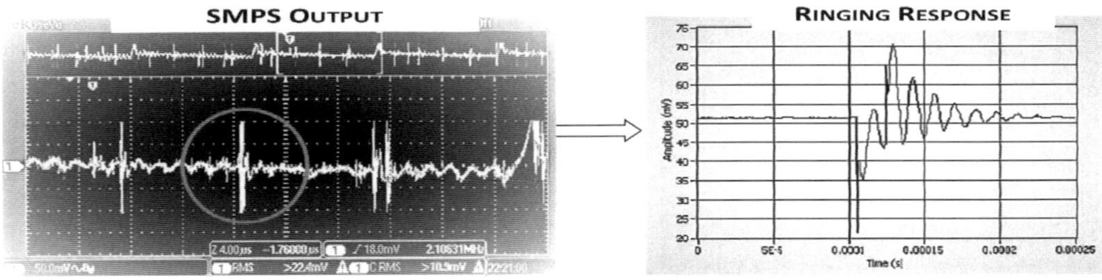

Figure 1. SMPS output voltage (left) and extracted damped-ringing response (right)

A damped-ringing response can be modeled as comprising many features such as a DC voltage, V_{DC}; a response amplitude, A_R; a dampening time constant, τ; a frequency, ω; and a phase shift, ϕ [4],[5]:

$$V_0 = V_{DC} + A_R\{\exp(-t/\tau)\}\{\cos(\omega t + \phi)\} \tag{1}$$

PoF analysis shows that as the filter capacitance of an SMPS degrades, the resonant frequency changes:

$$\omega \approx \omega_0\sqrt{C_0/(C_0 - \Delta C)} \tag{2}$$

Then from $f = \omega/2\pi = FD$,

$$FD = FD_0\sqrt{C_0/(C_0 - \Delta C)} \quad \text{or} \quad FD_i = FD_0\,(\sqrt{P_0/(P_0 - dP_i)}) \tag{3}$$

Substituting P_0 (nominal parameter value) for C_0 and dP (change in parameter) for ΔC, the radical can be replaced by $g(dP, P_0)$:

$$FD_i = FD_0\, g(dP, P_0) \tag{4}$$

Which is interpreted to mean the following: the measurement from a sensor is the nominal value of a parameter times a function that changes in correlation to degradation where dP is the change in value of that parameter. A fault-to-failure progression (FFP) data point is defined to be the following:

$$FFP_i = (FD_i - FD_0 - NM)/FD_0 \tag{5}$$

Where NM is a constant value chosen to mitigate the effects of noise so that an FFP data point > 0 indicative of a degraded state. By collecting terms, and letting $(-FD_0 - NM)/FD_0 = -\mathrm{C}$ (a constant), Eq. (5) becomes

$$FFP_i = FD_i/FD_0 - \mathrm{C}$$

And by substitution using Eq. (4)

$$FFP_i = g(dP, P_0) - \mathrm{C} \tag{7}$$

Fault-to-Failure Progression (FFP) Signature:

A set of FFP data points forms a collection of normalized, dimensionless data obtained by transforming FD points starting with a particular point, m, and ending at a final point, $m+n$, that forms a signature (see Figure 2):

$$\boldsymbol{FFP} = \{FFP_{m,} FFP_{m+1}, \ldots FFP_{m+n-1}, FFP_{m+n}\} \tag{8}$$

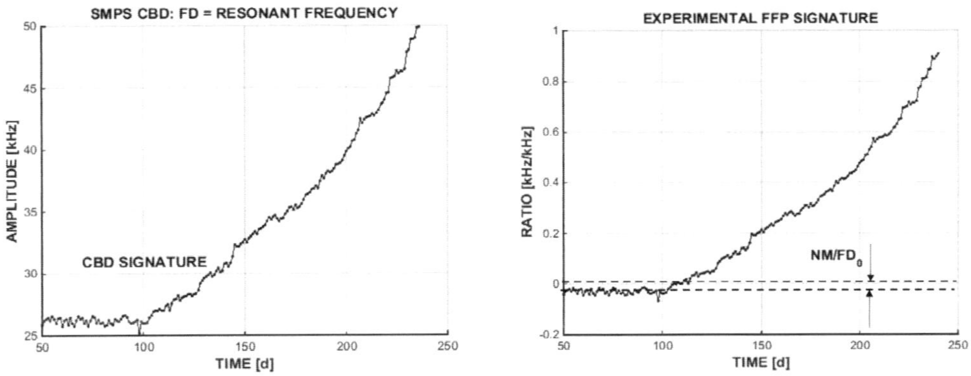

Figure 2. FD extracted from CBD (left) and transformed into FFP data (right)

Degradation Progression Signature (DPS)

A DPS is a collection of transformed FFP signature points: it is a function of a change in value, dP, of a parameter of interest P_0: (1) when there is no degradation, the value of the parameter is unchanged and $dP = 0$; (2) as degradation progresses, the magnitude of dP increases:

$$\boldsymbol{DPS} = \{DPS_{m\,,}\, DPS_{m+1}, \dots, DPS_{m+n}\} \tag{9}$$

$$DPS_i = dP_i/P_0 \tag{10}$$

A major advantage of DPS data is that its characteristic curve is defined by dP_i/P_0 and when that is linear, which is often the case, the DPS is linear. Even when dP_i/P_0 is not linear, a DPS signature curve is more linear than its underlying FFP signature.

2. Transforming FFP Signatures into DPS

Degradation-related signatures result from failure modes that can be modeled as power functions or exponential functions (Table 1). To transform signatures from one form to another, the authors developed seven sets of degradation functions (Table 2): five are power functions and two are exponential functions, which have either increasing or decreasing amplitudes – constant signatures are only applicable in the sense that they indicate the absence of degradation. Signatures having decreasing amplitudes are transformed to a complementary, decreasing amplitudes are transformed to a complementary, increasing amplitude signature and degradation model. Degradation signatures are further classified by whether the signature has a decreasing or increasing slope angle as degradation progresses: a linear, straight-line degradation signature is a special case that is classified as having a constant slope angle. The slope angle is the arc between a tangent to the curve and the horizontal axis. Sample plots of degradation functions and their DPS transforms are shown in Figure 3 through Figure 9.

Table 1. Degradation Functions

Function Set and Type	Decreasing Amplitude	Increasing Amplitude
Set 1: Power-function	$-(dP_i/P_0)^n$	$(dP_i/P_0)^n$
Set 2: Power-function	$1-[1/(1-dP_i/P_0)]^n$	$[1/(1-dP_i/P_0)]^n-1$
Set 3: Power-function	$[1/(1+dP_i/P_0)]^n-1$	$1-[1/(1+dP_i/P_0)]^n$
Set 4: Power-function	$1-(1+dP_i/P_0)^n$	$(1+dP_i/P_0)^n-1$
Set 5: Power-function	$(1-dP_i/P_0)^n-1$	$1-(1-dP_i/P_0)^n$
Set 6: Exponential-function	$1-\exp(dP_i/P_0)$	$\exp(dP_i/P_0)-1$
Set 7: Exponential-function	$\exp(-dP_i/P_0)-1$	$1-\exp(-dP_i/P_0)$

Table 2. Transform Models

Increasing FFP	Increasing DPS	DPS-based FFS
$\{\mathrm{FFP}\}=\{(\boldsymbol{FD_i}-\mathbf{FD_0})/\boldsymbol{FD_0})\}$	$\{\mathrm{DPS}\}=\{(\boldsymbol{dP_i/P_0})\}$	$\{\boldsymbol{FFS_{DPS}}\}=(\{\boldsymbol{DPS}\}/\boldsymbol{FL_{DPS}})\mathbf{100}$
$\boldsymbol{FD_i=FD_0\,g(dP_i,P_0)}$	$\boldsymbol{dP_i/P_0=f(FFP_i)}$	$\boldsymbol{FL_{DPS}=h(FL_{FFP})}$
$\boldsymbol{g(dP_i,P_0)}$	$\boldsymbol{f(FFP_i)}$	$\boldsymbol{h(FL_{FFP})}$
$(dP_i/P_0)^n$	$(FFP_i)^{1/n}$	$(FL_{FFP})^{1/n}$
$[1/(1-dP_i/P_0)]^n-1$	$1-1/(FFP_i+1)^{1/n}$	$1-1/(FL_{FFP}+1)^{1/n}$
$1-[1/(1+dP_i/P_0)]^n$	$1/(1-FFP_i)^{1/n}-1$	$1/(1-FL_{FFP})^{1/n}-1$
$(1+dP_i/P_0)^n-1$	$(FFP_i+1)^{1/n}-1$	$(FL_{FFP}+1)^{1/n}-1$
$1-(1-dP_i/P_0)^n$	$1-(1-FFP_i)^{1/n}$	$1-(1-FL_{FFP})^{1/n}$
$\exp(dP_i/P_0)-1$	$\ln(FFP_i+1)$	$\ln(FL_{FFP}+1)$
$1-\exp(-dP_i/P_0)$	$\ln(1/(1-FFP_i))$	$\ln(1/(1-FL_{FFP}))$

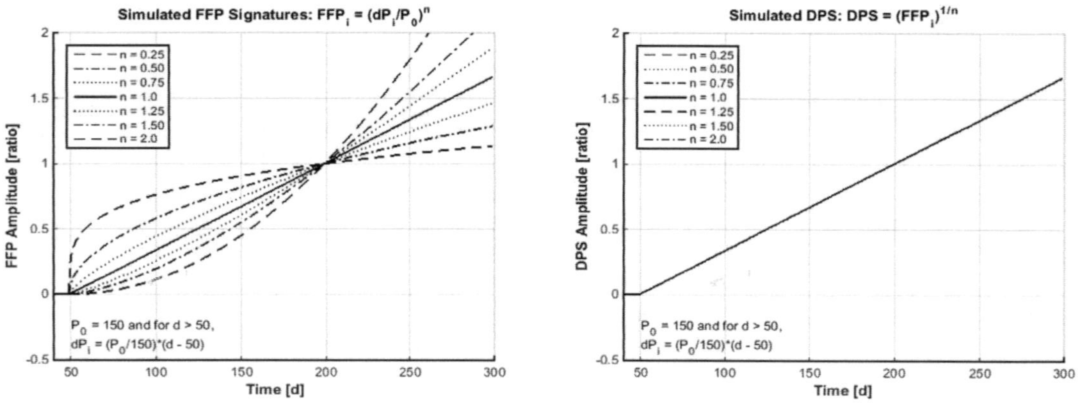

Figure 3. Example FFP and DPS plots for $\boldsymbol{f(dP_i,P_0)=(dP_i/P_0)^n}$

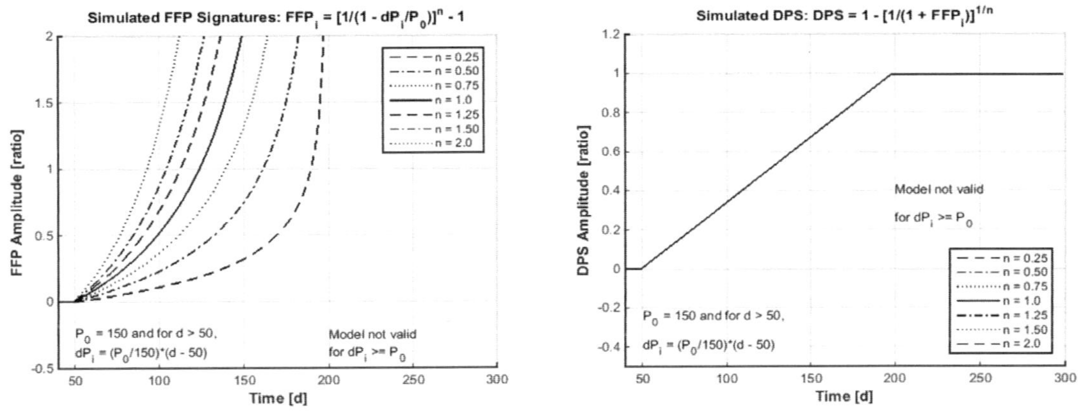

Figure 4. Example FFP and DPS plots for $f(dP_i, P_0) = [1/(1 - dP_i/P_0)]^n - 1$

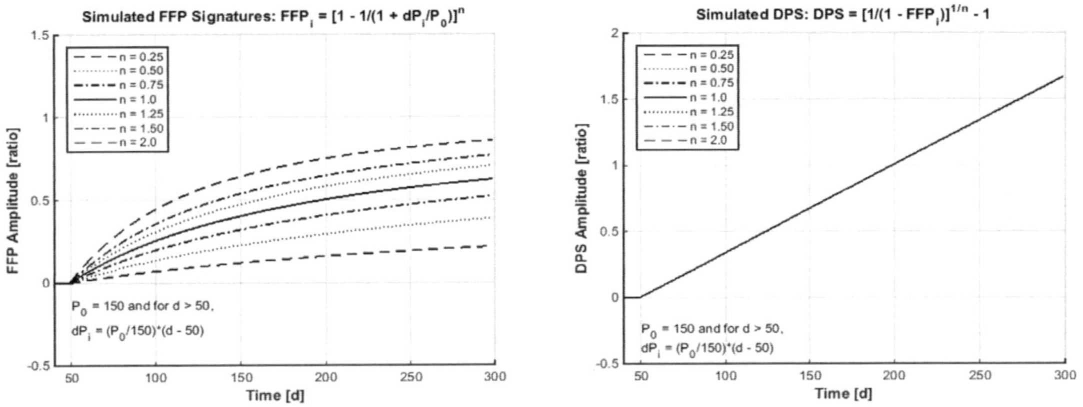

Figure 5. Example FFP and DPS plots for $f(dP_i, P_0) = 1 - [1/(1 + dP_i/P_0)]^n$

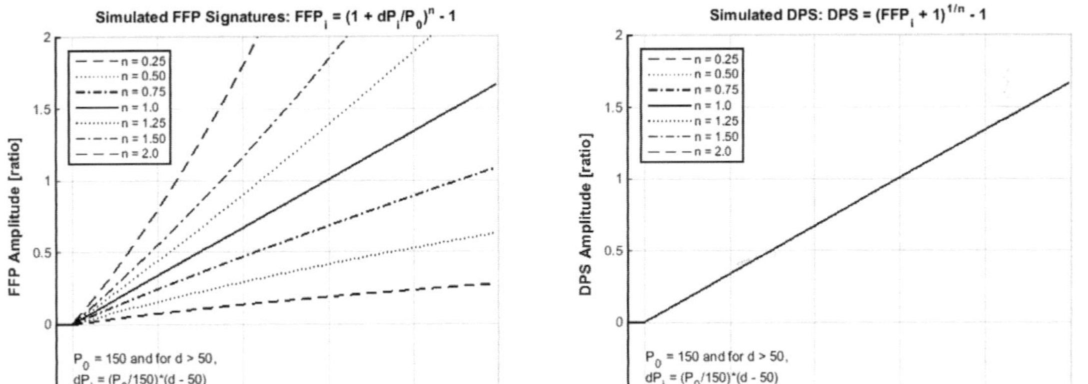

Figure 6. Example FFP and DPS plots for $f(dP_i, P_0) = (1 + dP_i/P_0)^n - 1$

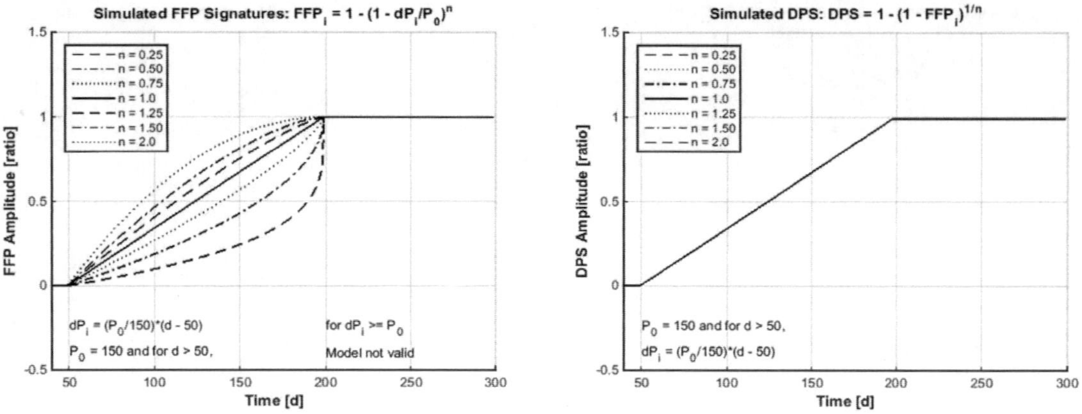

Figure 7. Example FFP and DPS plots for $f(dP_i, P_0) = 1 - (1 - dP_i/P_0)^n$

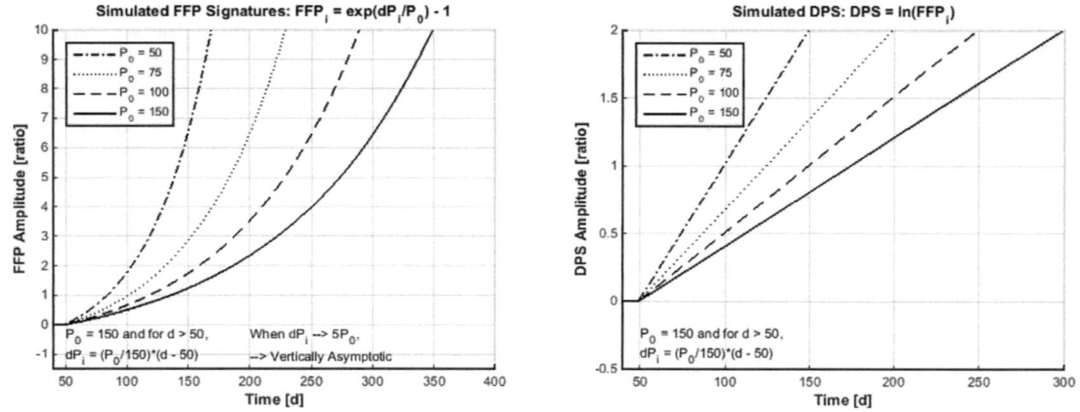

Figure 8. Example FFP and DPS plots for $f(dP_i, P_0) = \exp(dP_i/P_0) - 1$

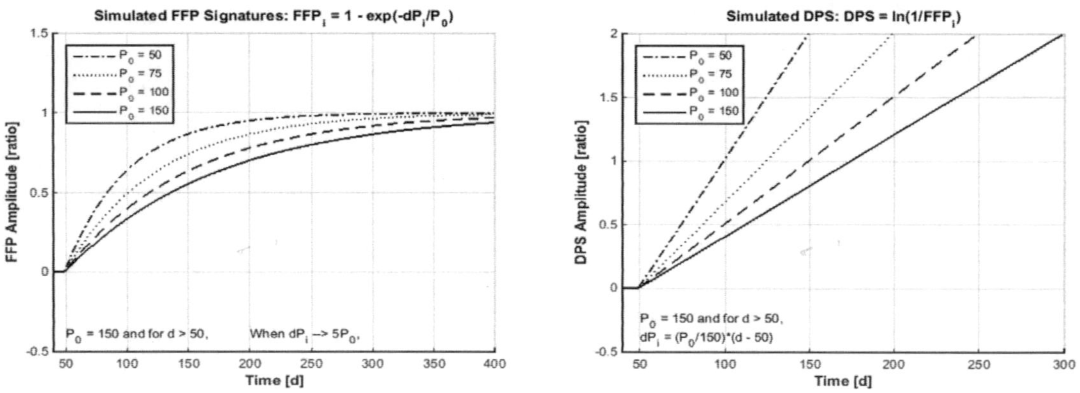

Figure 9. FFP and DPS plots for $f(dP_i, P_0) = \exp(dP_i/P_0) - 1$ and $f(dP_i, P_0) = 1 - \exp(-dP_i/P_0)$

3. TRANSFORMING DPS SIGNATURES INTO FUNCTIONAL FAILURE SIGNATURES (FFS)

The term 'functional failure' rather than simply 'failure' or 'physical failure' should be used in prognostics. Functional failure means reaching a level of degradation at which a device or component and the assembly in which it is located no longer operates within specifications:

functional failure is not absolute: it is dependent not only on the device, component, and assembly, but also on the application and the requirements for that application. A functional-failure signature (FFS) is obtained by doing the following:

1. For each failure mode of interest, collect historical data and/or experimental data and/or simulation data.
2. Transform that data (CBD) into FFP signature data using Eq. (5):

   $$FFP_i = (FD_i - FD_0 - NM)/FD_0 \quad .$$

   FFP signatures reduce modeling complexity because of the following: (1) common amplitude units-of-measure – a relative ratio, (2) ratio values of zero or less are indicative of no degradation – use a sufficiently large value of NM, and (3) values greater than zero are indicative of degradation – a value greater than 1.0 indicates the signal has more than doubled in magnitude.
3. Plot the FFP signature then compare and choose one, or maybe two, signature models from those shown in Figure 3 through Figure 9. For example, consider the example shown in Figure 10:

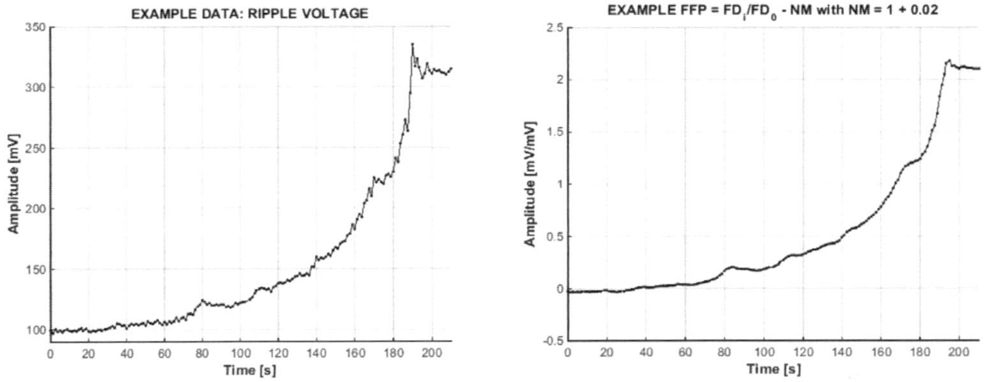

Figure 10: Example CBD-based and FFP signatures

Instead of using physics of failure (PoF) and failure-mode effects analyses (FMEA) to determine the effects of the failure on CBD and then solving for dP_i/P_0, use a heuristic approach: assume the solution is one of those listed in Table 2. For the example shown in Figure 10, by inspection and matching of curves, there are two candidate solutions:

$$DPS_i = 1 - 1/(FFP_i + 1)^{1/n} \tag{11}$$

and

$$DPS_i = \ln(FFP_i + 1) \tag{12}$$

Which results in the plots shown in Figure 11: the left-hand plots using n = 0.5 or n = 1.0 are more linear than the right-hand plot.

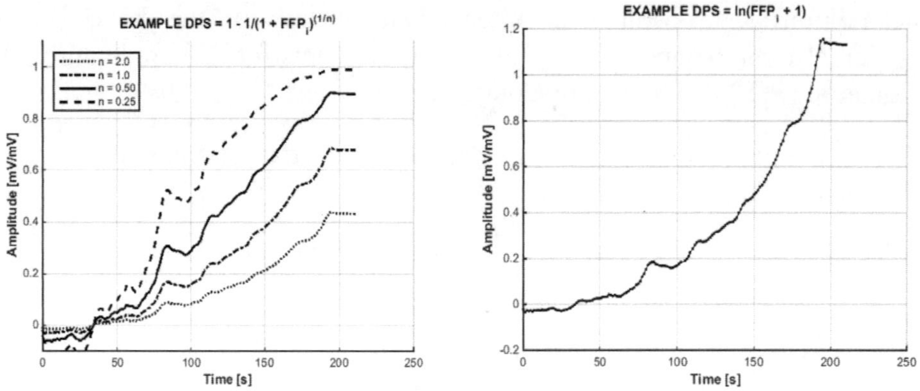

Figure 11. Plots from using Eq. 11 (left side) and Eq. 12 (right side)

4. Select or otherwise determine a value of FFP at which the prognostic-enabled device, component, or assembly is defined as functionally failed: FL_{FFP} = FFP value at failure. That defines the level of degradation at which the device or component is no longer capable of operating within specifications. Then calculate the equivalent failure DPS threshold. For example, suppose you decide to use $FL_{FFP} = 0.90$ and n = 0.5 or 1.0, then from Table 2:

$$FL_{DPS} = h(FL_{FFP}) = 1 - 1/(FL_{FFP} + 1)^{1/n} \tag{13}$$

$$FL_{DPS} = 0.72 \text{ for n} = 0.5 \text{ and } FL_{DPS} = 0.47 \text{ for n} = 1.0 \quad .$$

5. Then transform DPS data into FFS data, which results in the plots shown in Figure 12: note the two plots intersect each other at FFS amplitudes of 0 and 100%.

$$\boldsymbol{FFS}_{\text{DPS}} = (\boldsymbol{DPS}/FL_{DPS})100 \tag{14}$$

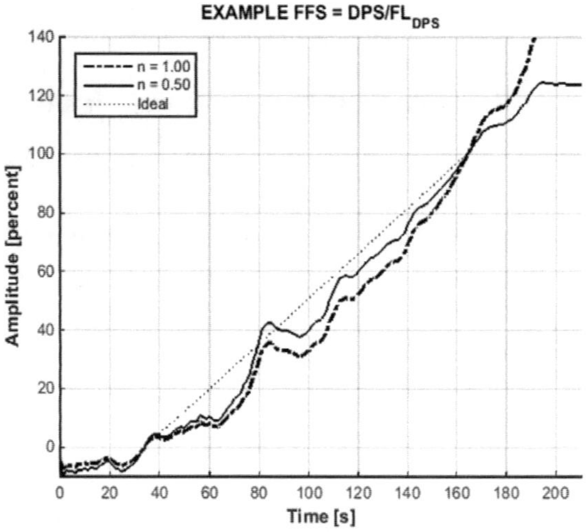

Figure 12. Example FFS plots

Evaluate the **FFS** plots and determine whether to slightly change the value of n: for n = 0.40 instead of 0.50, the **FFS** plot is shown in **Error! Reference source not found.**. Compare that to an ideal FFS transfer curve – the dotted straight line.

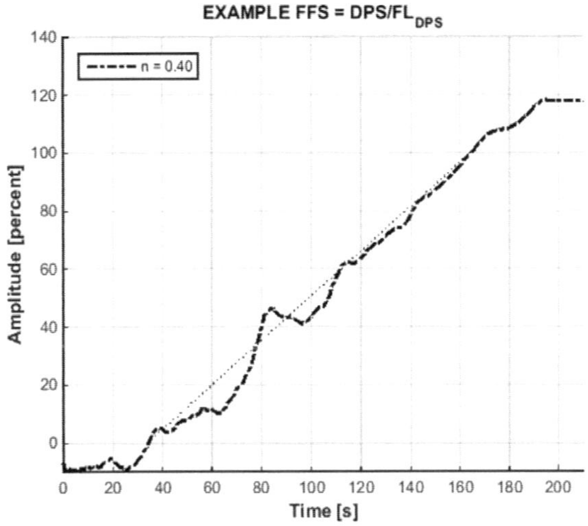

Figure 13. Final FFS plot using Eq. 14 with n = 0.40 in Eq. 11 and Eq. 14

4. VERIFY RESULTS

We define FFS nonlinearity (FNL) as a measure of the deviation between an actual FFS and an ideal FFS: very much like an integral nonlinearity (INL) assessment of an analog-to-digital data converter or transducer output transfer curve. A point-by-point FNL is a measurement of the error between an actual FFS value and the ideal FFS value at that point in time and is obtained using the following [6]-[9]:

$$TTF = t_{FAILURE} - t_{ONSET} \quad \text{calculate time-to-failure} \tag{15}$$

then for each point in the set $\{FFS_i\}$, we create an ideal point,

$$IDEAL\_FFS_i = 100\,(t_i - t_{ONSET})/TTF \tag{16}$$

and calculate point-by-point nonlinearity and total nonlinearity:

$$FNL_i = FFS_i - IDEAL\_FFS_i \tag{17}$$

Total FNL error (FNL_E) provides an assessment of the nonlinearity of the FFS curve:

$$FNL_E = \max(\text{abs}(\{FNL_i\})) \tag{18}$$

Figure 14 shows the non-linearity plot for transfer curve shown in Figure 13.

Despite the relatively large variation in the data between the approximate times of 60 and 90 seconds, the FNL_E is less than 13 percent, which in the assessment of the authors is quite remarkable. This data, when input to a prediction algorithm employing estimation techniques such as Kalman filtering, is very likely to produce very accurate prognostic information.

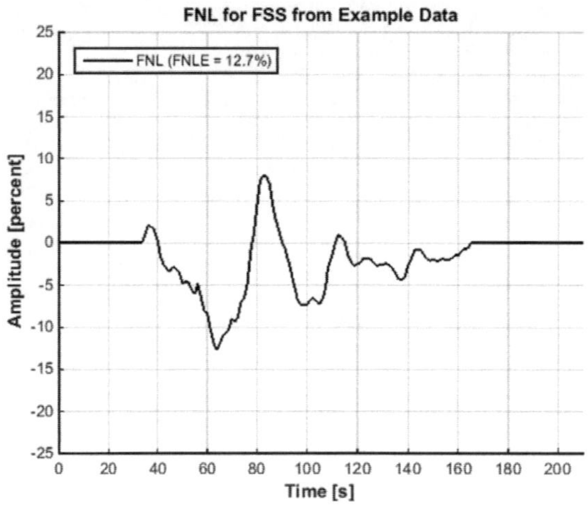

Figure 14. Non-linearity for FFS shown in Figure 13

5. SUMMARY

In this paper a concept of signatures is introduced and an example given to illustrate CBD features that can be extracted as FD to create an FFP signature, which can be further transformed into DPS data and then into FFS data – one data point at time. Instead of rigorous, time-consuming, and error-prone PoF and FMEA analysis to derive transformation models, the paper describes a heuristic-based approach for advantageous used of a set of already derived models (Table 1 and Table 2) to transform CBD into FFS data for use as a transfer curve for producing prognostic information. An FFS progresses from 0 percent or less (no degradation) to 100 percent or higher (functionally failed). Next a method is introduced to evaluate non-linearity of FFS data (FNL): that method is based one used for evaluating the nonlinearity of data converters.

Acknowledgements

The authors thank Naval Air, Naval Sea, U.S. Army, U.S. Air Force, and NASA research centers for their support and funding of multiple projects that led to the results described and shown in this paper.

References

[1] Hofmeister, J., Wagoner, R. and Goodman, D. (2013) "Prognostic Health Management (PHM) of Electrical Systems using Conditioned-based Data for Anomaly and Prognostic Reasoning," *Chemical Engineering Transactions*, Vol. 33, pp. 992-996.

[2] Hofmeister, J. Szidarovszky, F. and Goodman, D. (2017) "An Approach to Processing Condition-based Data for Use in Prognostic Algorithms," 2017 Machine Failure Prevention Technology, Virginia Beach, VA, 15-18 May 2017.

[3] Medjaher, K. and Zerhouni, N. (2013) "Framework for a Hybrid Prognostics," *Italian Association of Chemical Engineering, Chemical Engineering Transactions*, DOI: 10.3303/CET1333016, pp. 91-96, 2013.

[4] Judkins, J. B., Hofmeister, J. and Vohnout, S. (2007) "A Prognostic Sensor for Voltage Regulated Switch-Mode Power Supplies," IEEE Aerospace Conference 2007, Big Sky, MT, 4-9 Mar. 2007, Track 11-0804, pp. 1-8.

[5] Judkins, J. B. and Hofmeister, J. P. (2007) "Non-invasive Prognostication of Switch Mode Power Supplies with Feedback Loop Having Gain," IEEE Aerospace Conference 2007, Big Sky, MT, 4-9 Mar. 2007.

[6] Erickson, R. (1999), *Fundamentals of Power Electronics,* Norwell, MA: Kluwer Academic Publishers.

[7] TI (1995) "Understanding Data Converters," Application Report SLAA013, Texas Instruments, Inc., 1999.

[8] Carr, J.J. and Brown, J.M. (2000), *Introduction to Biomedical Equipment Technology,* 4th Ed., Upper Saddle River, New Jersey, Prentice Hall

[9] Jenq, Y.C. and Li, Qiong, (2002) "Differential Non-linearity, Integral Non-Linearity, and Signal to Noise Ratio of an Analog to Digital Converter," Department of Electrical and Computer Engineering, Portland State University, P.O. Box 751, Portland, OR. 97207, USA.

Degradation Signature Modeling

James P. Hofmeister
Ridgetop Group, Inc.
3580 West Ina Road
Tucson, AZ 86741
520-742-3300x107
hoffy@ridgetopgroup.com

Douglas L. Goodman
Ridgetop Group, Inc.
3580 West Ina Road
Tucson, AZ 86741
520-742-3300
doug@ridgetopgroup.com

Ferenc Szidarovszky
Ridgetop Group, Inc.
3580 West Ina Road
Tucson, AZ 86741
520-742-3300
szidarka@gmail.com

Abstract—**This paper presents models and methods to transform conditioned-based data (CBD) signatures into functional-failure signatures (FFS) that are particularly amenable to processing by state-estimator prediction algorithms. Failure modes generate characteristic CBD signatures that are correlated to changes in value of a parameter, such as stress or force, as degradation progresses. Features such as amplitude or frequency related to parameters that change are extracted from CBD signatures and transformed into fault-to-failure progression (FFP) signatures: a function of feature data (FD). FFP signatures are then transformed into degradation-progression signatures (DPS): a function of a change in the value of a parameter. A DPS, absent noise, has a characteristic linear curve: a straight line progressing from 0 or less (no degradation) to a value defined as a level of degradation at which the degrading component, and the assembly it is in, no longer functions within specification(s): functional failure occurs.**

1. Introduction

Prognostics includes the acquisition of data in support of reliable condition-based monitoring (RCM) using condition-based data (CBD) having features that can be used to create signatures that have characteristic curves related to changes in value(s) as a parameter degrades. A switched-mode power supply (SMPS) having a filter capacitance that degrades is used as an example to illustrate how a characteristic CBD signature is transformed into a fault-to-failure progression (FFP) signature; how that FFP signature is transformed into a degradation-progression signature (DPS); and how that DPS is transformed into a functional-failure signature (FFS) that is particularly amenable for processing by prediction algorithms to produce prognostic information such as remaining useful life (RUL) and state-of-health (SoH) estimates.

Seven sets of models are presented to generate CBD signatures, their characteristic curves, their related models for transforming CBD signature data into DPS data, and then into FFS data. The models are used to transform signatures from one form to other forms. Plots are included to illustrate the usefulness of such transforms.

2. Prognostics

Prognostics is an ability to accurately detect and report future failures in systems. The purpose of prognostics is to detect degradation and create prognostic information such as estimates of SoH and RUL of systems for the following benefits: (1) provide advance warning of failures; (2) minimize unscheduled maintenance; (3) predict the time to perform preventive replacement; (4) increase maintenance cycles and operational readiness; (5) reduce sustainment costs by decreasing inspection, inventory, and down-time costs; and (6) increase reliability by improving the design and logistic support of existing systems [1]-[3].

MFPT 2018, Virginia Beach, VA, 15-17 May 2018

Prognostics includes data acquisition by sensors and data manipulation by processing within a sensor framework (see Figure 1). Additional processing within a feature-vector framework to include detection, isolation, and identification is performed to extract feature data (FD) that are condition indicators that comprise leading indicators of failure (signatures). Health assessment and prognostic assessment are performed within a prediction framework. The sensor framework, feature-vector framework, prediction framework, and a control and data flow framework comprise a prognosis subsystem within a prognosis and health management (PHM) system [4],[5]. There are three classical approaches for handling data in PHM system: – model-driven, data-driven, and hybrid-driven – and a CBD approach that often employs analysis and modeling techniques such as reliability modeling, physics-of-failure (PoF) analysis, and failure-mode effects analysis (FMEA). The reliability of any approach for prognostics is dependent on reliable and accurate prognostic information, which depends on how accurately the sensing system collects and conditions data and how accurately that data is processed by prediction algorithms in producing prognostic information. Curvilinear CBD that is reliably and accurately transformed into linearized data is very useful for processing to produce reliable and accurate prognostic information.

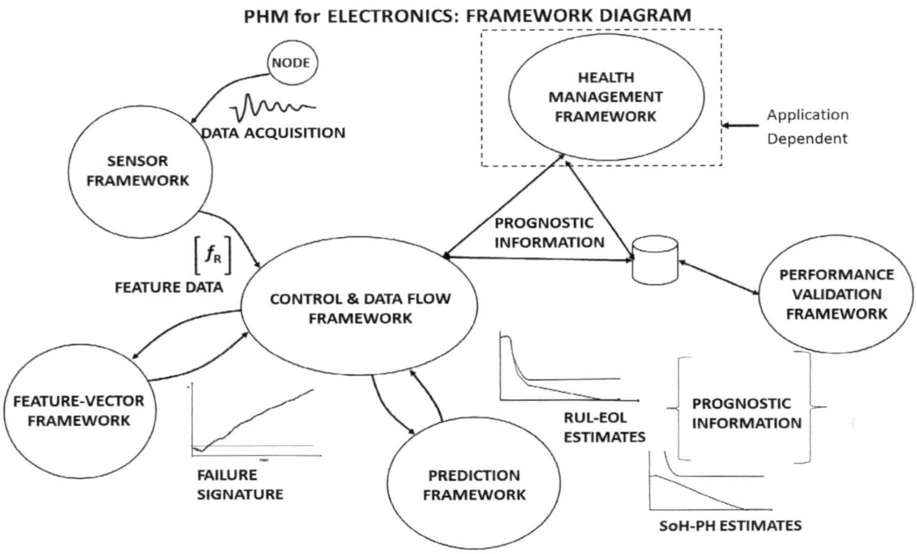

Figure 1. Example PHM system showing core frameworks (after [5])

The focus of this paper is transforming condition-based data (CBD) signatures into fault-to-failure progression (FFP) signatures, degradation-progression signatures (DPS), and functional-failure signatures (FFS) [6]-[8]. FFS data as input into a prediction framework, such as that shown in Figure 1, increases the reliability and accuracy of prognostic information such as RUL and SoH estimates.

3. SIGNATURES

A signature characterizes a feature of interest, such as feature data (*FD*) that changes in amplitude over a period of time: $T = \{t_m, t_{m+1}, \ldots t_{m+n}\}$. Amplitude refers to a characteristic value such as voltage, current, resistance, force, energy and so on that changes as the magnitude of damage/degradation increases over time:

$$\boldsymbol{FD} = \{FD_{m,}\, FD_{m+1}, \dots\, FD_{m+n-1}, FD_{m+n}\} \tag{1}$$

Four sets of signature data of interest related to signals and prognostic processing are identified: CBD (in terms of FD), FFP, DPS, and FFS. An FFP signature is a transform of a CBD signature to reduce modeling complexity and a DPS signature is a transform of an FFP signature to further reduce modeling complexity and to increase the reliability and accuracy of prognostic information. An FFS is a transform of either an FFP or a DPS signature that is particularly amenable to processing by prediction algorithms because of commonality of FFS signatures and values.

Condition-based Data (CBD) Signature

Referring to Figure 2, a sensor is located at a node of prognostic interest to monitor and collect data. An example is the output node of a switch-mode power supply (SMPS): the output voltage comprises multiple types of features and noise, including damped-ringing responses, switching noise from power transistors, ripple voltage, pulse-width modulator effects, effects of voltage and current regulation and feedback, and background noise. Sensors can be designed to support the isolation and extraction of leading indicators of failure, such as a damped-ringing response seen in the output of a SMPS [6],[7], [9]-[12].

$$\mathbf{V} = CBD_1 + CBD_2 + \cdots CBD_n + N_1 + N_2 + N_m \tag{2}$$

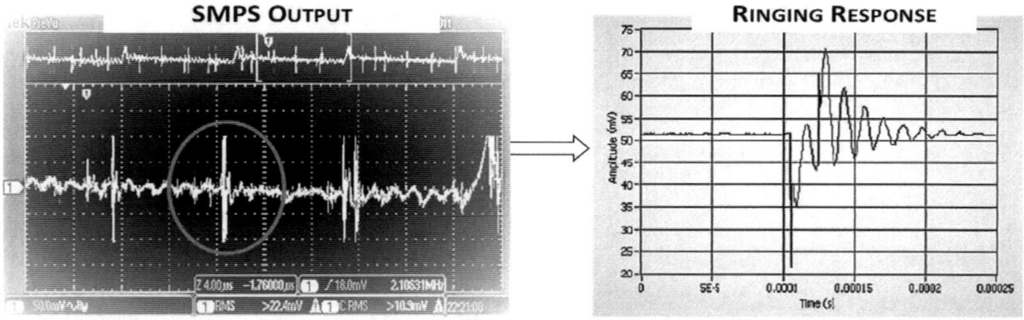

Figure 2. SMPS output voltage (left) and extracted damped-ringing response (right)

Signal processing isolates, extracts, conditions, and transforms data to create a particular *FD* value, such as the resonant frequency of a damped-ringing response. A collection of FD points over time is a CBD signature: an example of which is shown in Figure 3.

An idealized damped-ringing response can be modeled as multiple features of interest including the following: a DC voltage V_{DC}, a response amplitude A_R, a dampening time constant τ, a frequency ω, and a phase shift ϕ [10],[11]:

$$V_0 = V_{DC} + A_R\{exp(-t/\tau)\}\{cos(\omega t + \phi)\} \tag{3}$$

PoF analysis shows that as the filter capacitance of an SMPS degrades, the resonant frequency changes [10],[11]:

$$\omega \approx \omega_0 \sqrt{C_0/(C_0 - \Delta C)} \tag{4}$$

Then from $f = \omega/2\pi = FD$],

$$FD = FD_0 \sqrt{C_0/(C_0 - \Delta C)} \tag{5}$$

Fault-to-Failure Progression (FFP) Signature

Referring to Figure 3, an FFP signature is a collection of normalized, dimensionless data obtained by transforming FD points starting with a particular point, *m*, and ending at a final point, *m+n*:

$$\boldsymbol{FFP} = \{FFP_{m,} FFP_{m+1}, \ldots FFP_{m+n-1}, FFP_{m+n}\} \tag{6}$$

Where the FFP point for an *i*-th FD point is given by the following in which *NM* is a noise margin (NM/FD_0 in Figure 3) [7].

$$FFP_i = (FD_i - FD_0 - NM)/FD_0 \tag{7}$$

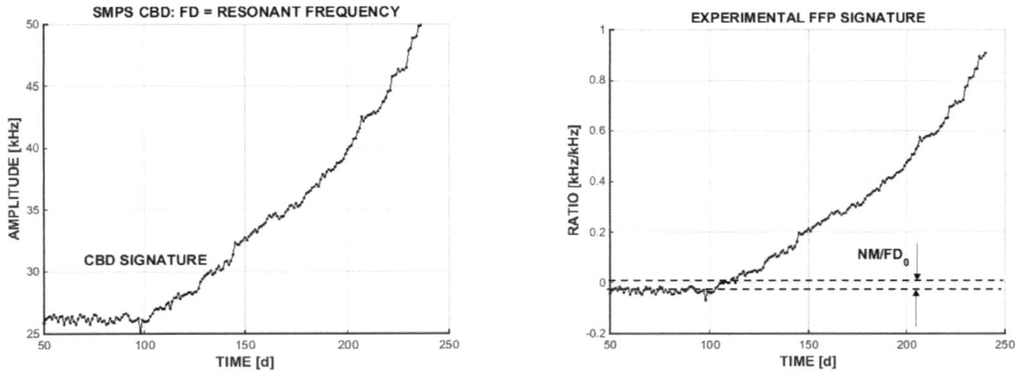

Figure 3. FD extracted from CBD (left) and transformed into FFP data (right)

Compared to CBD signatures, FFP signatures are more amenable to prognostic processing because normalized, dimensionless data simplifies modeling and processing. Values at or below zero indicate an absence of detectable degradation and values greater than zero indicate a presence of degradation. Functional failure occurs when degradation reaches a defined level: for example, 0.60 for one application and 0.70 for another application.

FFP-based Functional-failure signature (FFS)

Since functional failure occurs when degradation reaches a defined FFP value, a prediction algorithm that processes FFP signature data needs application-specific knowledge such as a failure threshold value and/or FFP model. To address that consequence, we introduce the FFS signature produced by dividing FFP signature data by a defined failure level (FL) and multiplying the result by 100:

$$\boldsymbol{FFS}_{\text{FFP}} = (\boldsymbol{FFP}/FL)100 \tag{8}$$

Referring to Figure 4, while such FFS data has the same, generally curvilinear, characteristic shape as the FFP signature it is based on, the signature simplifies processing by prediction algorithms: (1) $FFS_i \leq 0$ indicates no degradation; (2) $0 < FFS_i < 100$ indicates degradation; and (3) $FFS_i \geq 100$ indicates functionally failed.

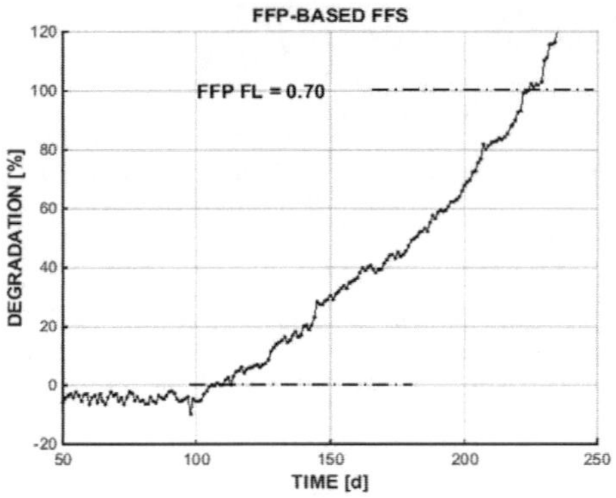

Figure 4 FFS based on the FFP in Figure 3: FL = 0.7

A disadvantage of FFP-based FFS data is the signature is generally curvilinear: it has the same characteristic curve as the original CBD signature (compare Figure 3 and Figure 4).

Degradation Progression Signature (DPS)

A DPS is a collection of transformed FFP signature points: it is a function of a change in value, dP, of a parameter of interest P_0: (1) when there is no degradation, the value of the parameter is unchanged and $dP = 0$; (2) as degradation progresses, the magnitude of dP increases:

$$\boldsymbol{DPS} = \{DPS_{m}, DPS_{m+1}, \ldots, DPS_{m+n}\} \tag{9}$$

$$DPS_i = dP_i/P_0 \tag{10}$$

From Eq. (5) and Eq. (7), for our exemplary SMPS, and using Eq. (9) and Eq. (10) as shown in Appendix A, we get the following:

$$FD_i = FD_0 \left(\sqrt{P_0/(P_0 - dP_i)}\right) \tag{10}$$

Solving for dP_i/P_0 in Eq. (10) leads to the following (see Appendix A):

$$DPS_i = dP_i/P_0 = 1 - (FD_0\,/FD_i)^2 \tag{11}$$

A major advantage of DPS data is that its characteristic curve is defined by dP_i/P_0 and when that is linear, which is often the case, the DPS is linear. Even when dP_i/P_0 is not linear, a DPS signature curve is more linear than its underlying FFP signature. Figure 5 (left) plots the ***DPS*** from using Eq. (11) to transform the ***FFP*** signature data.

DPS-based FFS

FFS data based on DPS data is functionally the same as FFS data based on FFP signature data:

$$\boldsymbol{FFS}_{\mathrm{DPS}} = (\boldsymbol{DPS}/FL_{DPS})100 \tag{12}$$

To ensure equivalency, the failure level for DPS data must be derived from the FFP-based failure level. For this example (see Appendix B),

$$FL_{DPS} = 1 - (1/(\,1 + FL_{FFP}))^2 \tag{13}$$

As example and referring back to Figure 3, suppose we defined functional failure to occur at FFP = 0.70, then from Eq. (13) we calculate the equivalent DPS-based failure level to be 0.65 with plotted FFS results shown in Figure 5: $FL_{DPS} = 1 - (1/1.70)^2 = 0.65$

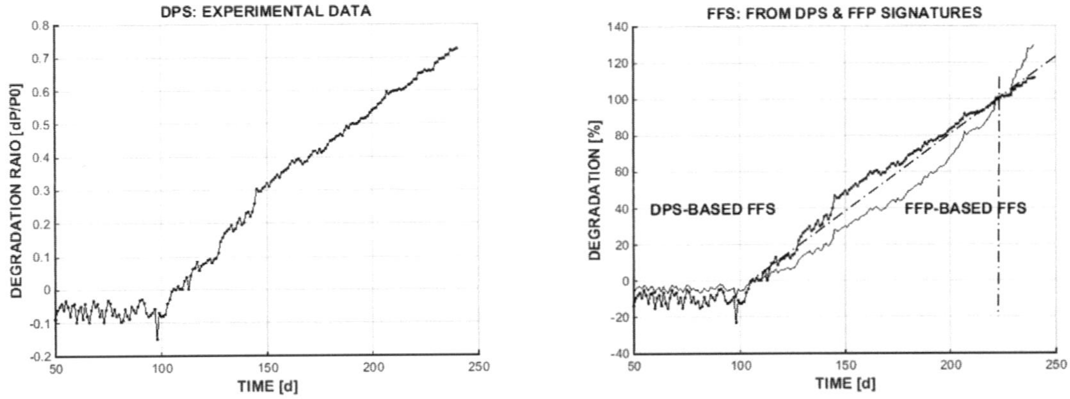

Figure 5. DPS plot (left) and FFS plots showing equivalent failure levels

Model Verification

It is important to verify models used in a PHM system: simulate models against experimental and/or actual fielded data. When FFS data is input to prediction algorithms in your PHM system, the resultant prognostic information needs to meet the accuracy, convergence, and reliability requirements. Model verification starts with a general form:

$$FD = FD_0 f(dP, P_0) \tag{14}$$

where the degradation function, $f(dP, P_0)$, depends on, for example, the component that is degrading and the failure mode. and the physics-of-failure. The degradation function for the example of Eq. (10) is the following: $f(dP, P_0) = \sqrt{P_0/(P_0 - dP_i)}$

Compare signatures—One verification step is to compare ideal model signatures to those of actual and/or experimental data. For example, Figure 6 shows there is a difference between the plots of the exemplary CBD signature and the plots of an ideal CBD from Eq. (10).

Explain variations—Another verification step is to explain variations/differences between what is obtained and what was expected: then either accept and/or remedy any variation. The major difference between the plots of the experimental data and the simulated (ideal) data was determined to be (A) the effects of the regulation feedback loop in the power supply and (B) step-like changes resulting from a test-bed method for injecting faults (loss of capacitance). This comparison explains the nonlinearity of the DPS-based FFS and an ideal, straight-line FFS – the 'dot-dash' straight line in Figure 5 (right).

Non-linearity—Yet another verification step is to evaluate the nonlinearity of an FFS signature compared to an ideal straight line. We define FFS nonlinearity (FNL) as a measure of the deviation between an actual FFS and an ideal FFS, a straight line: very much like an integral nonlinearity (INL) assessment of an analog-to-digital data converter (ADC) or transducer output transfer curve [13]-[15].

A point-by-point FNL is a measurement of the error between an actual FFS value and the ideal FFS value at that point in time and a total FNL error (FNL_E) provides an assessment of the nonlinearity of the FFS curve (see Figure 7). As seen, the maximum non-linearity error of the DPS is 13.1% - even before processing by a prediction algorithm employing, for example, Kalman or Extended Kalman filtering – and, the error is reduced to less than 10% within half of the prognostic distance: very accurate results.

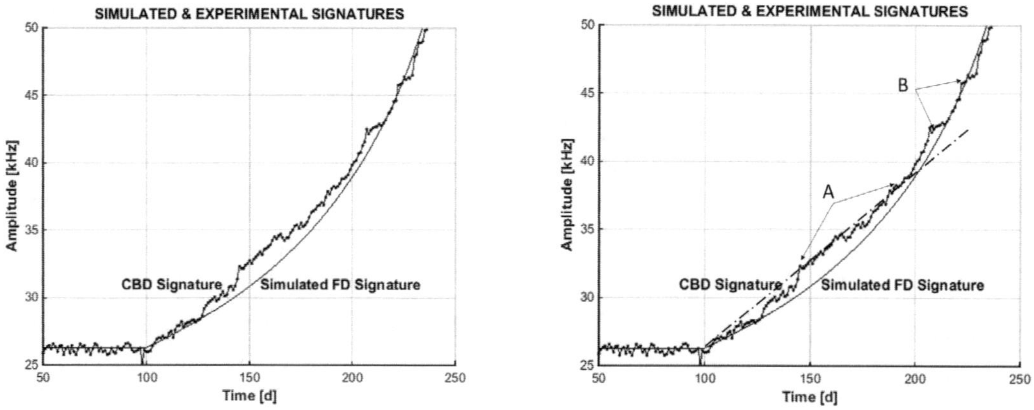

Figure 6. FFS comparison plots (left) with differences noted (right)

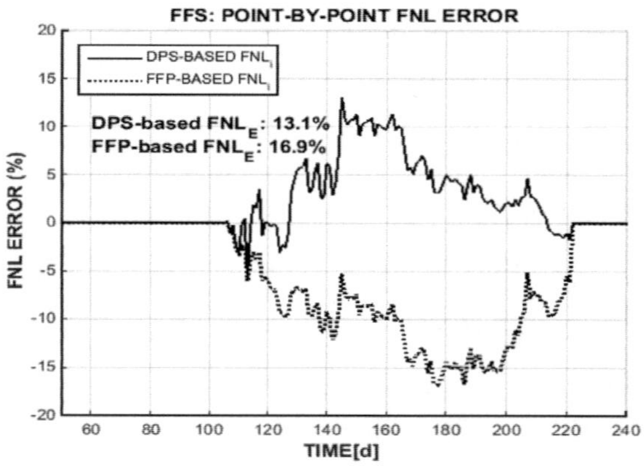

Figure 7. Example of point-by-point FNL for FFS data

4. DPS AND FFS MODELING

Degradation-related signatures result from failure modes that can be modeled as power functions or exponential functions.

Degradation Functions

To transform signatures from one form to another, the authors developed seven sets of degradation functions (see Table 1): five are power functions and two are exponential functions, which have either increasing or decreasing amplitudes – constant signatures are only applicable in the sense that they indicate the absence of degradation. Signatures having decreasing amplitudes are transformed to a complementary, decreasing amplitudes are transformed to a complementary, increasing amplitude signature and degradation model. Degradation signatures are further classified by whether the signature has a decreasing or increasing slope angle as degradation progresses: a linear, straight-line degradation signature is a special case that is classified as having a constant slope angle. The slope angle is the arc between a tangent to the curve and the horizontal axis. Sample plots of degradation functions and their DPS transforms are shown in Figure 8 through Figure 14.

Table 1. Degradation Functions

Function Set and Type	Decreasing Amplitude	Increasing Amplitude
Set 1: Power-function	$-(dP_i/P_0)^n$	$(dP_i/P_0)^n$
Set 2: Power-function	$1-[1/(1-dP_i/P_0)]^n$	$[1/(1-dP_i/P_0)]^n-1$
Set 3: Power-function	$[1/(1+dP_i/P_0)]^n-1$	$1-[1/(1+dP_i/P_0)]^n$
Set 4: Power-function	$1-(1+dP_i/P_0)^n$	$(1+dP_i/P_0)^n-1$
Set 5: Power-function	$(1-dP_i/P_0)^n-1$	$1-\ (1-dP_i/P_0)^n$
Set 6: Exponential-function	$1-\exp(dP_i/P_0)$	$\exp(dP_i/P_0)-1$
Set 7: Exponential-function	$\exp(-\,dP_i/P_0)-1$	$1-\exp(-dP_i/P_0)$

Table 2. Transform Models

Increasing FFP	Increasing DPS	DPS-based FFS
$\{\mathbf{FFP}\} = \{(\boldsymbol{FD_i} - \mathbf{FD_0})/\boldsymbol{FD_0})\}$	$\{\mathbf{DPS}\} = \{(\boldsymbol{dP_i/P_0})\}$	$\{\boldsymbol{FFS_{DPS}}\} = (\{\boldsymbol{DPS}\}/\boldsymbol{FL_{DPS}})\mathbf{100}$
$\boldsymbol{FD_i = FD_0\ g(dP_i, P_0)}$	$\boldsymbol{dP_i/P_0\ = f(FFP_i)}$	$\boldsymbol{FL_{DPS} = h(FL_{FFP})}$
$\boldsymbol{g(dP_i, P_0)}$	$\boldsymbol{f(FFP_i)}$	$\boldsymbol{h(FL_{FFP})}$
$(dP_i/P_0)^n$	$(FFP_i)^{1/n}$	$(FL_{FFP})^{1/n}$
$[1/(1-dP_i/P_0)]^n-1$	$1-1/(FFP_i+1)^{1/n}$	$1-1/(FL_{FFP}+1)^{1/n}$
$1-[1/(1+dP_i/P_0)]^n$	$1/(1-FFP_i)^{1/n}-1$	$1/(1-FL_{FFP})^{1/n}-1$
$(1+dP_i/P_0)^n-1$	$(FFP_i+1)^{1/n}-1$	$(FL_{FFP}+1)^{1/n}-1$
$1-\ (1-dP_i/P_0)^n$	$1-(1-FFP_i)^{1/n}$	$1-(1-FL_{FFP})^{1/n}$
$\exp(dP_i/P_0)-1$	$\ln(FFP_i+1)$	$\ln(FL_{FFP}+1)$
$1-\exp(-dP_i/P_0)$	$\ln(1/(1-FFP_i))$	$\ln(1/(1-FL_{FFP}))$

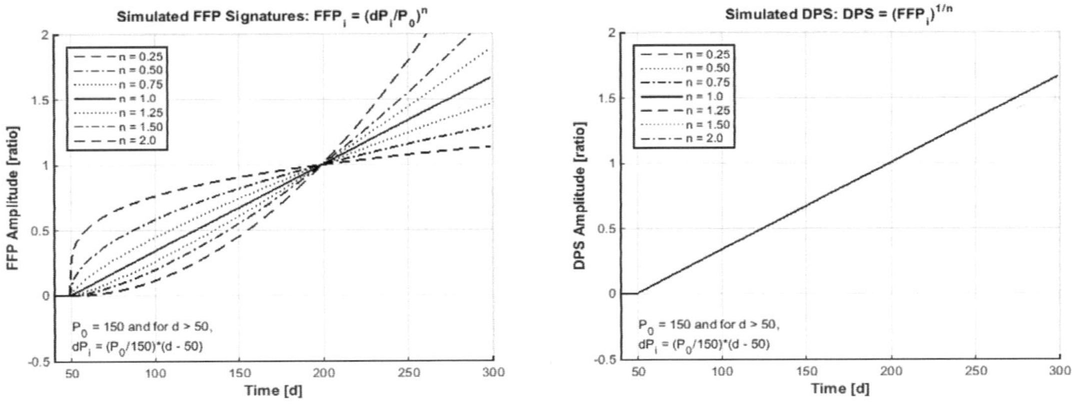

Figure 8. Example FFP and DPS plots for $\boldsymbol{f(dP_i, P_0) = (dP_i/P_0)^n}$

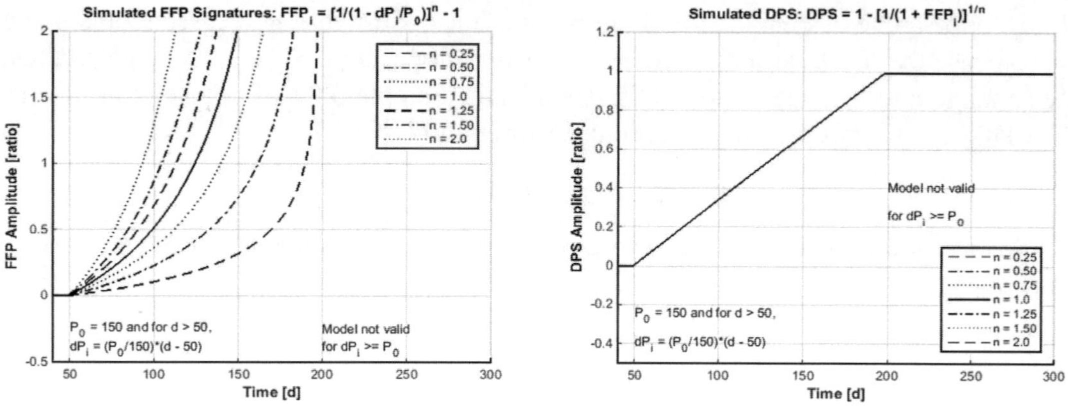

Figure 9. Example FFP and DPS plots for $f(dP_i, P_0) = [1/(1 - dP_i/P_0)]^n - 1$

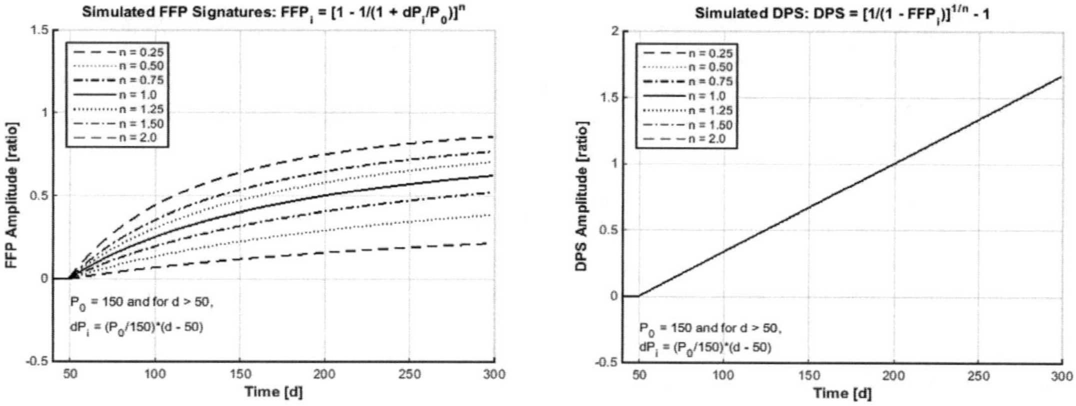

Figure 10. Example FFP and DPS plots for $f(dP_i, P_0) = 1 - [1/(1 + dP_i/P_0)]^n$

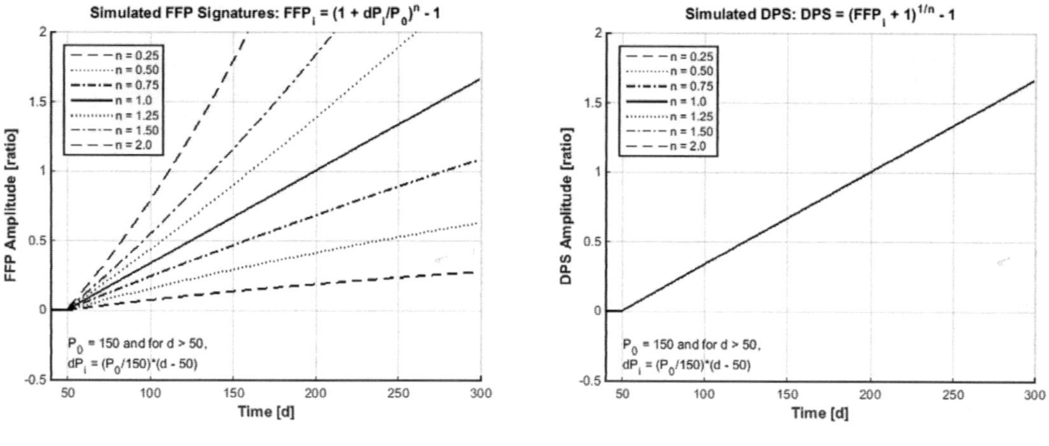

Figure 11. Example FFP and DPS plots for $f(dP_i, P_0) = (1 + dP_i/P_0)^n - 1$

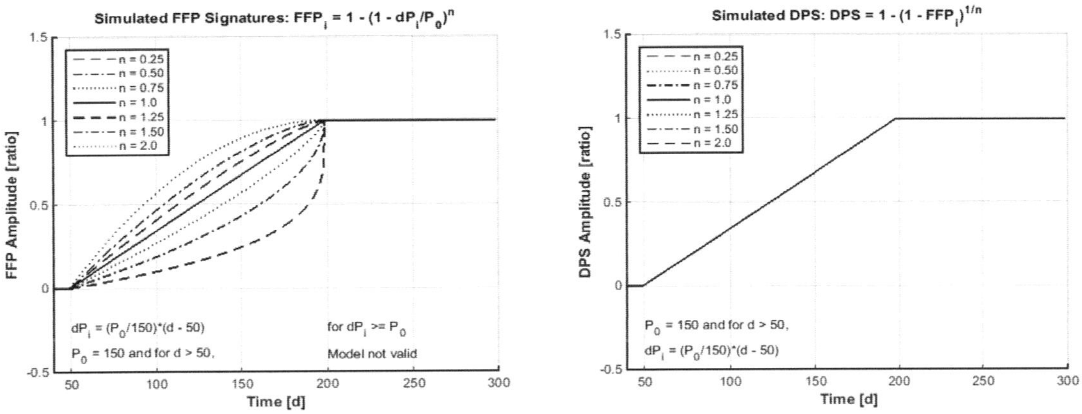

Figure 12. Example FFP and DPS plots for $f(dP_i, P_0) = 1 - (1 - dP_i/P_0)^n$

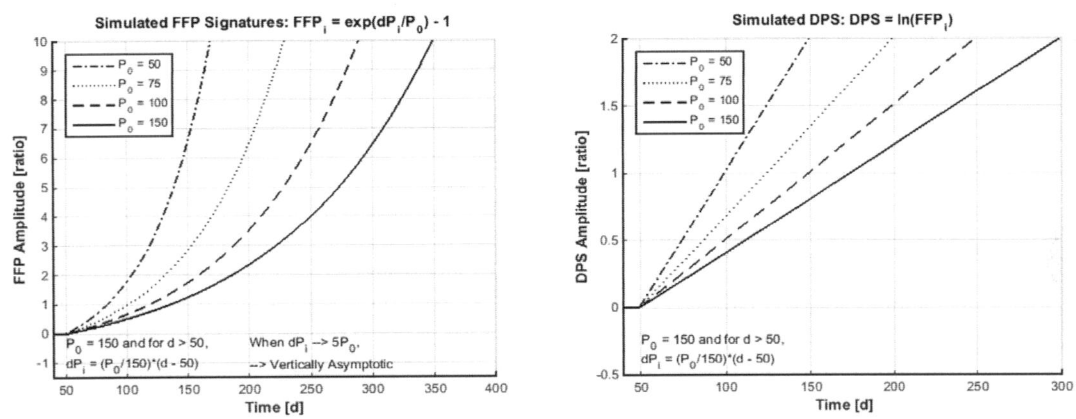

Figure 13 Example FFP and DPS plots for $f(dP_i, P_0) = \exp(dP_i/P_0) - 1$

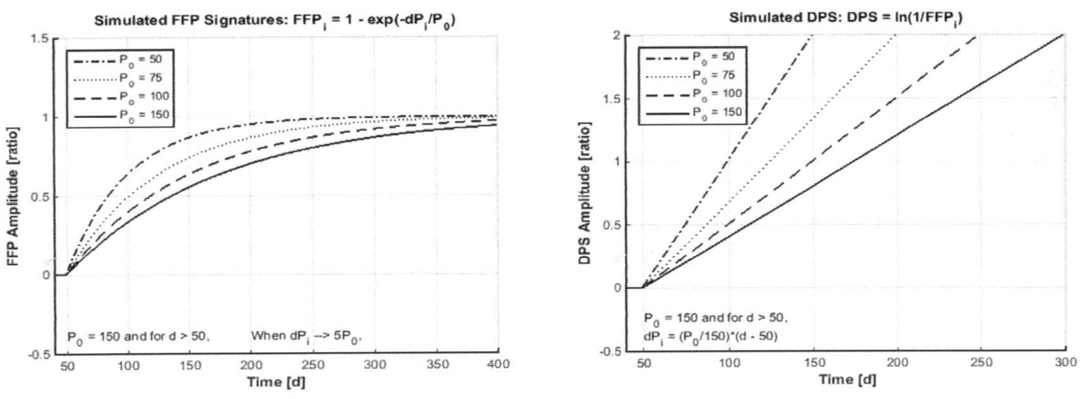

Figure 14. FFP and DPS plots for $f(dP_i, P_0) = \exp(dP_i/P_0) - 1$ and $f(dP_i, P_0) = 1 - \exp(-dP_i/P_0)$

Modeling CBD Signatures to FFS Data

A general procedure has been developed to use the seven sets of degradation functions to transform CBD signatures into DPS-based FFS data to be input to prediction algorithms (see Figure 15).

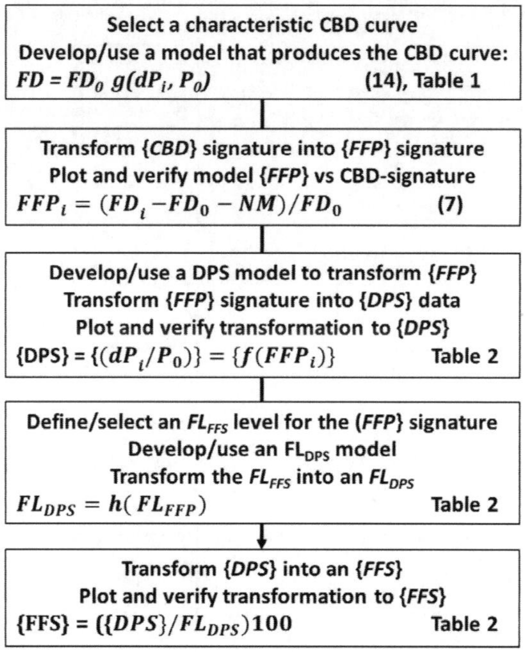

Figure 15. Procedure for transforming CBD signatures

Functional Failure Signature (FFS)

To create a set of DPS points, use a generalized form of Eq. (11) and a set of models from column 1 and column 2 of Table 1: $\{DPS\} = \{(dP_i/P_0)\}$ and $dP_i/P_0 = f(FFP_i)$. Then select/define a threshold FFP value that defines functional failure: a level at which the component, assembly, and/or system is no longer operating within specifications, $FL_{FFP} = FFP_i$ = threshold for functional failure.

Then transform FL_{FFP} into FL_{DPS} using a generalized form of Eq. (13) and a model from column 3 of Table 2: $FL_{DPS} = h(FL_{FFP})$. This transform is necessary to ensure that regardless of whether an FFS is based on an FFP signature or on a DPS, functional failure occurs at same point in time. For example, the plots in Figure 16 illustrate that regardless of whether the FFP results from a power or exponential function of degradation, functional failure occurs, as expected, at the same point in time.

A next step is to use Eq. (12) and FL_{DPS} to transform a DPS signature into an FFS signature: $\{FFS_{DPS}\} = (\{DPS\}/FL_{DPS})100$. Figure 17 shows the FFS plots for the two DPS and two FFP plots in Figure 16. Notable observations are the following: (1) the DPS-based FFS plots are more linear compared to the FFP-based FFS plots; (2) all four FFS plots have the same units of measure (percent in amplitude and days in time); (3) functional failure occurs when an FFS is at 100 percent; and (4) the two FFP-based FFS plots are similar, but not exactly the same.

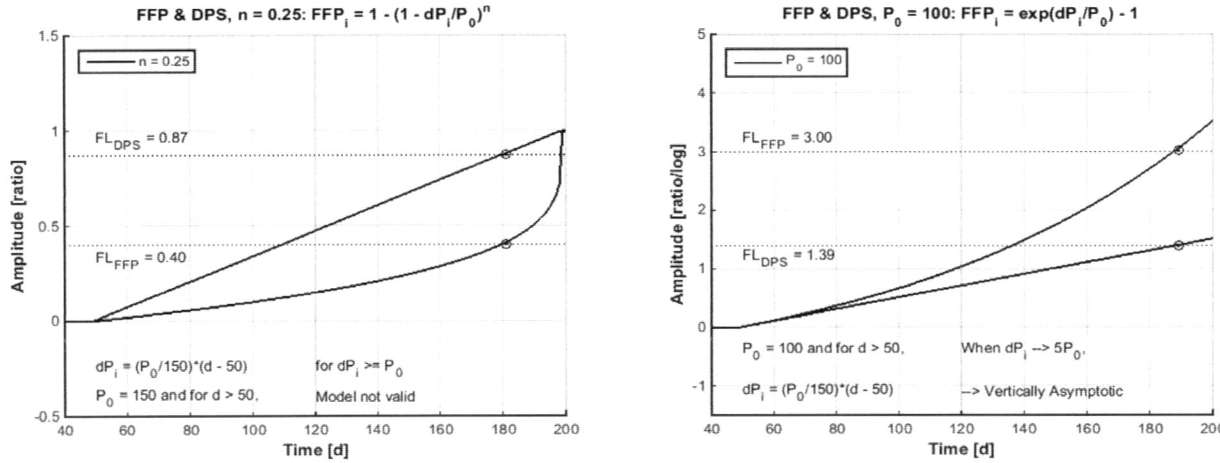

Figure 16. Plots for FL_{FFP} (selected) and FL_{DPS} (calculated): same time of failure

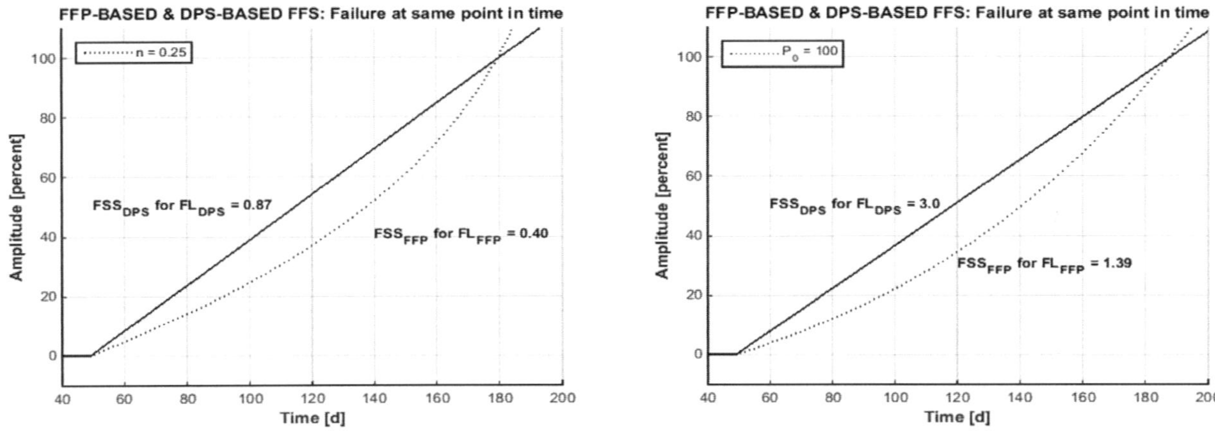

Figure 17. FFS plots for FL_{FFP} (selected) and FL_{DPS} (calculated): same time of failure

5. SUMMARY

In this paper an exemplary prognostic system is reviewed: a set of core frameworks, classical and CBD approaches to handle data, and control and data flow. Then a concept of signatures is introduced with a set of examples to illustrate that CBD contains features that can be extracted as FD to create an FFP signature, which can be further transformed into DPS data and then into FFS data – one data point at time. An ideal FFS, as a transfer curve for producing prognostic information, is a straight line progressing from 0 percent or less (no degradation) to 100 percent or higher (functionally failed). Next a method is introduced to evaluate non-linearity of FFS data (FNL): that method is based one used for evaluating the nonlinearity of data converters.

A procedure for transforming CBD signatures into FFS data is introduced: included is a set model functions developed for that purpose. Plots of the model functions showed that for ideal CBD (FFP) data, the resulting DPS and FFS curves are linear.

Acknowledgements

The authors thank Naval Air, Naval Sea, U.S. Army, U.S. Air Force, and NASA research centers for their support and funding of multiple projects that led to the results described and shown in this paper.

References

[1] Pecht, M. (2008) Prognostics and Health Management of Electronics. Hoboken, NJ: John Wiley & Sons.

[2] Kumar, S. and Pecht, M. (2010) "Modeling Approaches for Prognostics and Health Management of Electronics," *International Journal of Performability Engineering*, Vol. 6, No. 5, pp. 467-476.

[3] O'Connor, P. and Kleyner, A. (2012) *Practical Reliability Engineering*. Chichester, UK: John Wiley & Sons.

[4] IEEE 1856/D33 (2017), Draft Standard Framework for Prognosis and Health Management (PHM) of Electronic Systems, Mar 2017.

[5] CAVE3 (Accessed 2015, Nov.) Prognostics Health Management for Electronics, Auburn University, Auburn, Alabama. Web site: http://cave.auburn.edu/rsrch-thrusts/prognostic-health-management-for-electronics.html.

[6] Hofmeister, J., Wagoner, R. and Goodman, D. (2013) "Prognostic Health Management (PHM) of Electrical Systems using Conditioned-based Data for Anomaly and Prognostic Reasoning," *Chemical Engineering Transactions*, Vol. 33, pp. 992-996.

[7] Hofmeister, J. Szidarovszky, F. and Goodman, D. (2017) "An Approach to Processing Condition-based Data for Use in Prognostic Algorithms," 2017 Machine Failure Prevention Technology, Virginia Beach, VA, 15-18 May 2017.

[8] Medjaher, K. and Zerhouni, N. (2013) "Framework for a Hybrid Prognostics," *Italian Association of Chemical Engineering, Chemical Engineering Transactions*, DOI: 10.3303/CET1333016, pp. 91-96, 2013.

[9] Erickson, R. (1999), *Fundamentals of Power Electronics,* Norwell, MA: Kluwer Academic Publishers.

[10] Judkins, J. B., Hofmeister, J. and Vohnout, S. (2007) "A Prognostic Sensor for Voltage Regulated Switch-Mode Power Supplies," IEEE Aerospace Conference 2007*,* Big Sky, MT, 4-9 Mar. 2007, Track 11-0804, pp. 1-8.

[11] Judkins, J. B. and Hofmeister, J. P. (2007) "Non-invasive Prognostication of Switch Mode Power Supplies with Feedback Loop Having Gain," IEEE Aerospace Conference 2007*,* Big Sky, MT, 4-9 Mar. 2007.

[12] Hofmeister, J. Goodman, D. and Wagoner, R. (2016) "Advanced Anomaly Detection Method for Condition Monitoring of Complex Equipment and Systems," 2016 Machine Failure Prevention Technology, Dayton, OH, 24-26 May 2016.

[13] TI (1995) "Understanding Data Converters," Application Report SLAA013, Texas Instruments, Inc., 1999.

[14] Carr, J.J. and Brown, J.M. (2000), *Introduction to Biomedical Equipment Technology,* 4th Ed.*,* Upper Saddle River, New Jersey, Prentice Hall

[15] Jenq, Y.C. and Li, Qiong, (2002) "Differential Non-linearity, Integral Non-Linearity, and Signal to Noise Ratio of an Analog to Digital Converter," Department of Electrical and Computer Engineering, Portland State University, P.O. Box 751, Portland, OR. 97207, USA.

[16] Saxena, A., Celaya, J., Saha, B., Saha, S. and Goebel, K. (2009) "On Applying the Prognostic Performance Metrics," Annual Conference of the Prognostics and Health Management Society (PHM09), San Diego, CA, 27 Sep.-1 Oct. 2009.

Failure Analysis of a Primary Reactor Cooling Pump Using Modal and Vibration Analysis

Thomas J. Hazelwood and Blake W. Van Hoy

Oak Ridge National Laboratory

1 Bethel Valley Road

Oak Ridge, TN 37831

Telephone: (865) 540-3010

hazelwoodtj@ornl.gov

Abstract

Four pumps redundantly supply primary cooling water to the reactor of the High Flux Isotope Reactor (HFIR) at the Oak Ridge National Laboratory (ORNL). All four pumps underwent maintenance during a recently scheduled downtime. During preliminary evaluation following maintenance, one pump, PU-1A, exhibited higher vibration levels compared to the other three pumps. The pump, being a safety class item, underwent further performance testing to rule out potential damage resulting from the higher vibration. Vibration analysis and modal analysis including steady state spectrum, operational deflection shape, run up transients, and modal impact have been utilized to identify dynamic characteristics that could contribute to the increased vibration levels. This report will cover the testing setup, methodology, analysis results, and recommendations.

Keywords

Condition monitoring, fault analysis, failure prevention, signal analysis, diagnostics

Purpose of Data Acquisition and Analysis

Pump PU-1A, one of four original, primary cooling pumps for the High Flux Isotope Reactor (HFIR) at Oak Ridge National Laboratory (ORNL), continued to exhibit a vibration velocity greater than the other three pumps after maintenance. Due to the pump being a safety class item, several dynamic measurements were made to identify the root cause of the increased vibration and determine if this vibration might create a potential to cause damage to the pump, its motor, or its associated piping.

Data Acquisition Setup

An eight-channel data acquisition system was utilized for this measurement combined with a modal hammer and seven single axis laboratory grade accelerometers. Five of the accelerometers were epoxied in a single column and oriented radially along the pump's vertical axis. Two of the accelerometers where epoxied onto the attached piping a few feet from the pump. Table 1 lists these sensors, their location, and their software nomenclature.

Table 1 Acquisition Channels

Channel Number	*Sensor Type*	*Placement*	*Geometry Nomenclature*
1	Modal Hammer	Varied	Varied
2	Accelerometer	Top of Pony Motor, Radial	col.1.x
3	Accelerometer	Motor, Upper Bearings, Radial	col.2.x
4	Accelerometer	Motor, Lower Bearings, Radial	col.3.x
5	Accelerometer	Pump, Upper Flange, Radial	col.4.x
6	Accelerometer	Pump, Lower Flange, Radial	col.5.x
7	Accelerometer	Piping, Suction	suc.1.x
8	Accelerometer	Piping, Discharge	dis.1.x

Pump Information

The rated rotational speed of the 600-horsepower motor is 1775 rpm, but it runs at 1781 rpm or 29.68 Hz. Multiple orders of this rotational speed, noted in Table 2, are expected to be present within the measurement with amplitudes depending on system characteristics such as imbalance, misalignment, and looseness.

Table 2 Frequency of Orders

Order	**1**	**2**	**3**	**4**	**5**	**6**	
f (Hz)	29.688	59.376	89.064	118.752	148.44	178.128	
Order	**7**	**8**	**9**	**10**	**11**	**12**	**13**
f (Hz)	207.816	237.504	267.192	296.88	326.568	356.256	385.944

Some detectable vibration was expected originating from the rolling element bearings at frequencies noted in Table 3, but they were minimal due to the recent maintenance.

Table 3 Bearing Frequency

	Upper Guide Bearing SKF 6226C3		*Upper Thrust Bearing SKF 29326EJ*		*Lower Bearing 2x SKF 7222-BEGAM*	
	Order	f (Hz)	Order	f (Hz)	Order	f (Hz)
Rolling Element and Cage Assembly	0.41	12.17	0.45	13.36	0.43	12.77
Rolling Element About Own Axis	2.75	81.64	3.42	101.53	2.66	78.97
Over Rolling Inner	5.29	157.05	11.5	341.41	8.56	254.13
Over Rolling Outer	3.71	110.14	9.53	282.93	6.44	191.19
Over Rolling One Point On Ball	5.49	162.99	6.84	203.07	5.32	157.94

It was expected that an increased vibration would be noted at five times the rotational speed due to the pump's five vanes. The pump is positioned vertically in the following arrangement from top to bottom: pony motor, main motor, flange, pump housing. The motor and flange are depicted in Figure 1. Part of the pony motor can also be seen at the top of Figure 1. The pump assembly is anchored with four large bolts to an elevated steel platform. The flange, pump, and anchor bolts are depicted in Figure 2.

Figure 1 Motor and Flange

Figure 2 Flange and Pump

Modal Analysis

Modal analysis was performed on the pump using the single input multiple output (SIMO) roving modal hammer technique. This measurement captures system characteristics such as mass, stiffness, and damping to determine the natural frequencies and mode shapes of the system. Increased vibration amplitude is expected when these natural frequencies are excited due to motor rotation during normal operation. This interaction will be described in the steady state section of this document. It is worth noting that the pony motor was in operation, rotating the shaft and impeller, during this measurement. The system's driven rotation will distort the frequency response function (FRF) to some extent because force inputs are being provided by both the rotation and as the modal hammer.

Frequency Response Function

Four primary radial natural frequencies were identified at approximately 11, 88, 145, and 292 Hz and can be seen in Figure 3. The last three of these correlate with three, five, and ten times the rotational speed. The proximity of the fifth order, approximately 148 Hz, to the modal peak at 145 Hz should result in greater amplitudes due to the five vanes of the pump impeller. The top of the pony motor exhibited the highest amplitude throughout the modal analysis.

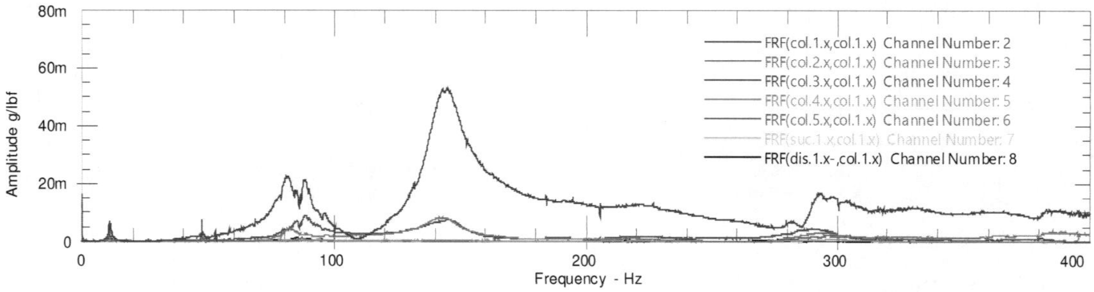

Figure 3 Modal Frequency Response Functions

Mode Shape

The following mode shapes were calculated using a single degree of freedom curve fitter. Mode shapes can be used to visualize the data from the FRF to depict the vibrations of the system. These shapes can be compared against the operational deflection shapes of the pump during normal operation to visualize possible interaction locations and intensities. Note that the amplitude of the first mode, in Figure 4, was quite low. The top of the pony motor exhibited the most motion in the remaining modes. Mode two, in Figure 5, includes motion at the lower bearings of the motor as well.

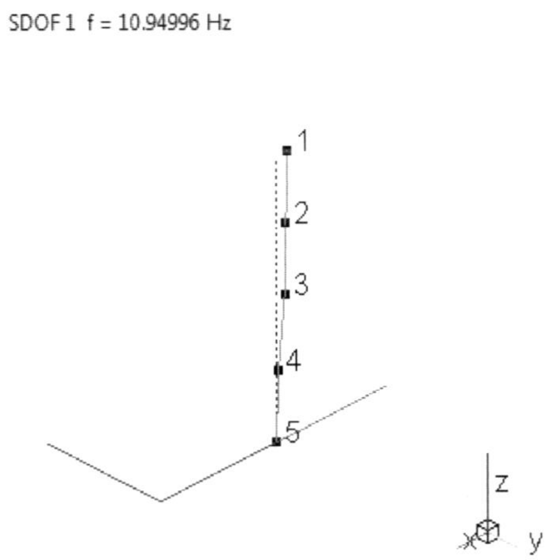

Figure 4 Mode Shape 1

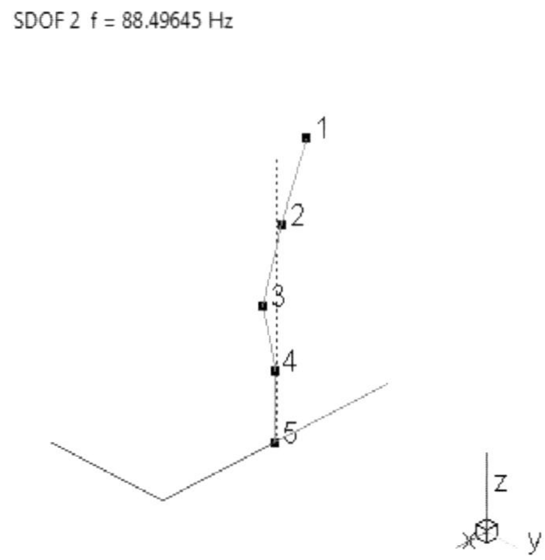

Figure 5 Mode Shape 2

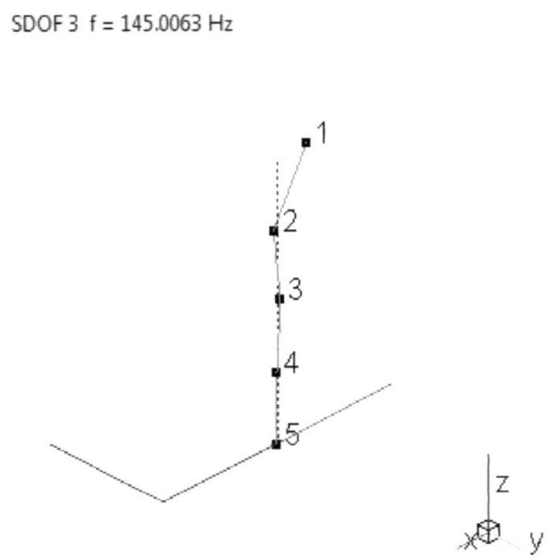

Figure 6 Mode Shape 3

Figure 7 Mode Shape 4

Steady State

About ten minutes of data, or 596.48 seconds, was taken with seven accelerometers while the pump was at its full operating speed. Two hundred and thirty-three fast Fourier transforms (FFT) were calculated and averaged into one spectrum data set for each of the sensors. This data is displayed within this section as a two-dimensional spectrum plot and operational deflection shapes.

Spectrum

Within the spectrum data displayed in Figure 8, five frequencies of interest exhibit elevated amplitudes of vibration. These frequencies closely correlate with the first, second, third, fifth, and tenth orders of rotational speed. The amplitudes of each of the sensors at these frequencies are presented in Table 4. As expected from the Modal Analysis section, the peak associated with the fifth order at 148.83 Hz exhibited the highest amplitude at 0.295 g rms.

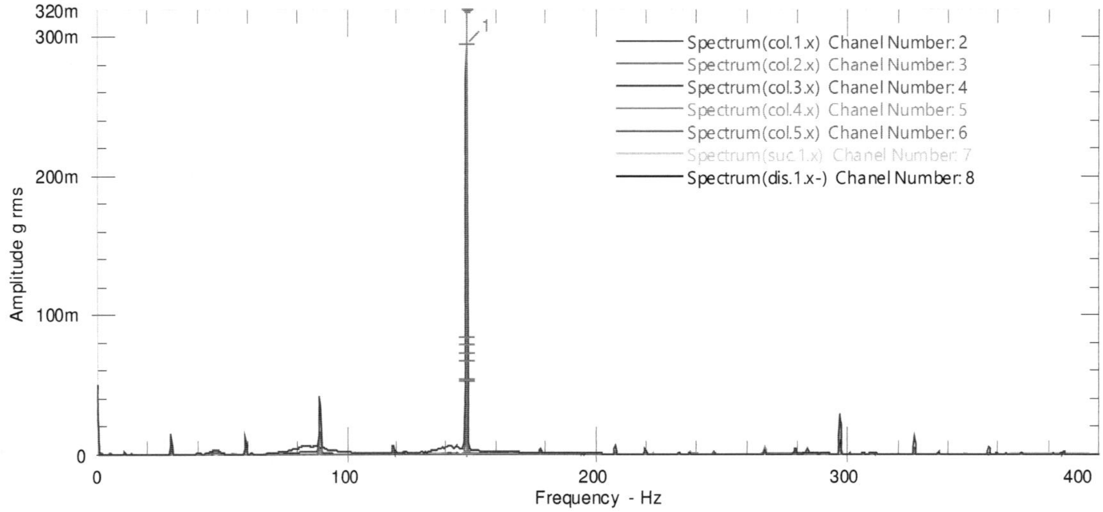

Figure 8 Steady State Spectrum, All Sensors

Table 4 Steady State Acceleration Ampitudes

Location	*29.688 Hz*	*59.375 Hz*	*89.063 Hz*	*148.83 Hz*	*297.27 Hz*
col.1.x (g rms)	0.016	0.013	0.041	0.295	0.030
col.2.x (g rms)	0.012	0.004	0.018	0.053	0.003
col.3.x (g rms)	0.006	0.005	0.043	0.055	0.007
col.4.x (g rms)	0.004	0.012	0.004	0.084	0.029
col.5.x (g rms)	0.002	0.009	0.005	0.068	0.027
suc.1.x (g rms)	0.004	0.009	0.006	0.073	0.010
dis.1.x (g rms)	0.000	0.006	0.002	0.080	0.015

The steady state spectrum data of the accelerometer on top of the pony motor is displayed in Figure 9. All orders of run speed within the range from 0 to 400 Hertz are present, but five times run speed stands out as the maximum excitation at almost 0.3 g rms.

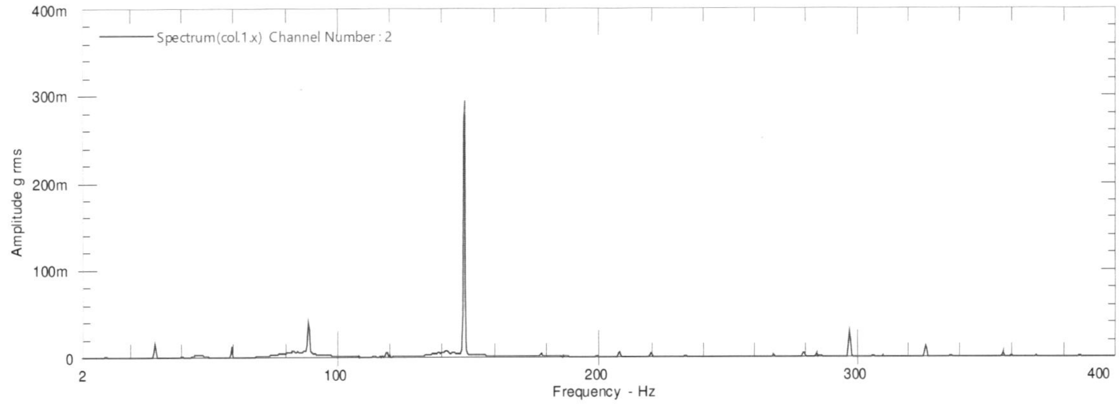

Figure 9 Steady State Spectrum, Top of Pony Motor

The steady state spectrum data of the accelerometer on the lower bearing area of the motor is displayed in Figure 10. Again, all orders of run speed are present. The five times run speed peak was still the maximum, but its amplitude was much lower at around 0.055 g rms. The next highest peak was at three times run speed with an amplitude of about 0.043 g rms, which is the maximum excitation for this frequency.

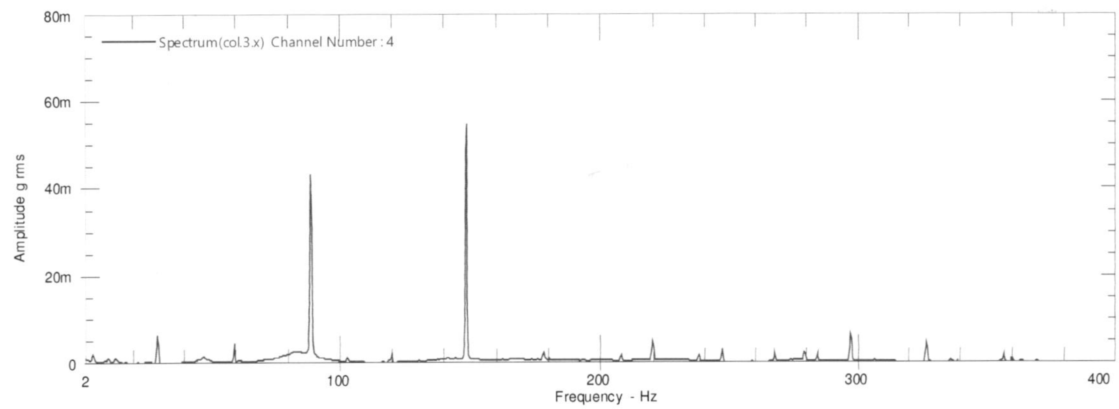

Figure 10 Steady State Spectrum, Motor Lower Bearings

The steady state spectrum data of the accelerometers of the suction and discharge piping are displayed in Figure 11 and Figure 12 respectively. As in the cases for the motor and pump accelerometers, all orders of run speed are present within the depicted range and five times run speed has the maximum amplitude.

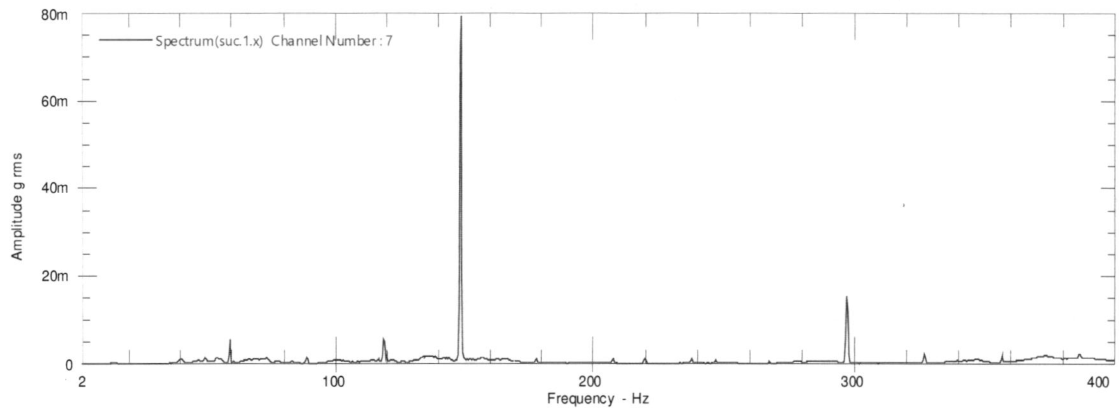

Figure 11 Steady State Spectrum, Suction piping

The amplitude of five times run speed of both piping accelerometers are both about 0.080 g rms. The next highest peak is at ten times run speed at about 0.015 g rms.

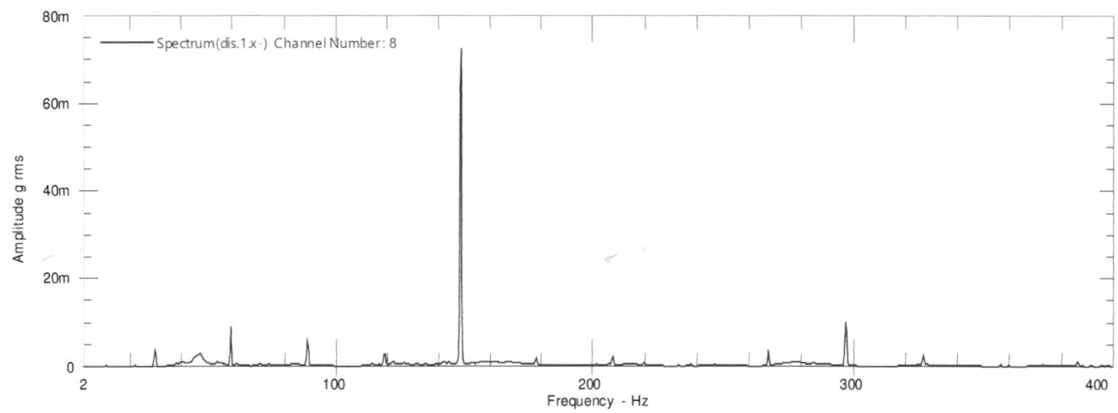

Figure 12 Steady State Spectrum, Discharge Piping

Operational Deflection Shape

The following operational deflection shapes (ODS), shown in Figures 13 though 17, depict the pump during normal operation. The deflections are normalized to the largest amplitude at that frequency.

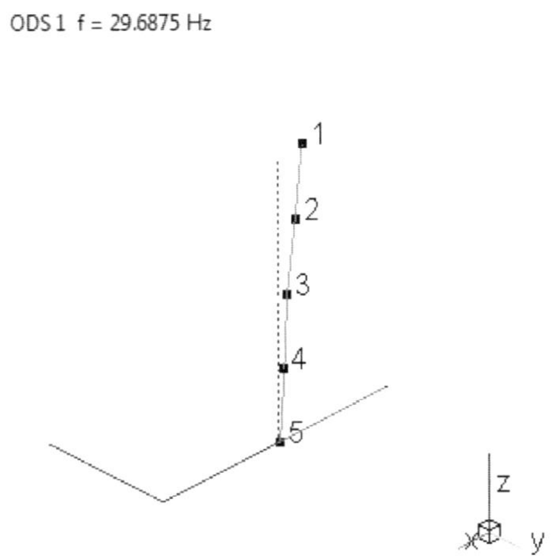

Figure 13 ODS 1

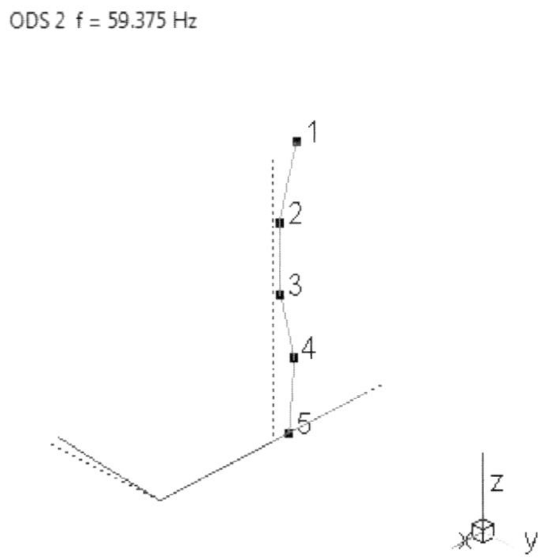

Figure 14 ODS 2

Operational deflection shapes three, four, and five, shown in Figures 15, 16, and 17, exhibit modal interaction with mode shapes two, three, and four respectively, shown in Figures 5, 6, and 7. Operational deflection shape four, shown in Figure 16, at five times run speed, exhibits the largest vibration amplitude.

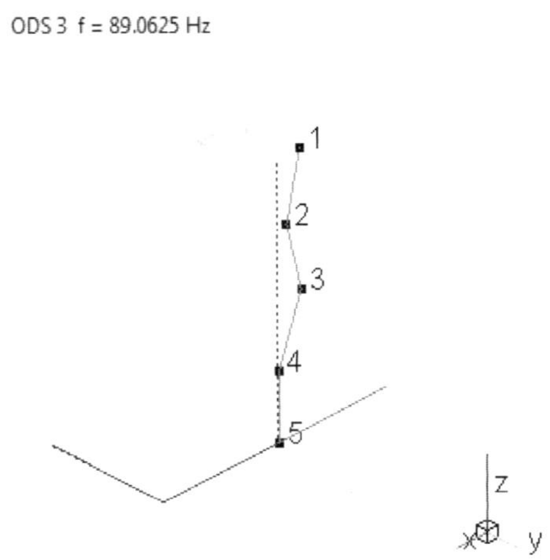

Figure 15 ODS 3

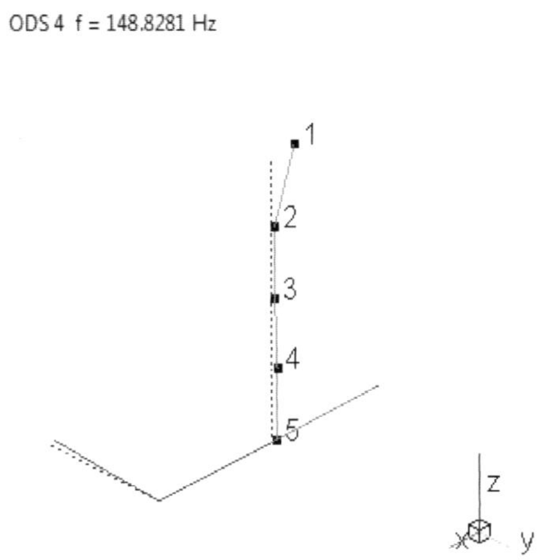

Figure 16 ODS 4

Operational deflection shape five, shown in Figure 17, at ten times run speed, differs from mode shape four, shown in Figure 7, by having a larger response from the accelerometers on the pump and flange as well as some motion in the piping.

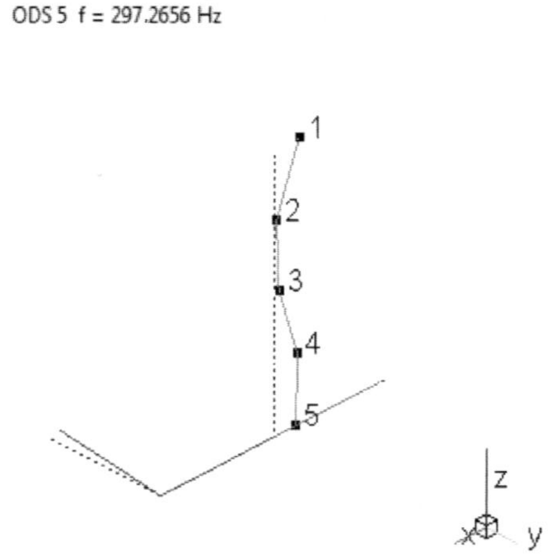

Figure 17 ODS 5

Transient Analysis

An additional ten minutes of data was taken for transient analysis. The main motor was off during the first two minutes of this acquisition and the pump was driven by the pony motor. The pump quickly came up to speed when the main motor powered on. No electrical grounding issues are apparent within the transient data. The portion of this data collected while the main motor was running closely matches the steady state spectrum section. The top of the pony motor, shown in Figure 18, and the suction piping, shown in Figure 19, were chosen as representative transient data.

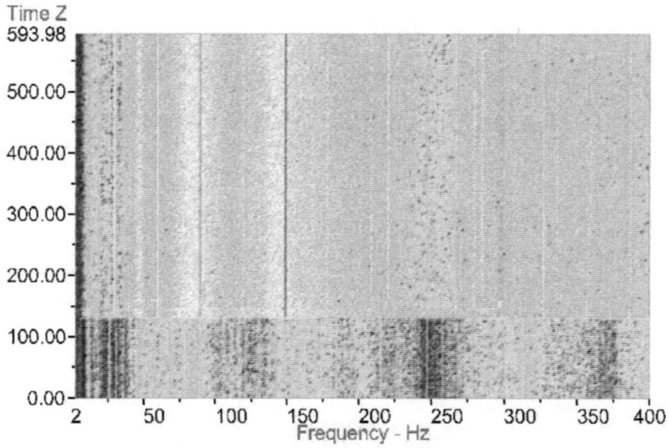

Figure 18 Transient, Top of Pony Motor

No broadband noise was present in the piping or pump; therefore, cavitation does not seem to be present in the system. Further, there were no indicators of looseness in the system.

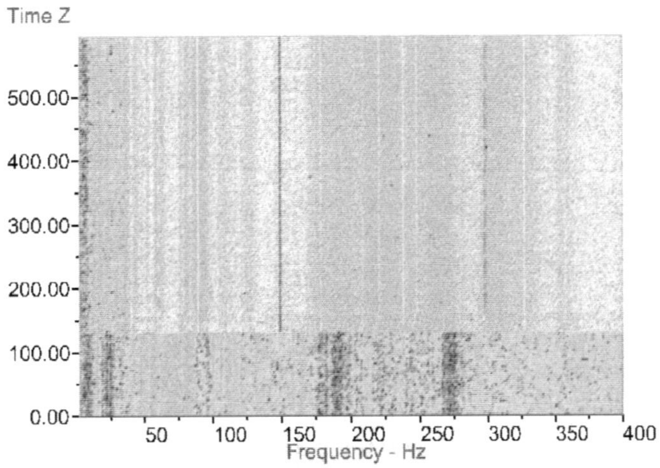

Figure 19 Transient, Suction Piping

Pony Motor Only Steady State Spectrum

About two minutes of data was taken with the main motor off while the pump is being driven by the pony motor. Fifty FFT where calculated and averaged into one spectrum data set for each of the sensors, shown in Figure 20. The top of the pony motor exhibited the most motion, just as with normal operation, but all the levels were expectedly lower. The largest amplitude was 0.0044 g rms at 78.9 Hz.

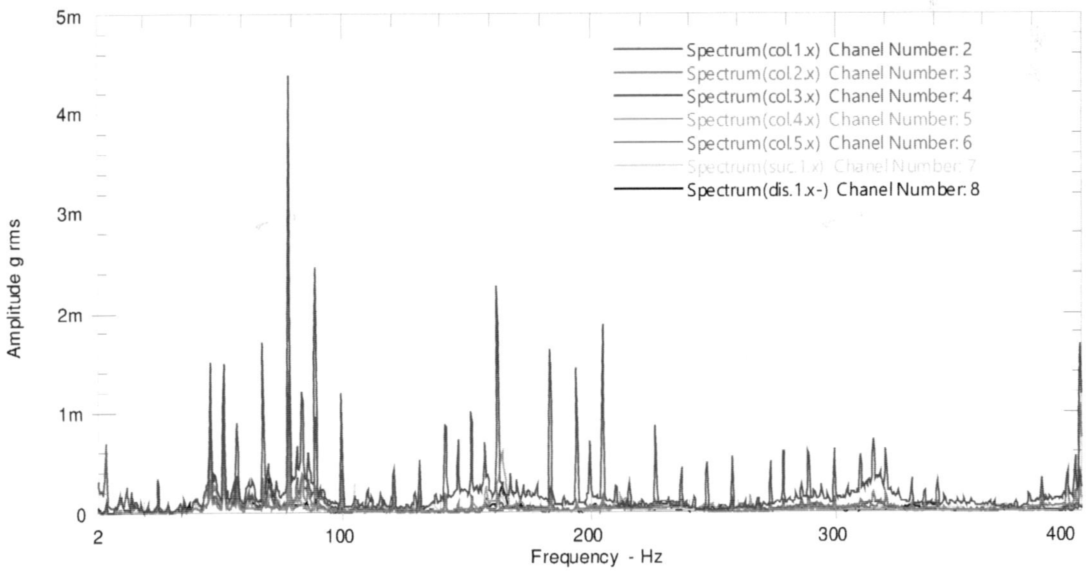

Figure 20 Pony Motor Only Steady State Spectrum, All Sensors

Analysis Results

Modal interaction was exhibited during the third, fifth, and tenth orders. The fifth order is excited further due to the five-vane pump. There is no evidence of electrical grounding, cavitation, looseness, misalignment, or imbalance within the data collected. Some bearing frequencies are present, but they are not elevated; therefore, it is concluded that the bearings have not experienced excessive wear or damage.

This pump is categorized as an integrated drive, rigid foundation, pump larger than 15 kW, or Group 4. Per ISO 10816 vibration standard, the motor falls within Zone C, which designates that the pump is considered unsatisfactory for long-term continuous operation. The standard states that the pump may be operated for a limited time until remedial action is performed. This categorization is a result of the high vibration found at the top of the pony motor of 0.295 g rms at 148.83 Hz for a 600 HP (447 kW) motor.

Using the same standard for the vibration measured from the piping at about 0.080 g rms at the same frequency categorizes the piping section within Zone A, which designates a newly commissioned machine. The piping accelerometers move a maximum distance of 0.125 mils peak to peak at the rotational speed of the pump. This indicates that piping damage due the measured vibration is not expected.

Recommendations

The rotational speed adjustment to decrease the excessive vibration would need to be significant due to the frequency width of the excited mode, which does not appear to be an option. It is suggested to perform vibration and modal analysis on the other three reactor cooling pumps to determine what differentiates PU-1A from the others. It is likely that a structural component is different between them, such as a brace, bolt, weld, flange, or joint. It is probable that with this information, PU-1A could be modified to increase the system's stiffness specifically at the top of the motor and pony motor to reduce the high vibrational amplitudes as compared to the other three pumps.

Reference

International Organization for Standardization. (2009). *Mechanical vibration -- Evaluation of machine vibration by measurements on non-rotating parts -- Part 3: Industrial machines with nominal power above 15 kW and nominal speeds between 120 r/min and 15 000 r/min when measured in situ (Standard No. 10816-3).* Retrieved from https://www.iso.org/standard/50528.html

Acknowledgments

This research used resources of the High Flux Isotope Reactor, which is a DOE Office of Science User Facility.

GENERATION OF TACHOMETER SIGNAL FROM A SMART VIBRATION SENSOR

Eric Bechhoefer
GPMS Inc
40 Ridge Rd
Cornwall, VT 05753
1 (802) 377-0160
eric@gpms-vt.com

Ed Spence
The Machine Instrumentation Group LLC
4605 Myra Glen Place
Durham, NC 27707
1 (781) 439-1277
ed.spence@machineinstrumentation.com

Abstract: The tachometer plays an important role in the quality of a vibration-based diagnostic. Because of the bandwidth limits of the machine controller, most machines have a slight change in speed over time. This change in speed necessitate the resampling of the data, based on a tachometer signal, to facilitate shaft, gear and bearing analysis of the machine. This is because a change in shaft rate changes the spectral content of the signal, upon which vibration analysis is dependent.

Unfortunately, there may be cases, such as glandless pumps, where due to heat and pressure it is impractical or infeasible to install a tachometer sensor. In other situation, such as monitoring gas turbine engine, interfacing with the existing tachometer for the power turbine or compressor turbine, may change certification requirements (adding cost) or increase system cost and weight. These issues may make adding a tachometer impractical.

Using a novel, two step process, we were able to generate a high quality tachometer signal from the vibration data. The first step uses an idealized bandpass filter to remove extraneous vibration signal such that the cyclic rate of the shaft can be constructed. The second step then removes any extraneous jitter in the tachometer signal. The resulting tachometer signal is usually of higher quality than that achievable from a traditional tachometer sensor. The tachometer signal is then used for the time synchronous average or time synchronous resampling algorithms, which are the basis for modern shaft, gear and bearing analysis. We demonstrate the efficacy of the technique on two data sets, showing that the resulting condition indicators are indistinguishable from a system using a traditional tachometer signal as an input.

Key words: CBM; TSA, Demodulation, Fourier Transform;

Introduction: R.M. Stewart (Ref. 1) introduced the time synchronous average (TSA) into gearbox analysis, and revolutionized condition based maintenance on helicopters and other rotating equipment. His analysis, such as FM0/1/2/3/4, the "difference signal", the envelope analysis, etc., are used by most, if not all, Health and Usage Monitoring Systems (HUMS) currently available. Stewart's design is heavily dependent on a tachometer input to allow resampling of vibration data for the analysis. Further, Stewart recognized the importance of tachometer jitter, noting that there was a limit to the number of revolutions to take on the TSA (e.g. at some point, the TSA was not consistent).

The need to interface with a tachometer adds cost and complexity for every application. Additionally, there may be cases, such as glandless pumps (Boiler Circulator Pumps), where due to heat and pressure it is impractical or infeasible to install a tachometer sensor. In other situation, such as monitoring a rotorcraft turbine engine, interfacing with the existing tachometer for the power turbine or compressor turbine, may change certification requirements (adding cost) and again increasing system cost and weight. These are just a few examples where adding a tachometer is impractical. Hence, the ability to derive a tachometer signal from vibration can reduce both cost and weight of a system, facilitating new applications.

For helicopter applications, weight and cost of HUMS integration can be reduced through the use of a bused, smart system (Ref 2). The use of a data bus greatly reduces interconnects, while smart sensor makes for a more scalable system. The smart vibration sensor also allows a path to locally processing vibration data to reconstruct a tachometer signal on the sensor, so that the TSA and other tachometer dependent analysis can be performed in situ.

Tachometer Signal From Vibration: This ability for a smart sensor to acquire vibration data, extract the shaft speed from the vibration data, and then process the data allows vibration based fault detection capability at a lower cost and weight than current systems. This also results in reduced installation complexity over traditional tachometer based system. Reducing cost, weight and installation complexity will expand the business case for condition monitoring, improving safety and reliability in many systems.

Cost is an important driver in the decision for operators to implement a condition monitoring system (CMS) such as HUMS. While it is well recognized that condition monitoring systems improves operational availability and reduces maintenance cost, the decision to install a CMS or HUMS for many operators is driven by cost. Removing a tachometer sensor and interface by adding this function to a smart sensor improves the value proposition of a system to a customer.

Spectral content of vibration is measured using the Fast Fourier Transform (FFT). The FFT is used in vibration-based diagnostics to: determine the magnitude and phase of component's vibration (such as shafts, gears or bearings), which can be indicative or wear and failure. Additionally, many common vibration analyses (Residual or Difference Analysis, Narrowband analysis (Ref 3, 4, 5)) use the FFT for ideal filtering of the signal, or to perform a Hilbert transform of the signal (Amplitude and Frequency Analysis).

The Fourier transform (and FFT) assumes that the signal under analysis is infinite in time. This assumption fails for real signals, but can be mitigated by the use of a window function (such as a Hamming or Hanning window). The other assumption is that the signal is stationary. Stationarity implies that the conditions of the signal do not change, e.g. the shaft rate is constant. This assumption in practice is violated: rotating machines have variation in the their shaft rate. This change in rotational rate is due to changing load and the limits of the feedback control bandwidth.

The failure of stationarity results in "spectral smearing" of energy associated with a shaft. This in turn results in not accurately measuring the energy associated with a particular fault frequency. In order to improve the performance of analysis using the FFT, Time Synchronous Averaging (the TSA, for shaft/gear analysis) and Time Synchronous Resampling (TSR) have been developed (Figure 1).

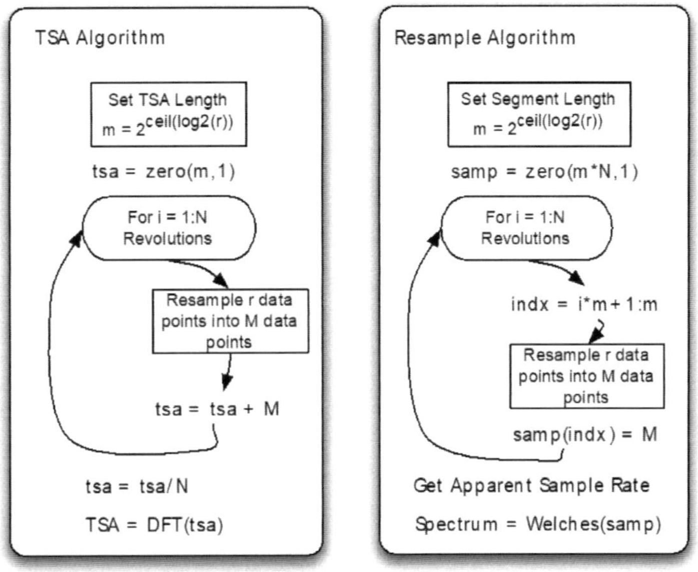

Figure 1: Resampling Algorithms for Vibration Based Diagnostics.

Resampling Algorithms: The TSA resamples the vibration associated with a shaft or gear, in the spatial domain, such that vibration associated with each shaft order, in the Fourier domain, represent one frequency bin. For example, consider a system in which the shaft rate is such that for a given vibration sample rate, the acquisition system on average collects 800 samples per revolution. The TSA would resample the 800 samples to the 1024 data points. The value 1024 is the next highest radix 2 value. Radix 2 values are used because the simplest implementation of the FFT is based on powers or 2, e.g radix 2 values. Consider that if the load on the shaft reduces, the shaft rate increases: the measured vibration data points decrease to 780 samples. It takes less time for the shaft to make one revolution, hence fewer samples. The 780 samples are resampled to 1024 points. At some future time, say that the load on the shaft may increase, slowing the shaft, resulting in measuring 820 samples, again the data is resample to 1024 points. For every revolution, the resampled data is summed point by point. After *n* revolutions, each of the 1024 points of is divided by *n*, essentially averaging the vibration data for that point in the TSA vector.

Assume in this case that there is a gear with 37-tooth gear on the shaft. The gear mesh energy of the gear would be a frequency of 37 x the shaft rate. In the Fourier domain, energy associated with the shaft rate would be in bin 2, and the gear mesh energy would be bin 38, and the second harmonic of that gear would be in bin 75 (37 x 2 + 1, bin 1 is the DC energy). The TSA also reduces non-synchronous vibration by 1/sqrt(n), where n is the total number of shaft revolutions which where used to construct the TSA. In this way, the TSA corrects for variation in the shaft rate and improves signal to noise.

The TSR similarly resamples (e.g. up samples) the vibration to correct variation in shaft speed. The apparent sample rate is the ratio of the total resampled time domain, vibration data set length divided by original data set length, multiplied by the original sample rate. Both the TSA and TSR use a tachometer signal to calculate the time over which a shaft completes one revolution. The time taken for any shaft to complete a rotation can be calculated even if the tachometer is not associated with a given shaft. This is done by taking into account the shaft ratio from the shaft with the tachometer, to the shaft under analysis, then interpolating based on that tachometer signal.

The Tachometer Signal: In implementation, the tachometer signal is the rising edge of a voltage trigger from the passing of a shaft key phasor (e.g. a stationary point of the shaft). The tachometer signal is then converted to time. This time is accrued for each pass of the key phasor. In an architecture where the tachometer signal is measured using an analog to digital converter (ADC), the resolution in time of the rising edge is 1 over the sample rate of the ADC. For condition monitoring purposes, the sample rate for a high-speed shaft could be 96,000 samples per second. In another architecture, the tachometer signal inputs into a voltage comparator. When the tachometer signal crosses zero (or some low voltage offset), the comparator goes high. The output of the comparator is monitored by the microcontroller using a general-purpose input/output (GPIO) pin. When the microcontroller senses the GPIO pin going high, it records the time. The resolution of time on the microcontroller is typically much higher than an ADC. For example, in a system using a 12 MHz clock, the microcontroller might run at 96 MHz, but the counter for time in the microcontroller would run at 48 MHz. The tachometer resolution in time would then be 2.0822e-8 seconds, or 500 x greater resolution than the ADC architecture.

Vibration signals from rotating equipment are sinusoidal and, by definition, synchronous with the shaft. However, the nature of vibration makes it impossible to use the vibration signal without significant signal processing. Measured vibration is the superposition (e.g. addition) of many signals in time domain. For example, consider a simple gearbox with an input shaft, and output shaft and a gear pair. The input shaft turns at 30 Hz, and has a 32 tooth gear, the output shaft has a 82 tooth gear with a speed of 11.707 Hz. The gear mesh frequency is 30 * 32 = 960 Hz. It is likely that the gear mesh frequency will have side bands as a result of any shaft imbalance being modulated onto the gear mesh. This can be proved using the trigonometric identity (Eq 1):

$$cos(a) * cos(b) = 1/2\,[cos(a+b) + cos(a-b)] \qquad (1)$$

Where in this example, cos(a) is 960 Hz, and cos (b) is 30 and/or 11.707 Hz shaft. Additionally, if the shaft is bent or bowed, there will be a 2x shaft vibration component. Other manufacturing defects such as the gear not being mounted perpendicular to the shaft, or not centering the shaft on the gear (e.g. eccentricity) adds additional tones.

Tachometer From Vibration: Bonnardot (6) in 2005 describes a method of using an acceleration signal of a gearbox to perform the TSA. This was further developed by Combet (7) in 2007. These methods call for band pass filtering of the signal around a gear mesh, then in (6) taking the Hilbert transform to estimate phase, and using phase to estimate zero crossing time. In (7), the signal was band pass filtered then demodulated, with the resulting time domain signal used for the zero crossing.

Both solutions are sub optimal, due to the nature of using the FIR (Finite Impulse Response) band pass filter. Additionally, because modeling errors are present even after band pass filtering, the tachometer signal is corrupted by jitter. In this paper, we present a two-step method to reconstruct a tachometer signal from vibration. Step one is to construct an ideal band pass filter and create an analytic signal in one functional procedure. Step two uses a jitter reduction model to remove noise (jitter) from the reconstructed tachometer signal not associated with changes in the shaft rate. While this system could be implement on any computer, it is ideally suited for implementation in a smart sensor, which distributes the processing cost and power within the condition monitoring system, reducing the overall cost of the system.

The pseudo code to recover a tachometer signal from vibration is:

Define the Sample rate = sr. The number of data points of vibration data, $n = sr$ x acquisition length in seconds, then:

1. Calculate the next larger radix-2 length for the FFT. nRadix = 2^ceil($\log_2(n)$)
2. Calculate the low and high bandwidth index (*bwlow, bwhigh*), which are centered are a know gear mesh
3. Take the zero padded FFT of the vibration data
4. Zero the FFT from zero to *bwlow*, and from *bwhigh* to nRadix
5. Take the inverse FFT
6. Calculate the unwrapped argument of the signal from 1 to n time series
7. Normalize the time series of radians by the number of teeth of the gear (assuming 1st harmonics)
8. Interpolate the number of indexes for every 2π radians
9. Normalized to tachometer zero crossing times by sr.

A band pass filter is the convolution of a low pass filter with a high pass filter. These filters are implemented as Finite Impulse Response (FIR) filters to improve their stability. Unfortunately, even rather large tap filters have poor filter response. Consider a case of a wind turbine gearbox, with an aforementioned approximately 29 Hz shaft and a 32 tooth gear, giving 928 Hz. The bandwidth of the filter is set from 910 Hz to 950 Hz, to exclude the 30 Hz of the high-speed shaft (e.g. 29 * 32 -29 = 899 Hz and e.g. 29 * 32 + 29 = 957Hz). The filter response for this case, using a 120 tap FIR filter is seen in Figure 2.

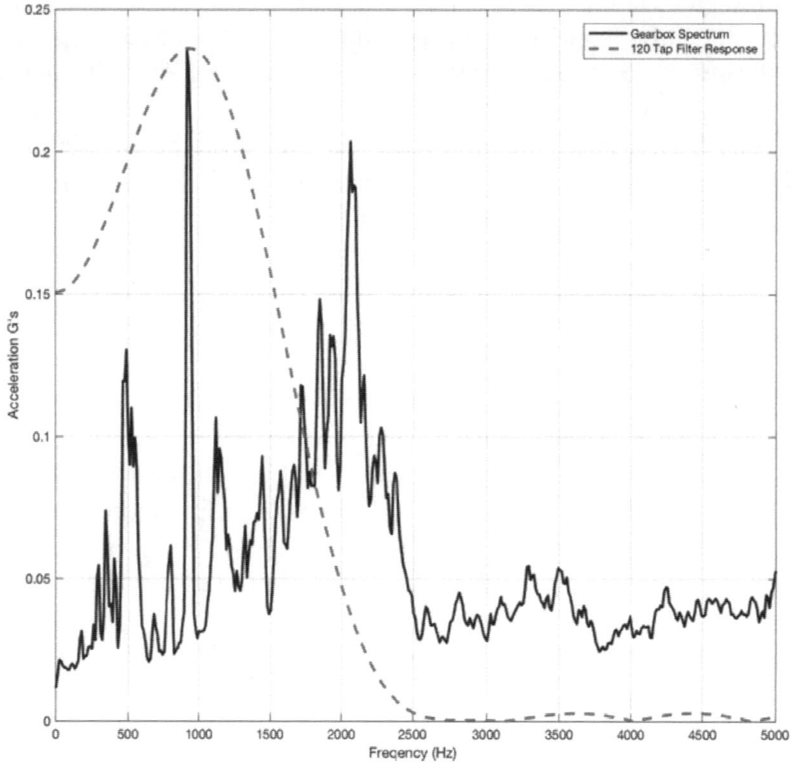

Figure 2 Spectrum from a wind turbine high-speed shaft, and a 120 tap FIR band pass filter

Note that the bandwidth (50% power, 3dB) of this filter is 1670 Hz. The filter does not reject the spectral content at 500 Hz, or at 1121 Hz. These additional tones distort the desired analytic signal, reducing the quality of the resulting tachometer signal.

Consider a process in which the developed analytic signal uses an ideal filter. The analytic signal is defined for real valued signal *s(t)* as Eq 2.

$$S(f) = F\{s(t)\} \tag{2}$$

Where *F* is the Fast Fourier Transform, then:

$$S_a(f) = S(f),\ \ f = 0$$
$$S_a(f) = 2S(f),\ \ f > 0$$
$$S_a(f) = 0,\ \ f < 0$$
$$s_a(t) = F^{-1}\big(S_a(f)\big)$$

where *S(f)* is as noted the Fourier transform of *s(t)*.

For a signal which is sampled at 97,656 samples per second, for six seconds, the total length of *s(t)* is *n*, 585,936 data points. As noted, there are advantages to using radix 2 lengths for the Fast Fourier Transform (FFT). By zero padding the FFT to next larger radix two value, 2^20 or 1,048,576, the index representing the cutoff frequency for the band pass frequencies are: *bwlow* = 910 Hz / 97,656 x 1,048,576 = 9771, and *bwhigh* = 960 / 97,656 * 1,048,576 = 10,308. Then one can define the band pass analytic signal in Eq 3 as:

$$S_a(f) = 2S(f), \;\; bwlow \leq f \leq bwhigh \qquad (3)$$
$$S_a(f) = 0, \;\; f < bwlow, f > bwhigh$$
$$s_a(t) = F^{-1}(S_a(f))$$

There is no need to multiply by 2, as the argument (e.g. angle) which is of interest, is the arctangent ratio of the imaginary parts of $s_a(t)$ and the real parts of of $s_a(t)$. The idealized band pass function reject all signals not associated with the desired pass band. Figure 3 is a zoomed view of the spectrum in Figure 2., comparing the pass band of the idealized filtered realized using Eq 3.

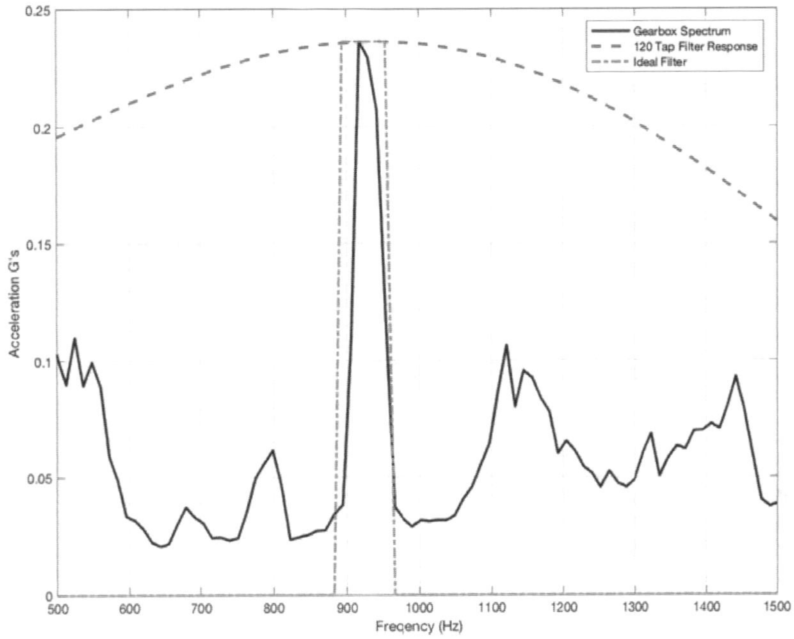

Figure 3. FIR band pass filter vs. the idealized filtered realized using Eq 3

The FIR filter does not reject the 500 Hz and 1150 Hz tone. The idealized filter captures only the signal associated with the desired gear mesh tone. This allows for higher signal to noise ratio and improved reconstruction of the tachometer signal from vibration. This tachometer signal is recovered from the arctangent of the analytic signal.

It is important to note that the arctangent function returns radians between 0 to π and -π to 0. But the evolution of the angle represents the incremental increase in phase for each sample in time. For example, the phase of the analytic signal for three cycles is: 2π x 3 or 6π. The arctangent of that signal will be -π to π for 3 cycles. The result of the arctangent must be unwrapped to capture these increases in phase vs. time. Unwrapping of the phase angle requires keeping track of the previous angle and current angle. The current angle is added to the previous angle, except when the returned arctangent goes from π to -π. In this case, π is added to the returned value to correct for the case when the returned value is between -π to 0.

After unwrapping the phase angle, the units are in radians per sample. While the FFT and inverse FFT operated on the radix 2 length (in this case 1,048,576), the argument and phase angle computation are performed only on the original sample length, *n*. Note that this time series of radians is for the gear mesh. To converts to radians per revolution of the shaft, the time series is divided by the number of teeth in the gear: in this example, 32.

However, it may be that the strongest gear mesh tone is the 2nd or 3rd harmonic, in which case the pass band is adjusted accordingly, and the time series of radian angle is divided by 2 x number of teeth for the 2nd harmonic, etc. The resulting time series represents the radian angle of the shaft, were each index advances the angle in time by *dt*, or 1/sample rate. Every 2π radians represents one shaft revolution. Because one is interested in the time, exactly every 2π, a form of interpolation is needed. For example, consider that the index just prior to 2π is 6.282780795474 (or 0.0004 less than 2π) at array index 3395, while at index 3396, the radian value is 6.284629142378, or 0.0014 greater than 2π. It is then a simple matter to interpolate between the index 3395 and 3396 for a radian value of 2π. In this case, the interpolated value is: 3395.21885053316.

This interpolation gives the number of index for each revolution. Note that this estimate of the tachometer zero cross signal is corrupted by noise. Combert (7) reports that the phase error standard deviation is related to the local signal to noise at the mesh harmonic *k* as:

$$\sigma[\delta] = {}^{1}\!/_{\sqrt{2}}\, 10^{-SNR/20} \tag{4}$$

Typical measured SNR are 6 to 8 dB. This suggests that the standard deviation of the phase error would be 6 to 10 degrees. This phase is zero mean, but as it is non-zero, it will add jitter to the reconstructed tachometer signal.

Controlling Jitter: In "*Improving Gear Fault Detection by Reducing Tachometer Jitter*" (8), it was shown that tachometer jitter contained a low a low frequency component associated with the engine control unit, and random, higher frequency components. Bechhoefer et. al. improved gear fault detection by using a zero phase, low order IIR, backward/forward filtering. As noted previously, both FIR and IIR filters bandwidth is defined by the 3dB reduction in signal energy. The filter does not remove all signals above the bandwidth, and in fact reduces some signal energy below cutoff (up to 50%).

The idealized filter using the FFT processing is zero phase as well. The improve analysis would proceed similar to the forward/backward process:

1. Take the pseudo derivative of the tachometer signal
2. Calculate the the radix-2 length of the pseudo derivative signal of length n
3. Zero pad the array from n to the radix-2 length
4. Calculate the bandwidth index of the FFT
5. Idx = floor(bandwidth * radix-2 length / 2);
6. Bandwidth is a normalized value, typically 0.12
7. Take the real FFT of the zero padded derivative signal
8. Set the real and imaginary parts of the FFT from Idx to the radix-2 length
9. Take the inverse real FFT.
10. Reconstruct the tachometer signal by taking the pseudo integral of the signal

Figure 4 demonstrates the effectiveness of the ideal low pass filter in removing jitter from the tachometer signal. The tachometer was from a wind turbine with 8 targets per revolution. The tachometer signal was resampled and the spectrum taken.

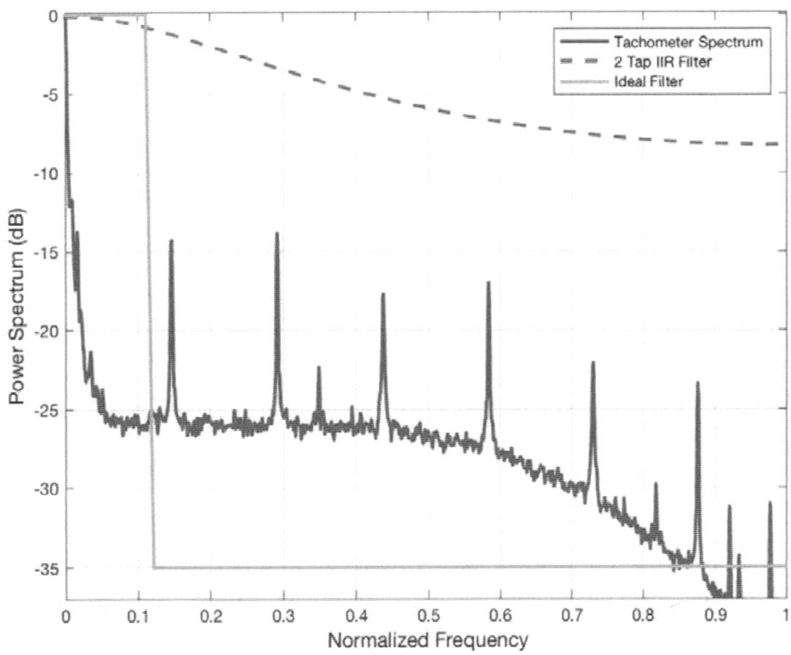

Figure 4: Tachometer pseudo derivative spectrum compared to the IIR and ideal low pass filter.

Note that the tachometer spectrum is primarily DC, with peaks associated with the unequal spacing of the targets. The idealized filter retains more low frequency response than the IIR filter, and effectively removes all spectral power (jitter) above the cutoff frequency.

Case 1: High Speed Pinion on a Wind Turbine: This machine was found, through vibration based diagnostics, to have a cracked tooth on the high-speed pinion. The acquisition length was six seconds, sampled at 97,656 samples per second. The high-speed shaft is approximately 30 Hz. This system was equipped with a tachometer, which is used as a reference to compare against the vibration based tachometer signal. The installed tachometer was a Hall sensor, using the GPIO architecture to capture zero crossing time. The tachometer target had eight pulses per revolution, but time was calculated using each 8th timing mark (e.g. one revolution). Figure 5 compares shaft speed derived from the Hall sensor base tachometer signal, vs. the vibration based tachometer signal, vs. the final processing use the new jitter reduction technique. Wind turbine analysis is particularly difficult in that there is a high degree of variability in the torque.

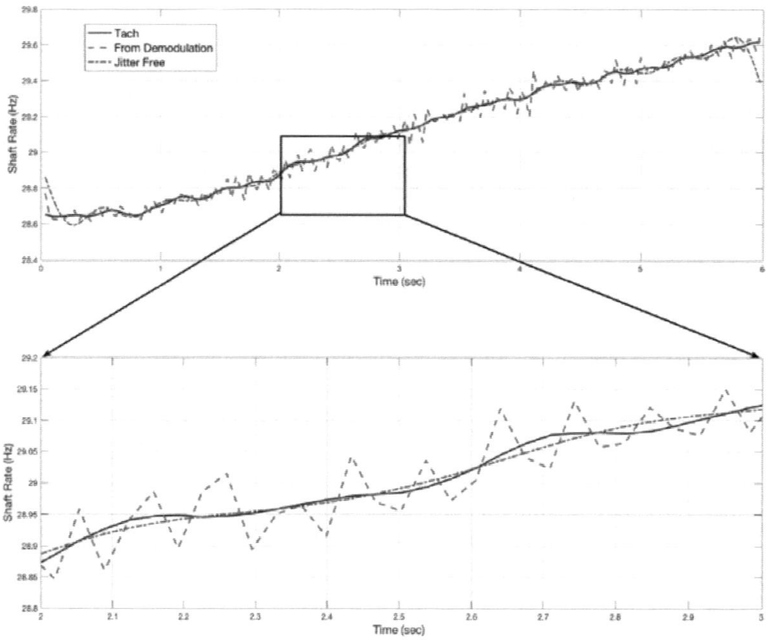

Figure 5. Shaft Rate from Tachometer signal in comparison with jitter free tachometer from vibration. Note that after reducing jitter, the two tachometer signals are virtually the same.

The bottom graph is a zoomed from the 2nd to 3rd second on the acquisition. In general, after removing jitter using the new process, the vibration based tachometer signal is not significantly different from the tachometer derived from the Hall sensor. Figure 6 depicts the TSA and the spectrum of the analysis. The difference between the TSA is effectively only phase angle. From the spectrum, we can calculate the signal to noise ratio to be approximately 6dB. Statistics derived from the TSA are given in table 1.

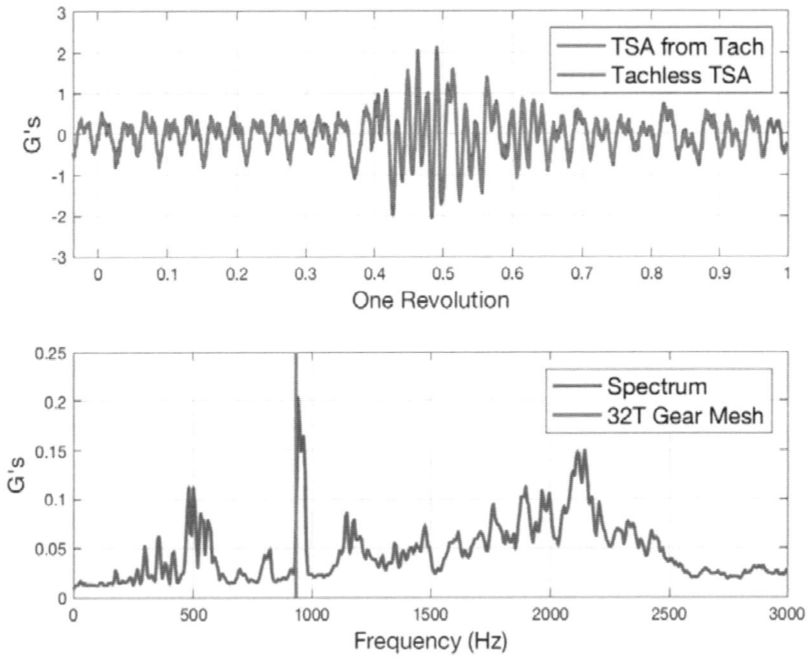

Figure 6: Comparison of the TSA derived from a tachometer and vibration sensor

Table 1 Tach vs. Tachless for Standard Vibe Statistics

Analysis	Tach	Tach from Vibe
SO1	0.0100 g	0.0104 g
SO2	0.0013 g	0.0016 g
SO3	0.0019 g	0.0018 g
TSA RMS	0.5091 g	0.4828 g
TSA P2P	2.0887 g	1.8430 g
FM0	4.278	4.00
AM RMS	0.100 g	0.099 g
AM Kurtosis	4.242	4.217
FM RMS	0.428 radians	0.426 radians
FM Kurtosis	4.995	4.844

Clearly, the gear fault was detectable with either a tachometer, or via vibration method.

Case 2: Tail Rotor Intermediate Gearbox: This was taken from public domain data set, donated by NAVAIR AIR 4.4.2 (Ref 9.). The sample rate was 100,000 samples per second, using the ADC architecture for recording zero crossing data. The tachometer sensor was a variable reluctance speed sensor, using a 22-tooth gear as a target. The shaft rate for the target gear was approximately 3000 rpm (e.g. 500 Hz), with a ratio from this shaft to the tail rotor drive shaft of 7.3:1. As seen in Figure 8, the tachometer signal is noisy (high jitter), with a mean shaft rate of 68.5 Hz.

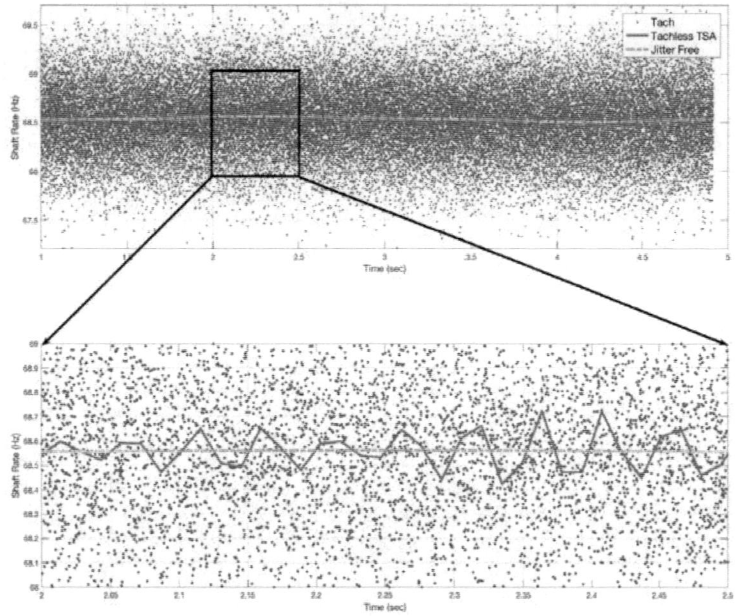

Figure 7: VR Tachometer vs. jitter free tach from vibration

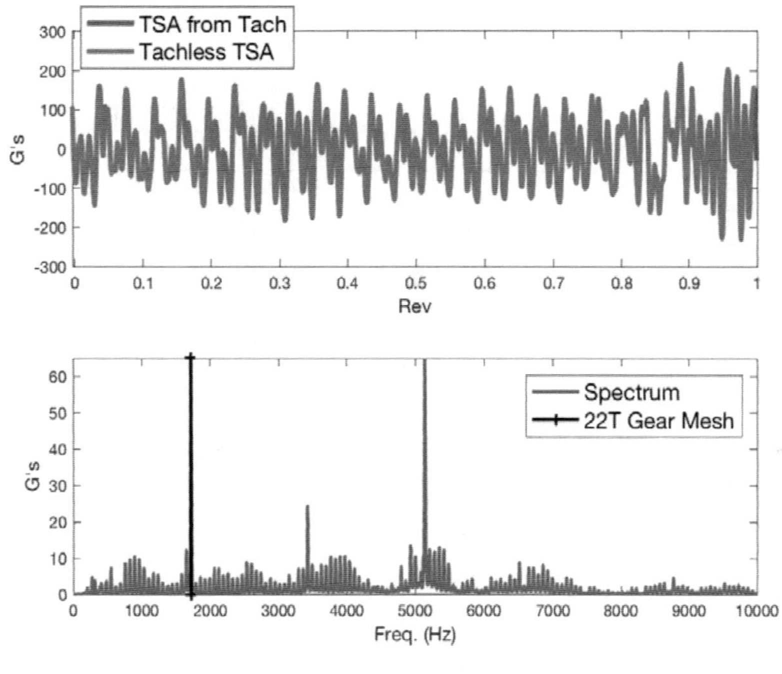

Figure 8: Comparison of the TSA derived from a tachometer vs. vibration sensor

The zoomed view shows the tachometer from vibration and with less jitter than the VR sensor tachometer signal, while the jitter free tachometer signal is smooth, capturing a slowly evolving change in shaft rate. The SNR for this example is approximately 12 dB, due to the 65g mesh frequency (Figure 8). The resulting TSA, aside from a difference is phase, is indistinguishable from the TSA using a tachometer from the VR sensor.

Statistics derived from the TSA for the tail rotor input gear are given in table 2:

Table 2 Tach vs. Tach-less for Standard Vibe Statistics

Analysis	Tach	Tach from Vibe
SO1	0.043g	0.043g
SO2	0.283 g	0.282 g
SO3	1.855 g	1.85g
TSA RMS	81.646 g	81.591g
TSA P2P	222.77g	223.02g
FM0	16.94	15.38
AM RMS	17.06g	17.06g
AM Kurtosis	4.126	4.126
FM RMS	3.037 radians	3.048 radians
FM Kurtosis	2.46	2.45

Conclusions: A method to reconstruct the tachometer signal from vibration was demonstrated using a two-step process. The first step uses an idealized band pass filter implemented with a complex FFT to create an analytic signal, which is needed to derive the tachometer signal for the shaft under analysis. The second step uses another idealized low pass filter on the pseudo derivative of the reconstructed tachometer signal to remove jitter. This processing is run on a smart vibration sensor, which facilitates improved vibration analysis on rotating equipment where in the past the addition of a tachometer would be prohibitive due to cost, weight, certification requirements or being physically impractical.

The quality of the resulting analysis compares favorably with traditional tachometers on real world fault data. In fact, the tachometer from vibration signal may have less noise, resulting in improved fault detection, than traditional tachometer sensor based system

References:

[1] Stewart, R.M., "Some Useful Data Analysis Techniques for Gearbox Diagnostics," Institute of Sound and Vibraiton Research, Proceeding of Meeting on Application of Time Series Analysis, Sept. 1977

[2] Bechhoefer, E., Augustin, M, Kingley, M., "Architecture for a Light Helicopter HUMS", AHS Forum 68, May 1, 2012.

[3] Bechhoefer, E., Van Hecke, B., He, D., Processing for Improved Spectral Analysis, PHM Society Annual Forum, 2013.

[4] M. Lebold, K. McClintic, R. Campbell, C. Byington, K. Maynard, "Review of Vibration Analysis Methods for Gearbox Diagnostics and Prognostics", Proceedings of the 54th Meeting of the Society for Machinery Failure Prevention Technology, 2000.

5S.A. McInerny, Q, Sun, B. Hardman, "Public Domain Algorithms for Bearing and Gear
[5] Fault Detection", Proceedings of the 11th International Conference on Structure Borne Sound and Vibration, 2004.
[6] Bonnardot, F., El Badaoui, M., Randal, R. B., Daniere, J., Guillet, J., "Use of the acceleration signal of a gearbox in order to perform angular resampling (with limited speed fluctuations)", Mechanical Systems and Signal Processing 19 (2005) 766-785.
[7] Combet, F., Gelman, L., "An automated methodology for performing time synchronous averaging of a gearbox signal without speed sensor", Mechanical Systems and Signal Processing 21 (2007) 2590-2606.
[8] Bechhoefer, E., He, D., "Improving Gear Fault Detection by Reducing Tachometer Jitter", AHS Forum 72, May 1, 2014.
[9] NAVAIR AIR 4.4.2, Helicopter Drivetrain Vibration Condition Monitoring Algorithms,
http://web.archive.org/web/20150514044141/http://qsun.eng.ua.edu:80/cpw_web_new/index.htm

Bibliography:

[1] **Eric Bechhoefer** received his B.S. in Biology from the University of Michigan, his M.S. in Operations Research from the Naval Postgraduate School, and a Ph.D. in General Engineering from Kennedy Western University. He is a former Naval Aviator who has worked extensively on condition based maintenance, rotor track and balance, vibration analysis of rotating machinery and fault detection in electronic systems. Dr. Bechhoefer is a board member of the Prognostics Health Management Society, and a member of the IEEE Reliability Society.

Improved RotoSense™ for Rolling Stock: Locomotives and Cars

James P. Hofmeister, Wyatt Pena, Robert Wagoner, and Matthew Nielsen
Ridgetop Group, Inc.
3580 West Ina Road
Tucson, AZ 86741
520-742-3300
hoffy@ridgetopgroup.com, wpena@ridgetopgroup.com, mhudgins@ridgetopgroup.com, rwagoner@ridgetopgroup.com, mnielsen@ridgetopgroup.com

Abstract— This paper describes work related to improving an accelerometer-based sensor, RotoSense™, used for monitoring rolling stock: the locomotives and cars used in trains. At the 2016 MFPT conference, the authors presented a paper, "Accurate Vibration and Speed Measurement on Rotating Shafts using MEMS and IoT Single Wireless Triaxial Sensor," That sensor was capable of measuring shaft speeds of up to 5500 RPM: two example applications were described: (1) a helicopter gearbox, and (2) conditioning monitoring of a railroad track. This paper describes subsequent improvements to that sensor, in terms of ruggedizing and signal quality, to meet requirements of a manufacturer of rolling stock. The sensor described in the previous paper was the first (and only known) to survive, intact, three days of testing at the National Test Track Center in Pueblo, Colorado, including a 10-hour, non-stop, 400-mile test run. Even so, a manufacturer of rolling stock wanted more ruggedizing and better signal quality. The rationale, the methods, and the results of those improvements are presented in the paper.

1. Introduction

Industrial equipment, robotics, and gear boxes that incorporate rotating shafts often need to monitor rotational vibration and shaft speed, as part of broader condition-based maintenance (CBM) systems. Fault detection equipment on drive systems, such as those use for land, sea, and air vehicles, typically use accelerometers mounted on transmission housing to capture, measure, and process vibration signals. The usefulness and flexibility of such detection equipment for applications involving rotating shafts, including pinion and planetary gears, have been limited by cabling, slip-ring approaches, and multiple sensors to obtain monitoring information. For more complex systems, especially those with poor signal transmission paths, a shaft-mounted, wireless solution based on a micro-electro-mechanical system (MEMS) is needed (see Figure 1) [1] [2].

A prototype solution was designed, developed, and embedded on a rotating shaft used in helicopter transmissions to capture and measure vibration signals, and from that, a MEMS version mounted on the wheel hubs of rolling stock [1] [2].

Helicopter Transmission: Pinion Gear—A specific application was the pinion gear of an OH-58C transmission (see Figure 2 and Figure 3). A NASA Glenn Research Center, Small Business Innovation Research (SBIR) award led to an experiment in which a tooth on the spiral bevel gear was pre-notched and then the transmission was run to tooth failure. The purpose was to prove the feasibility of shafted-mounted solutions and compare results with those obtained from traditional stationary, housing mounted accelerometer solutions.

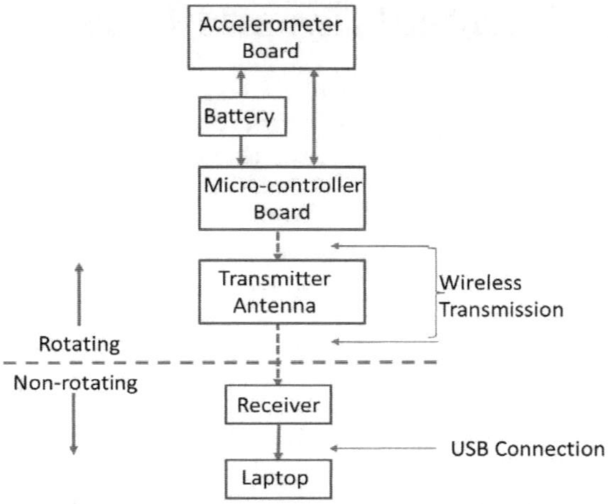

Figure 1. Simple MEMS block diagram [2]

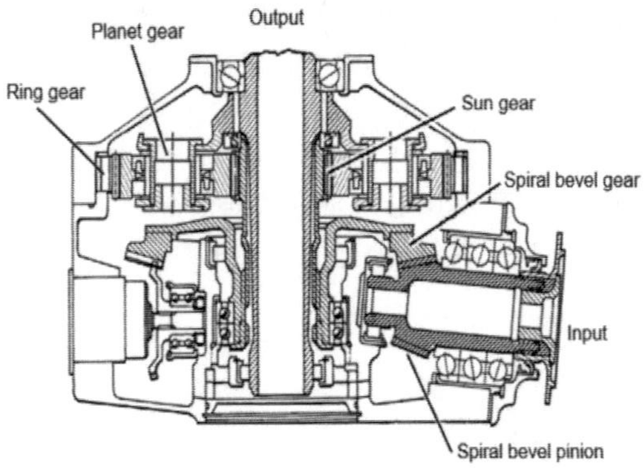

Figure 2. Illustration of an OH-58C transmission

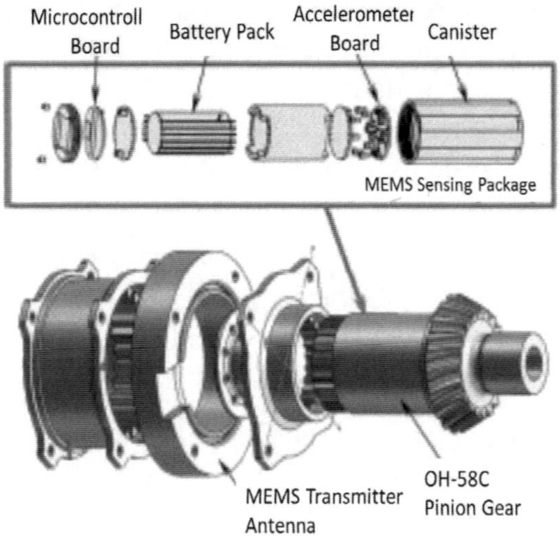

Figure 3. Diagram of MEMS sensor mounted on an OH-58C pinion gear

Figure 4 shows a partially disassembled MEMS sensor: the sensor comprises a microcontroller board, an accelerometer board, and a battery pack mounted in a cylindrical canister.

Figure 4. Canister, boards, and end cap

Wheel Hubs of Rolling Stock—A second example is the physical adaptation of the sensor for mounting on wheel hubs of the shafts of the trucks of train locomotives and cars (see Figure 5). This version of the MEMS sensor was designed and developed to prove feasibility for using such sensors to locate and identify anomalies related to railroad tracks.

Figure 5. Boards with batteries and housing assembly

Referring to Figure 6, the adapted system comprises a MEMs sensor and a gateway to collect data and write data to disk storage as files. The MEMS sensors were mounted on wheels of a train to produce shock data during test runs of a train over a High Tonnage Loop (HTL) test track (TT) to process the data, identify high-force events (HFEs), and locate the position on the HTL TT where HFEs occurred. The purpose was to demonstrate/validate an ability to support focused inspection of tracks to identify and locate anomalies requiring monitoring and service [2].

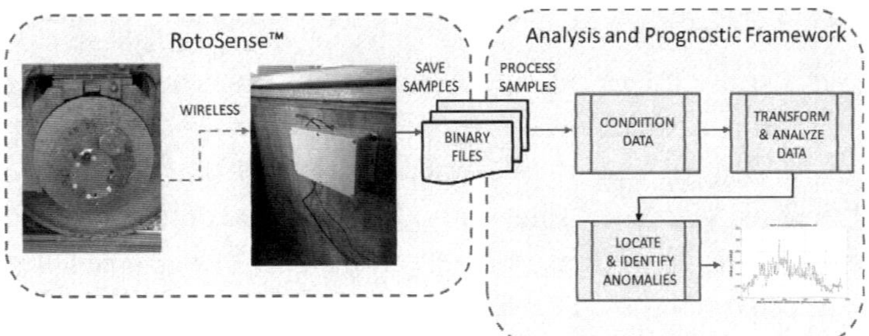

Figure 6. Block diagram of a rolling-stock sensor system

This paper presents initial results of applying the sensors to a helicopter transmission and to the wheel hub of a train car. Section three, Ruggedization, describes concerns and issues related to evaluation testing performed by an independent manufacturer of rolling stock. Those concerns and issues were fully addressed by solution approaches: physical potting of components, repositioning components, and updated software, firmware, and documentation, including new and improved functionality and descriptions. A description of the tests performed at the National Technical Systems (NTS) is described in section four, "NTS Test Description," and section five, "Improved Signal Qualify," presents the improvement in signal quality.

2. Initial Sensor Testing Results

This section presents the initial results of testing a MEMS-based, shaft mounted sensor for two applications are presented: (1) sensor mounted on the shaft of beveled pinion gear in a helicopter transmission, and (2) sensor mounted on the wheel hub of a truck axle of a freight car of a train.

Helicopter Transmission: Shaft of a Beveled Pinion Gear

Triaxle-shaft-mounted MEMS sensors, RotoSense™, were installed on pre-notched OH-58C spiral-bevel pinion gears and endurance tests at NASA's Glenn Research Center were performed and run to tooth fracture failure: the notch was extended at run time = 51.9 hours and widened at run time = 106 hours: see Figure 7.

Figure 7. Final notching of the gear tooth

The sensors performed well, lasted the entire test, and all MEMS accelerometers gave an indication of failure at the end of the test. The MEMS systems performed as well, if not better than, the existing stationary accelerometers mounted on the gear box housing with regards to gear tooth fault detection (Figure 8).

For both the MEMS sensors and stationary sensors, the fault detection time was not much sooner than the actual tooth fracture time. The MEMS sensor spectrum data showed large first order shaft frequency sidebands due to the measurement rotating frame of reference. The method of constructing a pseudo tach signal from periodic characteristics of the vibration data was successful in deriving time-synchronous-average (TSA) signals (Figure 9) without an actual tach: the method proved to be an effective way to improve fault detection for the MEMS [1].

Figure 8. Failed gear – fractured tooth.

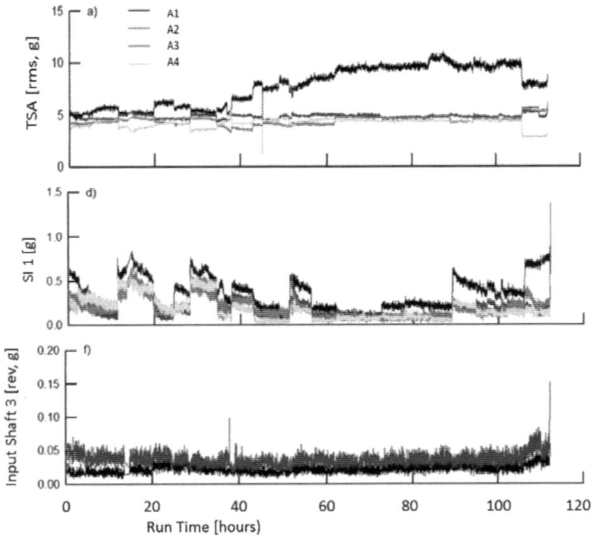

Figure 9. Example data: detection, condition indicators

Rolling Stock: Wheel Hubs of Truck Axles

Test Train—A test train at the National Test Track Center at Pueblo, Colorado, comprised three (3) locomotives and 110 freight cars and was run on a high-tonnage loop (HTL) test track (TT) used for research under heavy axle-loads to test track-component reliability, wear, and fatigue.

Test Track—The HTL track length is 2.7 miles divided into test sections that generally correspond to tangents, spirals, curves, and turnouts that are populated with features and test sections, as seen in Figure 11.

Table 1 lists the features of the TT, some of which are shown in Figure 12 and Figure 13. Fast-train operation is restricted to a maximum 40 miles per hour.

MEMS Configuration and Test Setup—The configuration for train applications includes a MEMS sensor, supporting firmware and software to support collecting, wireless transmitting of data to a gateway, and saving data in binary files. The assemblies are mounted to the wheel hubs to rotate with the axle so any anomalies in the wheels or track can be detected. The MEMS is configured as a three-axis accelerometer with 57mV/g sensitivity with a 161 Hz sampling rate: sensor was mounted concentrically on each end of a freight car axle and also on a locomotive axle of a train (see Figure 10).

Figure 10: MEMS sensor mounted on a wheel hub

Train Movement—The train was auto-controlled to run 15 laps per hour: 4 minutes per lap and 38,640 samples per lap. Four test runs were started on four days in 2015: May 11 – May 14.

Table 2 summarizes the May 14 test run from 2000 to 0632 the next morning: 10 hours, 32 minutes and over 4 million sets of six-byte data. The train started to move about 30 minutes after the sensors were turned on: the train was moved to the test track and two laps of test conditioning were run. After that, the train was kept at a constant speed of 15 laps per hour (4-minute laps).

Table 1. Features of the HTL TT – refer to Figure 11

1. Lubricator	2. Steel Bridge	3. Crib Ties
4. 405 turnout and frog	5. Thermite Welds	6. Rail Temperature
7. Rail Performance Test	8. Concrete Bridge	9. LTM Tests
10. Machine Vision	11. Fiber Optic cable	

Table 2. Summarization of the test run started on May 14, 2015

Description			**Comments**
Train	3 locomotives 110 cars		6,780' long (1.3 miles). Hopper car lengths, coupler to coupler, range from ~58.5 to 60.5 feet: used 60-foot length.
Build	92 minutes		2000 start; 2132 completed build
Run	540 minutes	132 laps	2132: started test conditioning run (TCR)
	17 minutes	2 laps	2149: completed TCR
	518 minutes	129 laps	2149 – 0627: testing
	5 minutes	1 lap	0632: end of test
Wheel 1	20645 files		
Wheel 4	20565 files		

Test Data—Data was collected and buffered for each axis at a sampling rate of 161 Hz and the buffered data was transmitted to a collection hub and saved in output files about once every

1.11 seconds. The data was analyzed, nominal values determined for zero-force conditions, and transformed into +/- values with respect to zero-force.

Data Analysis—Data analysis was hampered by the loss of Global Positioning Data (GPS) caused by a broken antenna. Consequently, raw data (see Figure 14) was analyzed by binning the data in terms of magnitude (xy-plane, the z-plane, and both planes) and relative laps (see Figure 15), examining the peaks, and comparing those peaks to features of the test track: pattern matching. We successfully proved that, even without GPS, we were able to synchronize the data to the start of the test track and then locate those features that could be located: as seen indicated in Figure 16 and Table 3.

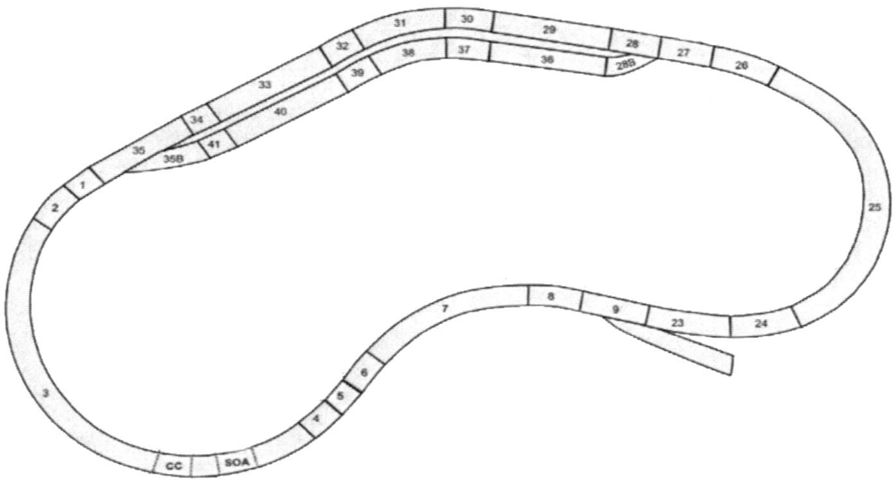

Figure 11. Layout of the Heavy Tonnage Loop Test Track.

[A] [B] [C]

Figure 12. Turnout and frog, switch left [A] and switch Right [B]; steel bridge [C].

[A] [B]

Figure 13: Concrete bridges [A]; crib ties, top of [B]; concrete bridges, bottom of [B].

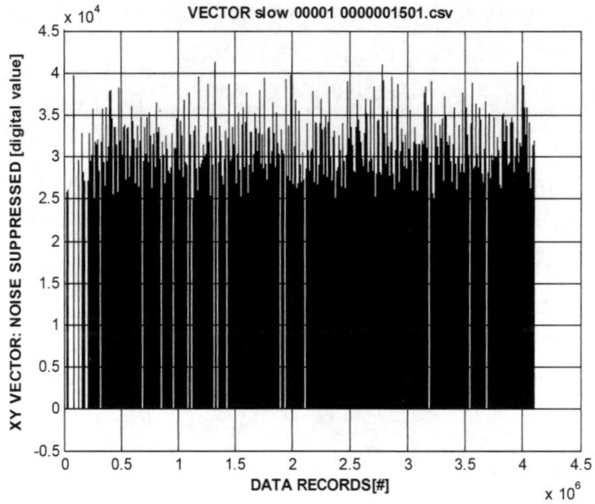

Figure 14. Raw data, xy-plane

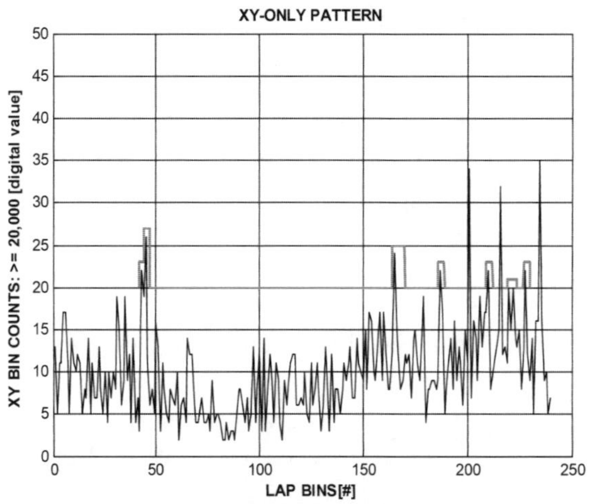

Figure 15. Binned data, xy-plane, by lap-bin number

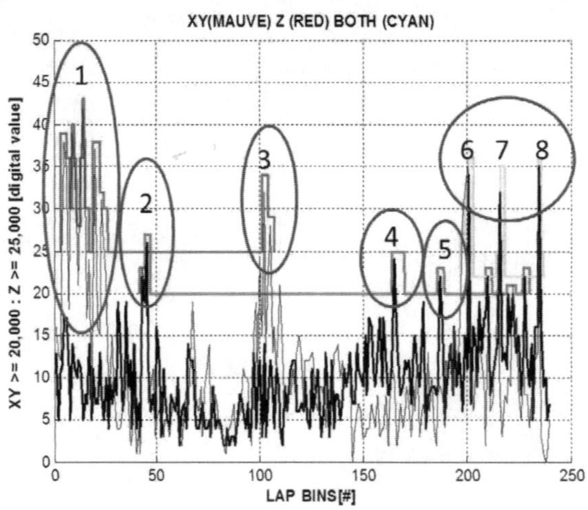

Figure 16. Data and identified features

Table 3. Test track description and detection evaluation

Track Sections	Feature Sections	TT ID	Track Feature	Detection Evaluation	
				XY-vector	Z-vector
1 – 3		S1	Lubricator	ND	ND
4 – 5		S2			
6 - 62	5-26	S3	Repair/overlay welds		Yes (1)
	30-40	S3	Concrete bridge	maybe	maybe
	42-46	S3	Concrete bridge	maybe	Yes (2)
63 – 66		S4	Steel bridges	ND	ND
67 – 69		S5	Bridge deflection	ND	ND
70 – 73		S6	Steel bridges	ND	ND
74 – 92		S7	Rail performance	ND	ND
93 – 97		S8	Fiber optic cable	ND	ND
98 – 108		S9	405 turnout/frog		Yes (3)
109 - 117		S23	405 turnout/frog		Yes (3)
118 – 125		S24	Lubricator	ND	ND
126 – 163		S25	TPO, Tie and fastener, performance	Yes (4)	
164 – 170		S26			
171 – 175		S27	Lubricator	ND	ND
176 – 180		S28	Turn out, steering switch, foundation	Yes (5)	
181 – 193		S29	LTM Tests	ND	ND
194 – 198		S30			
199 – 208		S31	FRA: Rail-seat deterioration, Thermite welds	Yes (6)	Yes (6)
209 – 212		S32			
213 – 225		S33	Crib ties	Yes (7)	Yes (7)
226 – 229		S34			
230 - 240		S35	407 turnout	Yes (8)	Yes (8)

Note: ND means Not Detectable.

3. Ruggedization

Subsequently, a manufacturer of rolling stock obtained sensor units and software and performed additional testing and evaluation. The units were found deficient because of the following: (1) physical failure at high-force testing of up to 100 g vibration and shock leading to board flexing and subsequent solder joint failures and battery displacement; (2) inaccurate vibrational readings caused by flexing of the printed circuit boards.

Operational concerns were reported in the following areas: (1) setting up communications between the Sentinel Gateway and Laptop; (2) setting up communications between RotoSense sensor and Sentinel Gateway; (3) sending commands to the RotoSense sensor; and (4) setting a synchronized time on a RotoSense sensor.

Solution Approaches

The following solution approaches were employed to address the issues and concerns: (1) employ potting; (2) reposition the PCB board and the battery; and (3) improve the quality of the built assembly to ensure an ability to withstand stresses and flexing due to high-g vibration and shock. Quality improvements also included inspections and procedures regarding use, assembly, and testing. In addition to hardware and software solutions, documentation improvements were made.

Final testing of the improved MEMS-based sensor, see Figure 17, was performed at the National Technical Systems (NTS) test facilities in Tempe, Arizona and at Ridgetop Group laboratories in Tucson, Arizona.

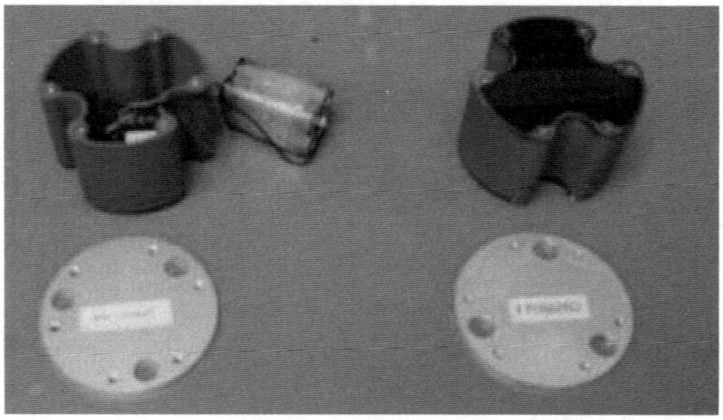

Figure 17. Original (left) and improved (right) sensors

Epoxy-based Potting—An epoxy-based potting was employed: (1) prevented battery displacement; (2) PCB protection from internal vibration and shock forces; (3) increased accuracy in sensor readings; and reduction of additional internal forces. Experiments were performed on sensor units with no potting, partial potting, and full potting. Partial and full potting addressed battery displacement, but partial potting did not fully address board flexing: full potting yielded the best results: zero defects related to shock and vibration occurred during initial and final testing conducted at Ridgetop Group and at NTS.

Component Repositioning—Experiments were performed using variations of component placement: the highest quality signals were obtained using the component placement shown in Figure 18.

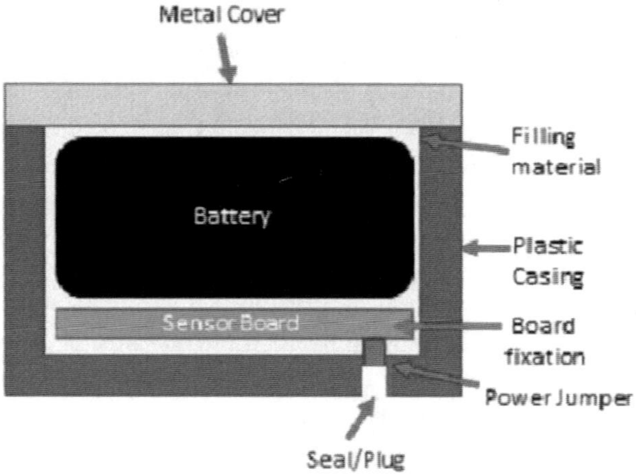

Figure 18. Component placement diagram

Software and Firmware Improvement—Supporting software updates include an updated file structure for more flexible control of future parameter changes. Firmware updates include the following:

1. Updated sensor-gateway communications
2. Updated gateway firmware to reduce timeouts
3. Sensor firmware, functional improvements:
    - Ability to change sensor node address
    - Ability to change the computer IP address for which the data is transmitted to
    - Set time on sensors
    - Read sensor temperature
    - Read radio frequency
    - Improved the wireless speed capabilities from 5.5KB/S to 11KB/S.

Documentation Improvement—The documentation was improved as follows:

1. Improved description of software commands
2. How to change sampling rates
3. How to change node addresses
4. Improved description of use and operation:

- Procedure to turn off the firewall
- Memory map functions
- Operating as an administrator (Admin)

4. NTS Test Description and Results

Test plans were updated from one to three types of vibration test: (1) fixed frequencies, (2) sweep frequency, and (3) pulse (shock). Testing included those run at NTS in June of 2017.

Fixed Vibration Frequencies—Three vibration frequencies at seven levels of force defined and tested: Table 4. The sequences were run at four sampling rates: 160 s/s, 250 s/s, 500 s/s, and 1000 s/s.

No physical damage or anomalies were found. There was no loss of signal.

Table 4. Fixed vibration frequency

Vibration Frequency: Force
Dwell @ 50 Hz: 5g, 10g, 20, 30g, 50g, 70g, 100g
Dwell @ 100 Hz: 5g, 10g, 20, 30g, 50g, 70g, 100g
Dwell @ 250 Hz: 5g, 10g, 20, 30g, 50g, 70g, 100g

Sweep Vibration Frequency—The sweep frequency as defined as 10 Hz to 500 Hz at two different levels of force: 10g and 40g. The sequences were run at the defined four sampling rates.

No physical damage or anomalies were found. There was no loss of signal.

Pulse Vibration (Shock)—The pulse was defined as a six-millisecond, positive half-sine wave repeated five times at the six different magnitudes of force used for fixed-frequency testing. The sequence was repeated for each of four different sampling rates.

No physical damage or anomalies were found. There was no loss of signal.

5. Improved Signal Quality

After potting and repositioning of components, the signal quality is significantly improved as seen in Figure 19 through Figure 22.

The output signals from the MEMS-based sensor after repositioning and potting of the boards and battery pack no longer exhibit extreme harmonic distortion and the amplitudes more accurately reflect the magnitude of the vibration force.

The bandwidth has also been improved from 4-6 Kb/s to 20-22 Kb/s.

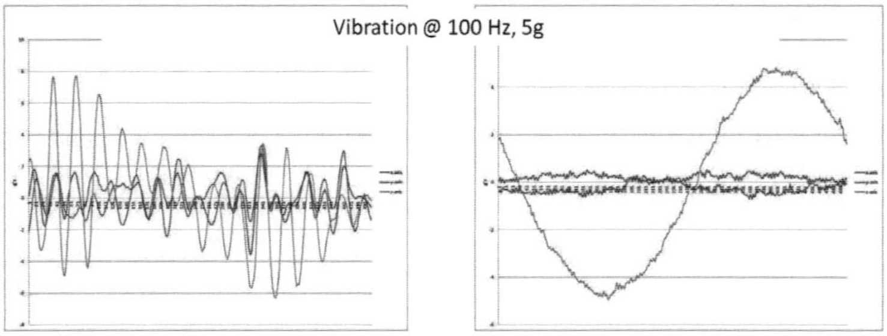

Figure 19. Vibration @ 100 Hz, 5g – before (left) and after (right) repositioning and potting

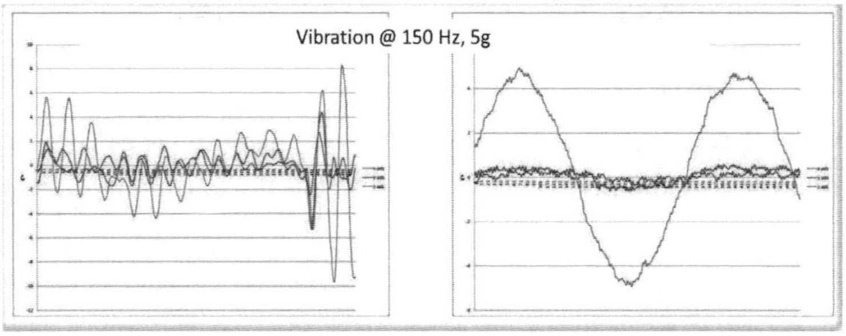

Figure 20. Vibration @ 150 Hz, 5g – before (left) and after (right) repositioning and potting

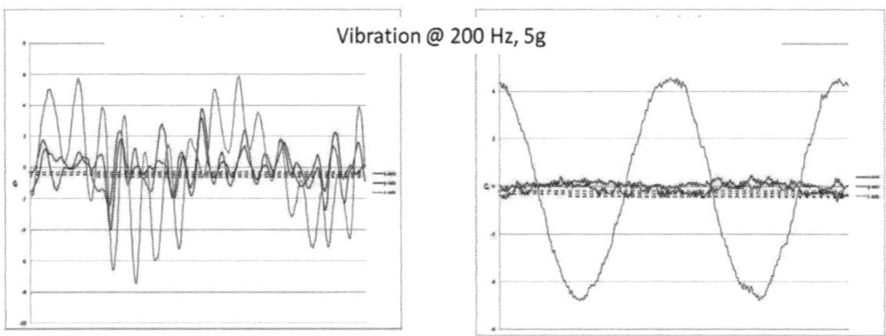

Figure 21. Vibration @ 200 Hz, 5g – before (left) and after (right) repositioning and potting

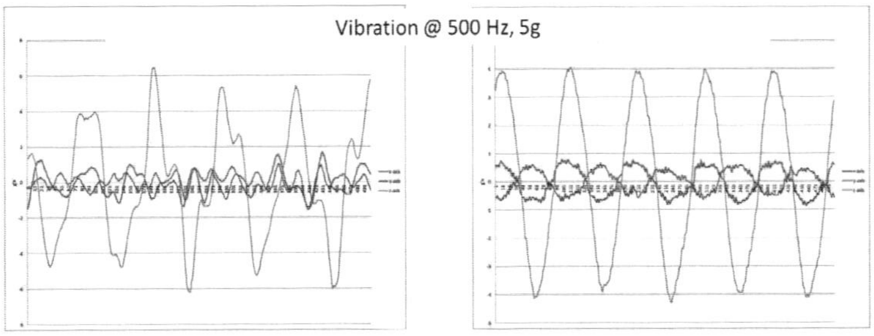

Figure 22. Vibration @ 500 Hz, 5g – before (left) and after (right) repositioning and potting

6. Gateway Improvements

A new version of the Gateway for helicopter use is significantly more rugged: (1) it uses super strong Vicor power supply, that went through extensive power cycle testing on the Apache helicopter; and (2) all internal components are compact, soldered to a carrier board, and/or immobilized.

Also, the new Gateway is compatible with IoT sensors that can communicate through a Wi-Fi connection: at speeds up to 300 MB/s.

7. Summary

Two versions of MEMS-bases sensors were designed and developed: one version for use in gear boxes of helicopters and the other version for hub mounting on axles of rolling stock. Each version of the sensors, their application, and their testing and results were described and the results evaluated: both versions functioned and operated successfully.

Subsequently, it was discovered the version for rolling stock was not rugged enough to pass 100 g vibration testing, and that level of testing was a customer required. After analysis and research, that version of the sensor was ruggedized by repositioning and potting of the components: the boards and the battery pack. The ruggedized version of the sensor was tested at the National Technical Systems at various regimes: fixed frequency, sweep frequency, and pulse mode vibration at various frequencies and levels of force - no physical failure or anomalies occurred.

The quality (level of harmonic distortion) and the accuracy of measured force were proved by comparing signals from non-ruggedized sensors to those from ruggedized sensors. In addition to ruggedization, improvements were made to the hardware to increase sensor bandwidth and to improve the Gateway with respect to ruggedness, to IoT compatibility, and to communication speeds. The software, firmware, and documentation were updated to increase communication, to reduce timeouts, and to increase functionality and usability.

Plans are to apply the potting approach to increase the ruggedization of the helicopter version of the sensor. That version has already been improved by increasing the sensor bandwidth, making the Gateway more rugged, and improving Gateway communications. We anticipate further improvements in the future, especially with regards to extending battery life.

ACKNOWLEDGEMENTS

The authors thank Naval Air, Naval Sea, U.S. Army, U.S. Air Force, and NASA research centers for their support and funding of multiple projects that led to the results described and shown in this paper.

REFERENCES

[1] Lewicki, D.G., Lambert, N.A., and Wagoner, R.S., (2015) "Evaluation of MEMS-Based Wireless Accelerometer Sensors in Detecting Gear Tooth Faults in Helicopter Transmissions," NASA/TM-2015-218722.

[2] Hofmeister, J.P., Goodman, D., and Wagoner, R. (2016), "Accurate Vibration and Speed Measurement on Rotating Shafts Using MEMS and IoT Single Wireless Triaxle Sensor," 2016 Machine Failure Prevention Technology, Dayton, OH, 24-26 May, 2016.

Modeling and Simulation Analysis of Dual-Rotor Vibration System with Multiple Faults

Yizhou Yang, Chao Liu, Dongxiang Jiang and Wenguang Yang
State Key Lab of Control and Simulation of Power Systems and Generation Equipment,
Department of Thermal Engineering, Tsinghua University
Beijing 100084, China
Telephone: +86 15201522492
yangyz16@mails.tsinghua.edu.cn

Abstract: Dual-rotor system is an important rotor form in rotating machinery like gas turbine engine. Its complex structure results in rich dynamical behaviors and more probability to failure. The modeling and simulation of the dual-rotor system can help to understand its dynamic characteristics and provide theoretical support for the design, operation and maintenance. In this paper, a dynamic model of a dual-rotor system with multiple rotor faults is established. The dual-rotor vibration model without any fault is built by finite element method, where the two shafts are connected by an inter-shaft bearing and nonlinear models of rolling element bearing and squeeze film damper are considered. The numerical integration method of Newmark-β is used to obtain the steady-state vibration response of the system. Then rotor faults are introduced to the system model, including unbalance, misalignment, looseness and rub-impact. The steady-state responses of single faults in the dual-rotor system are analyzed and typical fault features are obtained. Then coupling characteristics between different rotor faults, and the influences of the squeeze film dampers on the dynamic characteristics and fault features of the system are studied.

Keywords: dual-rotor system, modeling and simulation, rotor faults, rolling element bearing, squeeze film damper.

Introduction: Dual-rotor is a common form of rotor system in rotating machinery like aero-engines. The high-pressure rotor and the low-pressure rotor are coupled via an inter-shaft bearing, and the speed of the rotors are different and variable, making the vibration signals more complex than general rotating machinery. The modeling and simulation of the dual-rotor system and rotor faults can help to understand their dynamic characteristics and provide theoretical support for the design, operation and maintenance of the rotor. The finite element model turn the continuous flexible structure into finite discrete elements, obtains rich internal information, and are also convenient to add nonlinear forces, which is suitable for numerical calculation [1, 2]. Processing the flexible parts as finite elements and the parts of large stiffness as discrete lumped masses could reduce the complexity of the system model and ensure the accuracy of simulation results at the same time, making it a more efficient modeling approach [3, 4].

Rolling element bearings and squeeze film dampers are introduced in rotor models in forms of nonlinear external forces. The earliest complete rolling element bearing model was established by Gupta in 1975, which had 2 degrees of freedoms (DOFs) and Hertzian contact, the non-linearity and time-varying characteristics are considered in. Later more nonlinear factors had been taken into consideration in this bearing force model by other researchers, like bearing race clearance and the rolling element slippage. Squeeze film damper is a type of fluid-film bearing whose simplest form consists of an oil-filled annular cavity surrounding the outer race of a rolling element bearing. In aero-engines where rolling element bearings provide little damping, squeeze film dampers are widely used to introduce additional damping to attenuate vibration. The cavitated short bearing theory is employed for calculating the fluid-film forces in a damper, as the outer race of the bearing acts as the damper's journal and is prevented from rotating but allowed to whirl.

In this paper, a finite element approximate model of an aero-engine dual-rotor system which includes nonlinear models of rolling element bearings and squeeze film dampers are established. Steady-state vibration response of the dual-rotor system is obtained by solving the dynamic equation of the system. Then models of typical rotor faults are introduced to the system by adding nonlinear forces to the generalized external force vector. Simulated faults include rotor unbalance, rotor misalignment, support looseness, and dynamic-static rub-impact. Vibration responses of single rotor faults are calculated and fault features are obtained. Those fault features are compared with situations with the absence of squeeze film dampers. Vibration responses of situations with more than one rotor fault are calculated, whose fault features are also compared with ones of single fault situations.

Dynamic model of the dual-rotor system: The dual-rotor system consists of two rotors connected by an inter-shaft bearing. Each rotor is composed of a flexible shaft, rigid discs and supports. The schematic of the rotor system is shown in Figure 1. The support arrangement of the rotor is similar to the actual aero-engine rotor.

The shafts are modelled with finite element method by using the 3-D elastic beam element. The 3-D elastic beam element has 2 nodes and each node has 6 degree of freedoms (DOFs), which are translations in the X, Y and Z directions and rotations around the X, Y and Z directions. The DOFs and the motion equation of the element are:

$$q^{\mathrm{e}} = \left[u_{x1}, u_{y1}, u_{z1}, \theta_{x1}, \theta_{y1}, \theta_{z1}, u_{x2}, u_{y2}, u_{z2}, \theta_{x2}, \theta_{y2}, \theta_{z2}\right]^{T} \quad (1)$$

$$M^{e}\ddot{q}^{e} + (-\omega G^{e})\dot{q}^{e} + K^{e}q^{e} = 0 \quad (2)$$

M^{e} is the consistent mass matrix, G^{e} is the gyroscopic matrix, K^{e} is the stiffness matrix and ω is the rotational speed.

The rigid discs are modelled as discrete lumped masses using the 3-D mass element which has one node with 6 DOFs. The DOFs and the motion equation of the element are:

$$q^e = \left[u_x, u_y, u_z, \theta_x, \theta_y, \theta_z\right]^T \tag{3}$$

$$M^e \ddot{q}^e + (-\omega G^e)\dot{q}^e = Q^e \tag{4}$$

Q^e is the external force vector and could be the unbalance force or other types of excitation on the disc.

As shown in the Figure 1, the shaft of the low-pressure (inner) rotor has been divided into 17 elements and has 18 nodes, and the shaft of the high-pressure (outer) rotor has been divided into 18 elements and has 19 nodes. The shafts lie along the X direction. Combine all the elements of a shaft and discs together, the motion equation of one rotor is:

$$M_i^r \ddot{q}_i^r + (C_i^r - \omega_i G_i^r)\dot{q}_i^r + K_i^r q_i^r = Q_i^r, \qquad i = 1,2 \tag{5}$$

C^r is the damping matrix which is using Rayleigh damping for simplicity and is in the form of proportional damping [5] as:

$$C^r = \alpha M^r + \beta K^r \tag{6}$$

The rotor system model is established by assembling all the finite elements and add the external forces. The motion equation could be written as:

$$M^s \ddot{q}^s + (C^s - \omega G^s)\dot{q}^s + K^s q^s = Q^s \tag{7}$$

Q^s is the generalized external force vector of the system, which contains the unbalance forces, the weight forces, and the nonlinear forces of the rolling element bearings and the squeeze film dampers:

$$Q^s = F_u + W - F_B + F_D \tag{8}$$

The unbalance force on the disc is the main excitation of this dynamic model. The nonlinear equation of the rotor system model is solved by the explicit Newmark-β method. This numerical integration approach is suitable for obtaining the dynamic response of a nonlinear system. Then the displacement, velocity and acceleration at each node of the finite element model can be obtained.

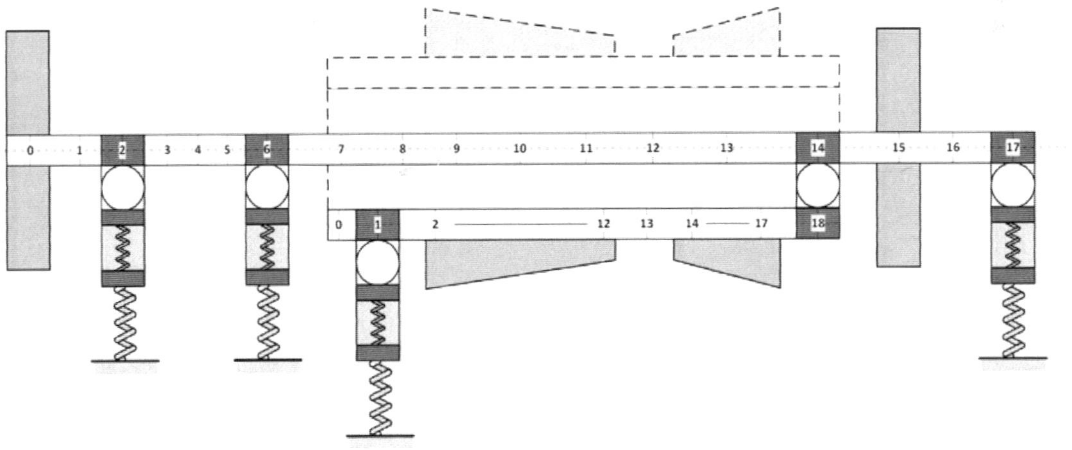

Figure 1. Dual-rotor system model.

Rolling element bearing model: The rotor is supported by rolling element bearings. Here a nonlinear multi-body dynamic model is used [6, 7], which is shown in Figure 2. The rolling element bearing model is assumed as nonlinear springs with no mass and evenly arranged along the rolling element raceway. The inner race of the bearing is assumed to be fixed to the shaft and the outer race is modelled as a lumped mass.

When the rotor rotates at constant speed of ω, the cage speed is ω_c. The angular position of the ith rolling element at time t will be θ_i.

$$\omega_c = \frac{\omega}{2}\left(1 - \frac{d}{D}\cos\alpha\right) \tag{9}$$

$$\theta_i = \frac{2\pi(i-1)}{Z} + \omega_c t + \theta_0 + (0.5 - rand) \times \theta_{slip} \quad i = 1, 2, \dots Z \tag{10}$$

θ_0 is the initial position of the first rolling element. $(0.5 - rand) \times \theta_{slip}$ is the phase deviation caused by slippage. The value of phase variation θ_{slip} could be 0.01 ~ 0.02 rad. If the displacement of the shaft (x_s, y_s) and the outer race (x_o, y_o) is known, the contact deformation for the ith rolling element is given as:

$$\gamma_i = (x_s - x_o)\cos\theta_i + (y_s - y_o)\sin\theta_i - \delta \quad i = 1, 2, \dots Z \tag{11}$$

δ is the initial clearance between the rolling elements and the bearing races. The rolling elements are compressed only when γ_i is positive, so the contact deformation will be multiplied by a factor as:

$$\lambda_i = \begin{cases} 1 & \gamma_i > 0 \\ 0 & \gamma_i \le 0 \end{cases} \tag{12}$$

According to Hertzian contact theory, the contact force of the ith rolling element is:

$$f_i = k_b(\lambda_i \gamma_i)^n \tag{13}$$

k_b is the contact stiffness factor and can be obtained from geometric and material properties of the contact between the rolling element and the inner race or the outer race. The exponent n is 3/2 when the bearing is a ball bearing and 10/9 when a roller bearing. The total bearing force can be calculated by summing the contact force of each rolling element in the X and Y directions:

$$\begin{cases} F_{Bx} = k_b \sum_{i=1}^{Z} (\lambda_i \gamma_i)^n \cos\theta_i \\ F_{By} = k_b \sum_{i=1}^{Z} (\lambda_i \gamma_i)^n \sin\theta_i \end{cases} \tag{14}$$

Squeeze film damper model: The squeeze film damper (SFD) is mounted between the outer race of a rolling element bearing and a bearing pedestal. The outer race forms the non-rotating journal of the squeeze film damper, and are supported by centering strings [8-10]. The model of SFD is shown in Figure 2.

Theoretical oil-film forces in squeeze film dampers are obtained by integrating the pressure distribution over the entire damper surface. Based on the Reynolds equation and the short bearing approximation, the pressure distribution in a π-film damper may be derived as:

$$p(\theta, z) = -\frac{6\mu}{c^2}\left(\frac{L^2}{4} - z^2\right)\frac{\varepsilon\dot{\phi}\sin\theta + \dot{\varepsilon}\cos\theta}{(1+\varepsilon\cos\theta)^3} \tag{15}$$

In the equation, ε and $\dot{\varepsilon}$ respectively are the eccentricity ratio and velocity in the radial direction. $\dot{\phi}$ is the angular velocity. θ is the angular position measured from the maximum film thickness end of the line of centers. L is the length of the damper, z the position in the axial direction, c the damper's radial clearance, and μ the dynamic viscosity of the oil.

Positive pressures occur for $\theta_1 < \theta < \theta_1 + \pi$ when

$$\varepsilon\dot{\phi}\sin\theta + \dot{\varepsilon}\cos\theta < 0 \ \ and \ \tan\theta_1 = \frac{\dot{\varepsilon}}{-\varepsilon\dot{\phi}} \tag{16}$$

Then the oil film forces can be integrated as:

$$F_r = -\frac{\mu R L^3}{c^2}\left(\dot{\varepsilon}I_1 + \varepsilon\dot{\phi}I_2\right) \tag{17}$$

$$F_t = -\frac{\mu R L^3}{c^2}\left(\dot{\varepsilon}I_2 + \varepsilon\dot{\phi}I_3\right) \tag{18}$$

where

$$I_1 = \int_{\theta_1}^{\theta_1+\pi}\frac{\cos^2\theta}{(1+\varepsilon\cos\theta)^3}d\theta \tag{19}$$

$$I_2 = \int_{\theta_1}^{\theta_1+\pi}\frac{\cos\theta\sin\theta}{(1+\varepsilon\cos\theta)^3}d\theta \tag{20}$$

$$I_3 = \int_{\theta_1}^{\theta_1+\pi}\frac{\sin^2\theta}{(1+\varepsilon\cos\theta)^3}d\theta \tag{21}$$

The closed form expressions of the integrals have been evaluated analytically as:

$$I_1 = \frac{\varepsilon\sin\theta_1\left(3 + (2-5\varepsilon^2)\cos^2\theta_1\right)}{(1-\varepsilon^2)^2(1-\varepsilon^2\cos^2\theta_1)^2} + \frac{\alpha(1+2\varepsilon^2)}{(1-\varepsilon^2)^{2.5}} \tag{22}$$

$$I_2 = -\frac{2\varepsilon\cos^3\theta_1}{(1-\varepsilon^2\cos^2\theta_1)^2} \tag{23}$$

$$I_3 = \frac{\varepsilon\sin\theta_1\left(1 + (\varepsilon^2-2)\cos^2\theta_1\right)}{(1-\varepsilon^2)(1-\varepsilon^2\cos^2\theta_1)^2} + \frac{\alpha}{(1-\varepsilon^2)^{1.5}} \tag{24}$$

$$\alpha = \frac{\pi}{2} + \arctan\left[\frac{\varepsilon\sin\theta_1}{(1-\varepsilon^2)^{0.5}}\right] \tag{25}$$

Suppose the displacement of the outer race (x_o, y_o) and the bearing pedestal (x_p, y_p), and the velocity of the outer race $(\dot{x}_o, \dot{y}_o)$ and the bearing pedestal $(\dot{x}_p, \dot{y}_p)$, so

$$x = x_o - x_p, \qquad y = y_o - y_p \tag{26}$$

$$\dot{x} = \dot{x}_o - \dot{x}_p, \qquad \dot{y} = \dot{y}_o - \dot{y}_p \tag{27}$$

$$\varepsilon = \frac{e}{c} = \frac{\sqrt{x^2 + y^2}}{c}, \qquad \dot{\varepsilon} = \frac{x \cdot \dot{x} + y \cdot \dot{y}}{c\sqrt{x^2 + y^2}} \tag{28}$$

$$\phi = \arctan\frac{y}{x}, \qquad \dot{\phi} = \frac{x \cdot \dot{y} - y \cdot \dot{x}}{\sqrt{x^2 + y^2}} \tag{29}$$

The oil-film forces in polar coordinates can be recast in Cartesian coordinates like:

$$\begin{cases} F_{Dx} = F_r \cos\phi - F_t \sin\phi \\ F_{Dy} = F_r \sin\phi + F_t \cos\phi \end{cases} \tag{30}$$

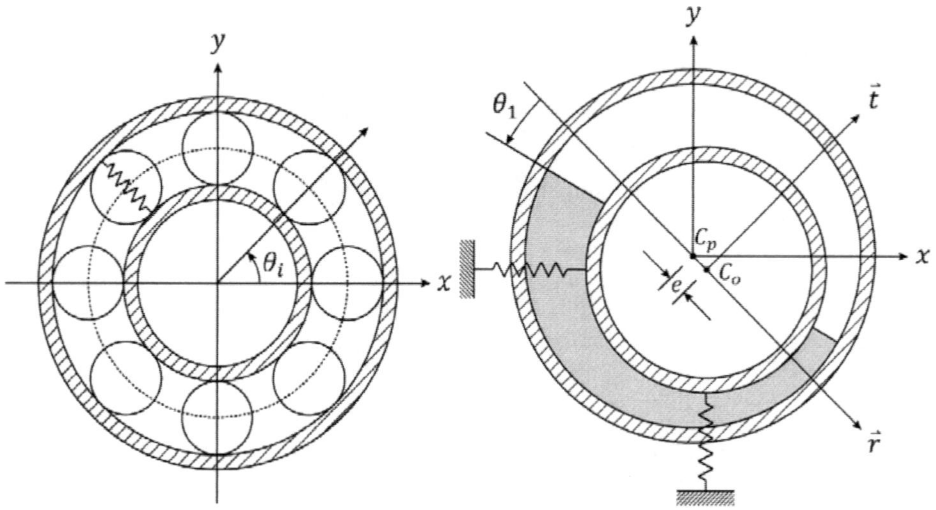

Figure 2. Rolling element bearing and squeeze film damper model.

Table 1. Parameters of rolling element bearing and squeeze film damper model.

Model	Notation	Description	Value	
Rolling element bearing	D	Pitch diameter	306 mm	
	d	Rolling element diameter	17 mm	
	Z	Rolling element number	30	
	α	Contact angle	0	
	k_b	Contact stiffness factor	7×10^8 N·m$^{-3/2}$	6×10^7 N·m$^{-10/9}$
Squeeze film damper	μ	Dynamic viscosity of oil	5 mPa·s	
	R	Damper radius	180 mm	
	L	Damper length	42 mm	
	c	Damper radial clearance	0.36 mm	

Normal state vibration response: Vibration response of the dual-rotor system without any rotor fault is calculated. The rotating speed is set to a fixed value of (70, 90) Hz. The displacement waveform, spectrum and shaft centerline orbit of node 0 and node 15 on the inner rotor are in Figure 3. Due to the coupling effect of the inter-shaft bearing, both frequencies of the rotating speed are reflected on the same node. Node 15 is near the inter-shaft bearing than node 0, so the spectral line of the outer shaft frequency is higher than the one of the inner shaft frequency, and vice versa. Besides the two main frequencies, no other frequency components is clear in the vibration signals.

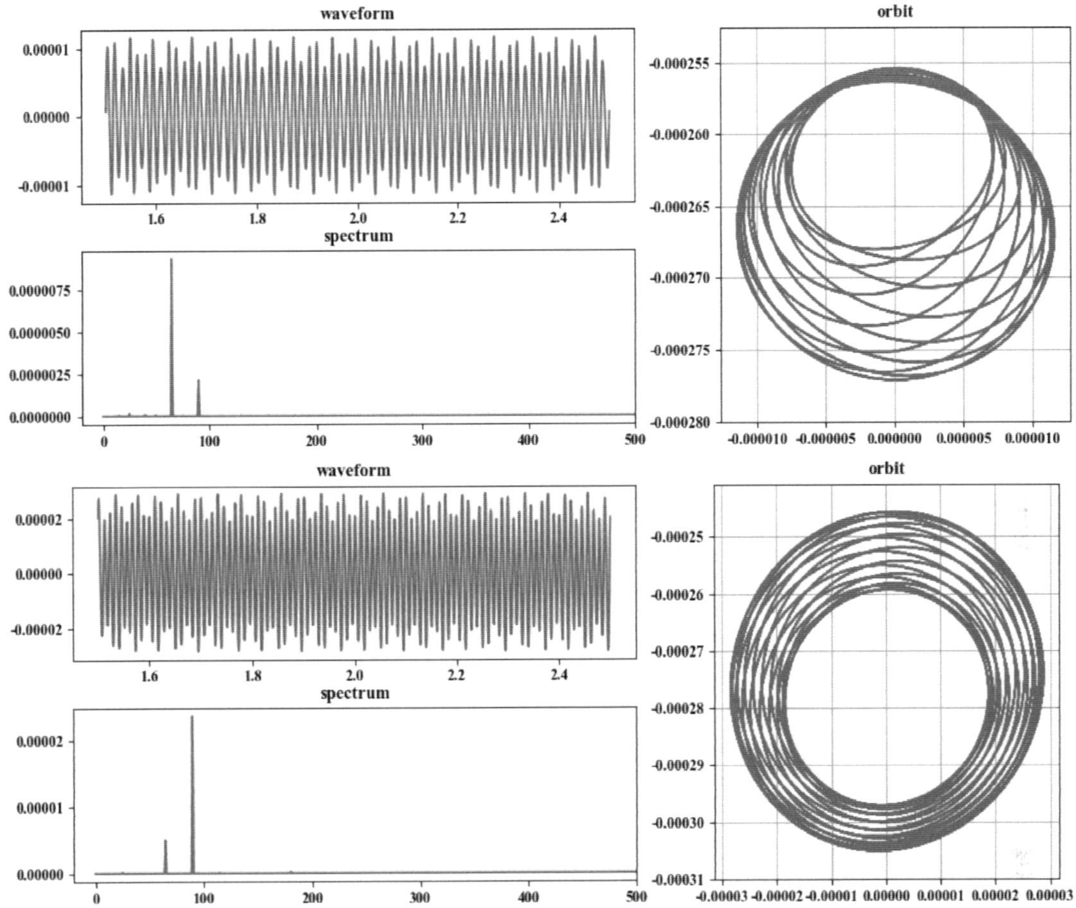

Figure 3. Waveform, spectrum and orbit of node 0 and node 15 on the inner rotor.

Some states are considered including: (1) no gravity effect, (2) no squeeze film damper, (3) SFD has no centering string, (4) centering string has have the stiffness. Vibration responses of those states are calculated and their shaft centerline orbits of the outer race (journal of SFD) of support 3 and node 3 at the outer rotor are plotted in Figure 4, comparing with the normal state. If the gravity effect is neglected, the shaft centerline orbit will be centered on the origin. When there is gravity or the stiffness of the SFD centering string is weakened, the center of orbit declines. And if there are no SFD in the dual-rotor system, the trajectory of the rotor will be relatively messy, and the inner rotor frequency become more prominent because there are three SFDs on the inner rotor but only one on the outer rotor. It can be seen that the damping effect of the squeeze film damper is very significant.

Calculate vibration responses at different rotating speed with a speed ratio of 1.5. All the spectra are plotted in the 3D spectrum of Figure 5. As the rotating speed increases, the vibration amplitude will increase significantly as the two rotational frequencies pass through the natural frequencies of the dual-rotor system. When the rotating speed is about (20, 30) and (56, 84) Hz the outer rotor has a high vibration amplitude. When the rotating speed is about (32, 48) and (84, 126) Hz the inner rotor has a high vibration amplitude. Thus it can be inferred that the first two natural frequencies is around 31 and 84 Hz.

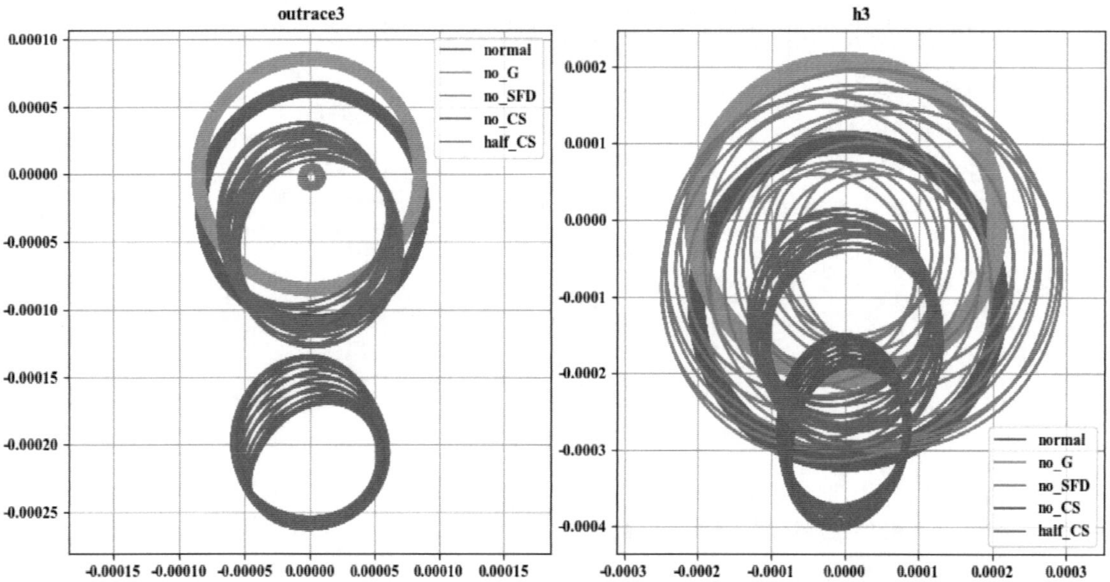

Figure 4. Shaft centerline orbits of the outer race 3 and node 3 at the outer rotor.

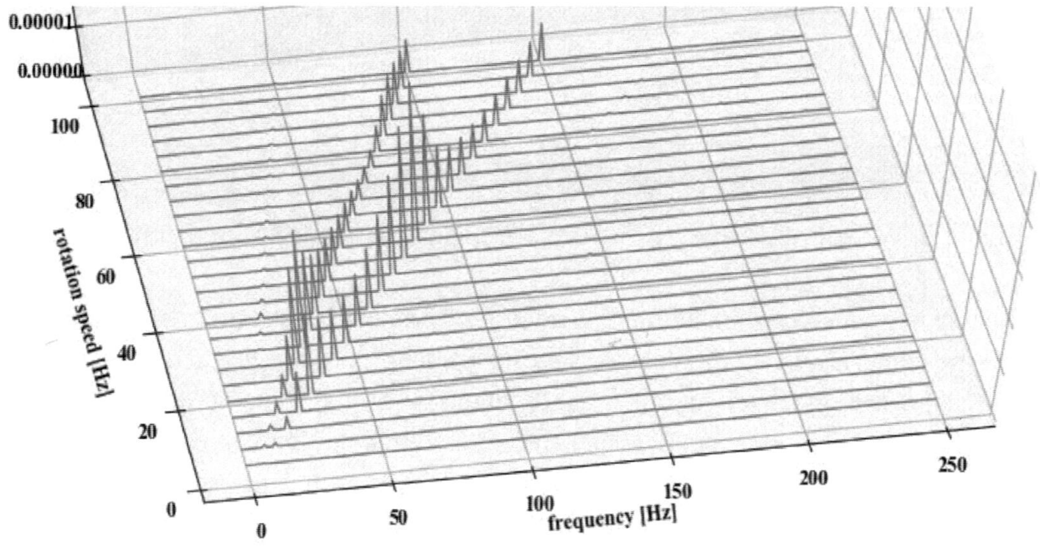

Figure 5. 3D spectrum at different rotating speed with a speed ratio of 1.5.

Models of rotor faults: Typical rotor faults include rotor unbalance, rotor misalignment, support looseness, and dynamic-static rub-impact. In the dual-rotor system dynamics model, a typical rotor fault is achieved by adding a corresponding load in the generalized external force vector.

Rotor unbalance manifests as centrifugal forces of the eccentric mass. Unbalance force is often the main excitation of a rotor dynamic system, but large eccentric mass will cause excessive vibration when the rotor rotates at a high speed. The external force of the unbalance fault model is:

$$\begin{cases} F_{uiy} = m_{di} e \omega^2 \cos(\omega t) \\ F_{uiz} = m_{di} e \omega^2 \sin(\omega t) \end{cases} \tag{31}$$

Rotor misalignment happens when two rotors connected by a coupling do not have their axes on the same line. The coupling will rotate as an eccentric mass at twice the rotor speed, adding a misalignment force to the rotor [11]. The external force of the misalignment fault model is:

$$\begin{cases} F_{miy} = -m_{ci} e \omega^2 \sin(2\omega t) \\ F_{miz} = m_{ci} e \omega^2 \cos(2\omega t) \end{cases} \tag{32}$$

Support looseness leads to a reduction in the stiffness of the connection and even creates an unconstrained gap. So the looseness fault manifests as piecewise linear stiffness of the foundation support [12]. The external force of the looseness fault model is:

$$F_{li} = \begin{cases} (k_0 - k_{l1}) \cdot x_i & x_i > d_l \\ (k_0 - k_{l2}) \cdot x_i & x_i \leq 0 \\ k_0 \cdot x_i & 0 < x_i \leq d_l \end{cases} \tag{33}$$

Dynamic-static rub-impact is the friction and collision between the rotor and the stator due to the reduction of the clearance between them. When a rub-impact occurs, the moving part will come into contact with the static part and then spring back. The part with a lower material hardness will suffer greater wear as the rubbing progresses [13]. The external force of the rub-impact fault model is:

$$\begin{cases} F_{riy} = k_r(1 - d_r/r_i)(-y_i + f \cdot z_i) \\ F_{riz} = k_r(1 - d_r/r_i)(-z_i - f \cdot y_i) \end{cases} \quad r_i = \sqrt{y_i^2 + z_i^2} \geq d_r \tag{34}$$

Table 2. Parameters of rotor faults.

Fault	Parameter	Description	Value
Normal	e	eccentricity	0.02 mm / 0.01 mm
Unbalance	e	eccentricity	0.02 mm / 0.03 mm
Misalignment	$m_c e$	eccentric mass	0.004 mm·kg
	d_l	loose clearance	0.05 mm
Looseness	k_{l1}, k_{l2}	looseness stiffness	5×10^8 N/m
	k_0	clearance stiffness	0
	d_r	rotor-stator clearance	0.15 mm
Rub-impact	k_r	contact stiffness	7.5×10^7 N/m
	f	friction factor	0.1

Fault simulation results: Adding a single typical rotor fault in the dual-rotor system, the vibration response of each fault above is calculated. Parameters of the rotor faults used in the calculation are listed in Table 2.

The displacement waveform, spectrum and shaft centerline orbit of node 15 on the inner rotor at the state of unbalance fault is in Figure 6. The eccentric mass of the inner rotor disc increase, thus the amplitude of its rotation frequency increase. No other frequency component is added to the vibration signal. The shaft centerline orbit still looks smooth.

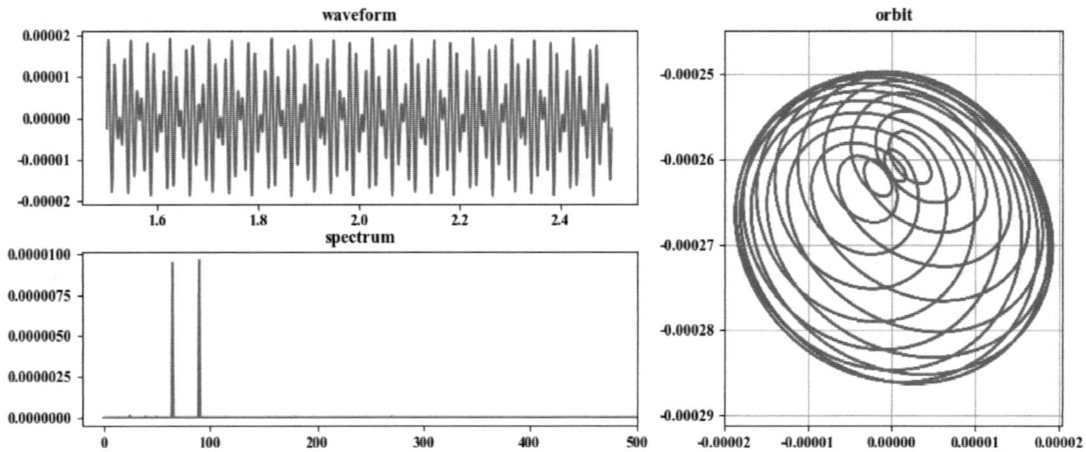

Figure 6. Waveform, spectrum and orbit of unbalance fault.

The displacement waveform, spectrum and shaft centerline orbit of node 15 on the inner rotor at the state of misalignment fault is in Figure 7. Due to the excitation of misalignment on the inner rotor, the frequency component of double the inner rotor rotation frequency becomes prominent.

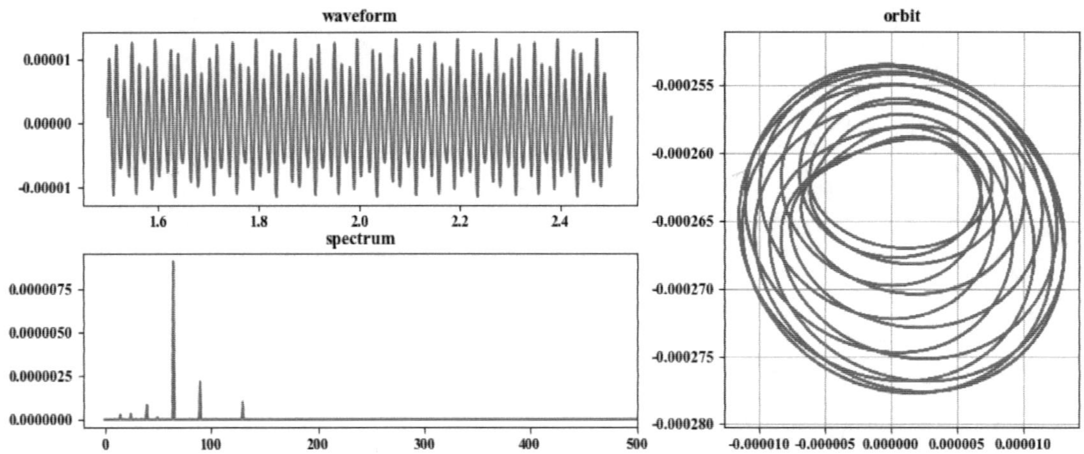

Figure 7. Waveform, spectrum and orbit of misalignment fault.

The displacement waveform, spectrum and shaft centerline orbit of node 15 on the inner rotor at the state of looseness fault is in Figure 8. The loose clearance is set in the horizon direction, so the constraint in this direction weakens. Low frequency components below the rotation frequency appear in the spectrum.

When there is no squeeze film damper, fractional frequency components under the rotation frequency become more abundant. The shaft centerline orbit becomes less smooth. Looseness will cause a certain degree of impact vibration, but the presence of SFDs weakens the impact. This is also an important manifestation of the shock absorption of SFDs.

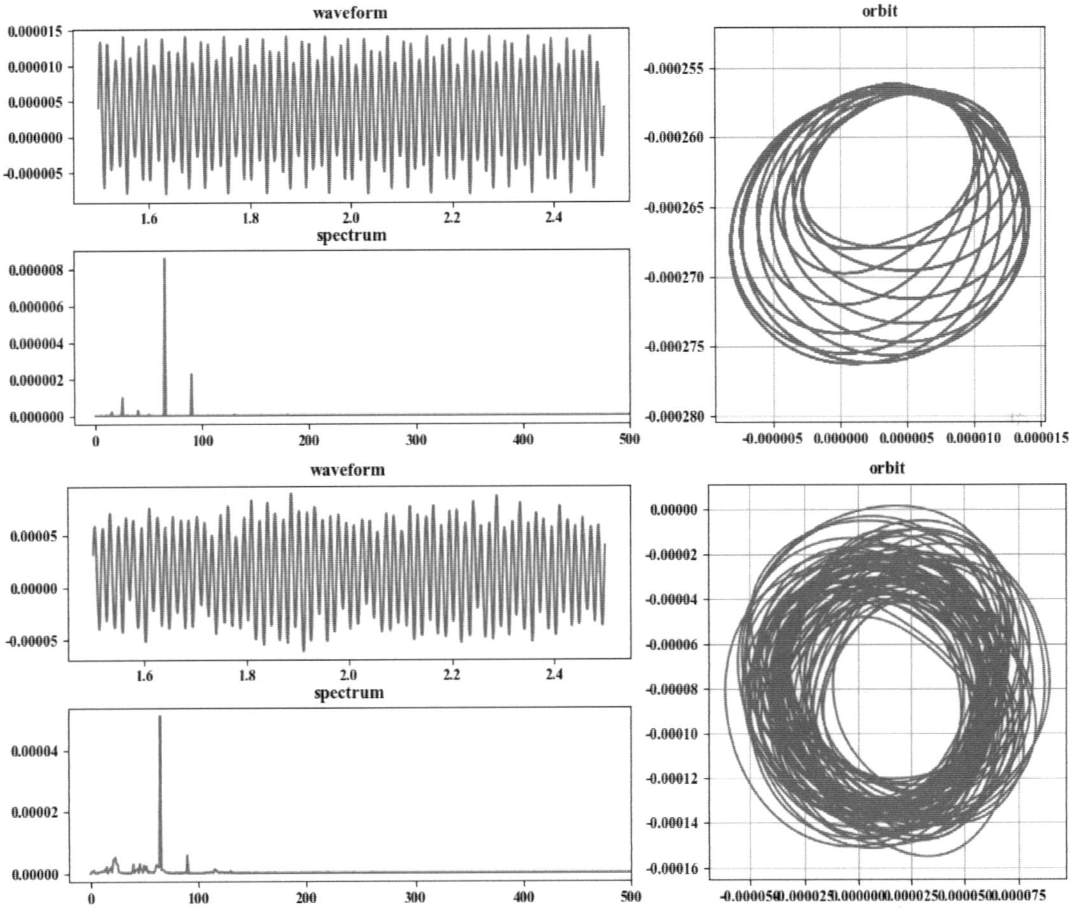

Figure 8. Waveform, spectrum and orbit of looseness fault, with or without SFD.

The displacement waveform, spectrum and shaft centerline orbit of node 15 on the inner rotor at the state of rub-impact fault is in Figure 9. When the clearance outside the inner rotor disc decrease, the disc will touch the bottom first because of the gravity effect, thus forms a state of single point rub-impact. Contact impacts make the shape of the shaft centerline orbit no longer close to a circle, and the amplitude of the inner rotor rotation frequency becomes more prominent. There are some frequency components in both low and high frequency areas due to the fault.

When there is no squeeze film damper, those low and high frequency components become more abundant. The shaft centerline orbit becomes less smooth. Like the situation of looseness, SFDs weaken the impacts caused by the rub-impact fault.

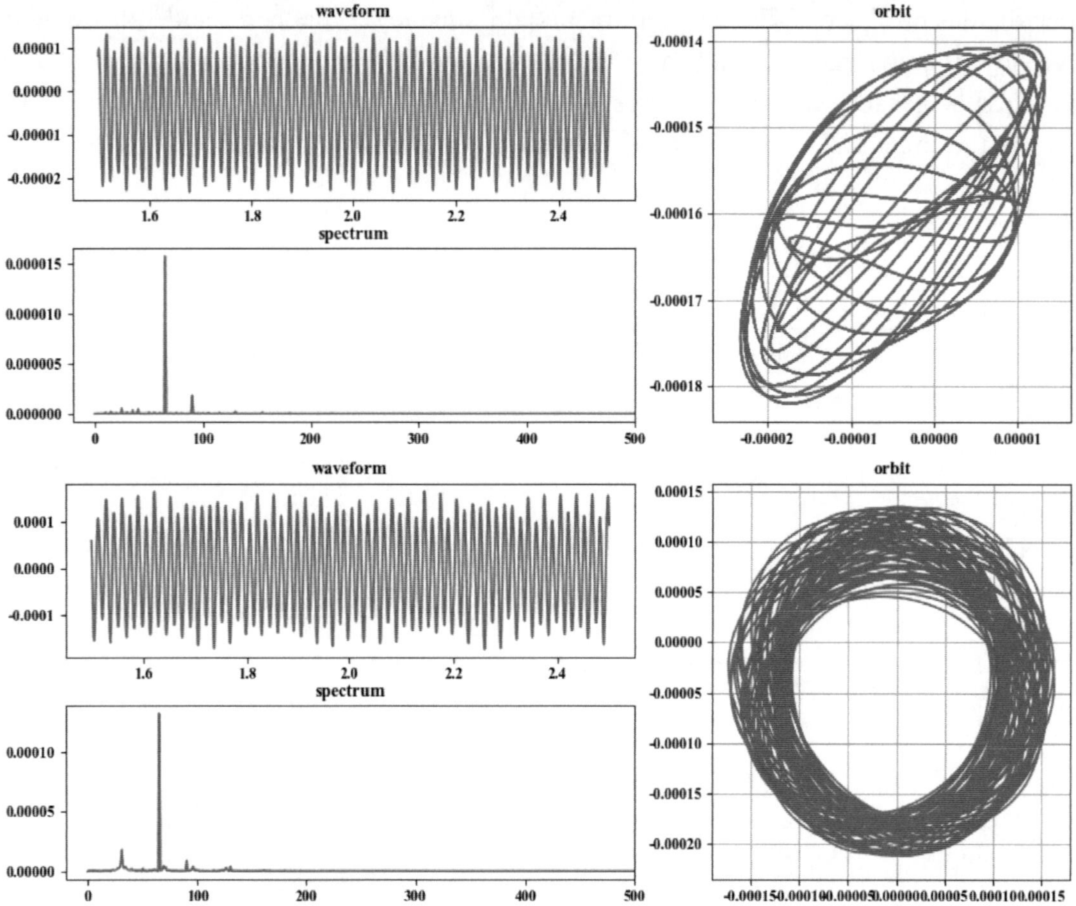

Figure 9. Waveform, spectrum and orbit of rub-impact fault, with or without SFD.

Figure 10 shows the situation when the unbalance fault and the looseness fault both exist, and the situation when the unbalance fault and the rub-impact fault both exist. The location and parameter of the faults are the same as single faults. The increase in the amount of unbalance increases the overall vibration amplitude of the rotor, thus enhances the nonlinear phenomena of the looseness or rub-impact fault. The fractional frequency components in the looseness fault and the low-frequency and high-frequency components in the rub-impact fault all become more significant.

Figure 11 shows the situation when the looseness fault and the rub-impact fault both exist, and the location and parameter of the faults are the same as single faults. The half rotational frequency component is much higher than single fault state of looseness or rub-impact. It can be said that in the presence of looseness, the constraint of the rotor is weakened and the rub-impact fault will cause more pronounced nonlinear phenomenon.

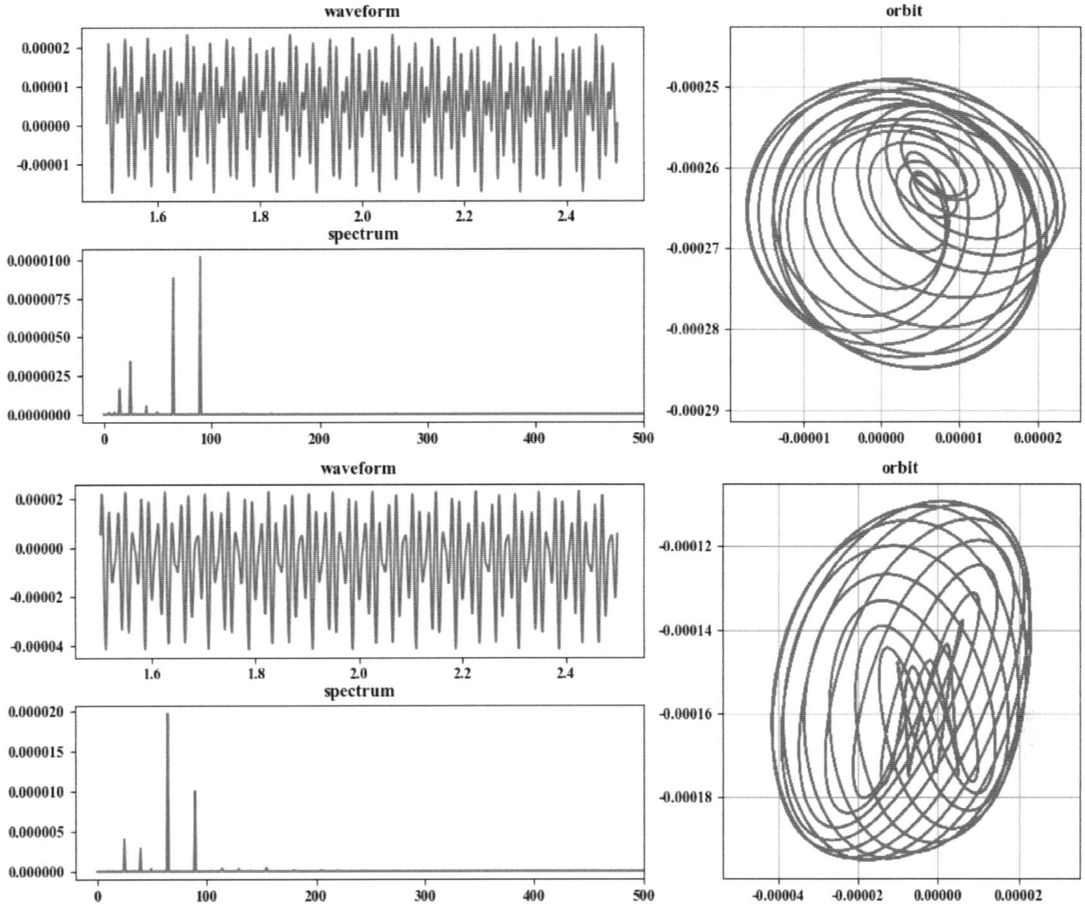

Figure 10. Waveform, spectrum and orbit of unbalance-looseness fault and unbalance-rub-impact fault.

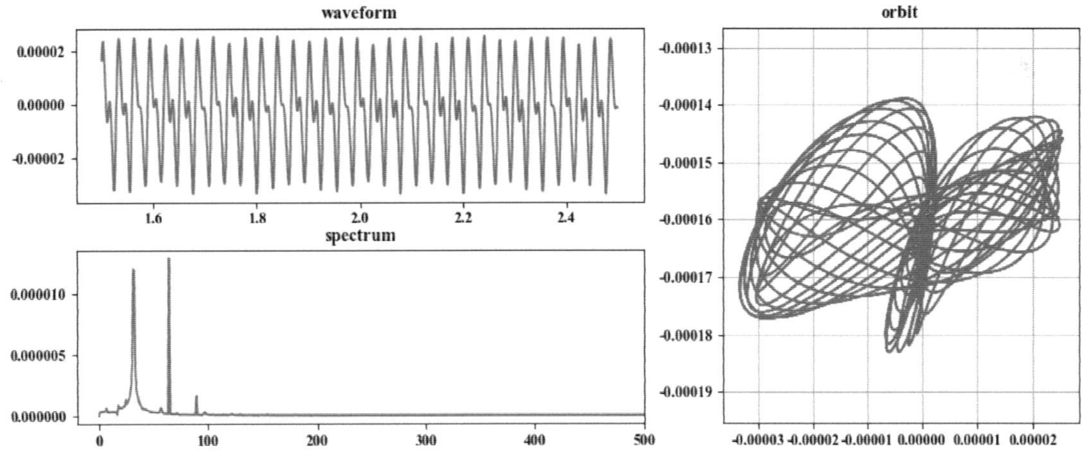

Figure 11. Waveform, spectrum and orbit of looseness-rub-impact fault.

Conclusion: This paper presents a finite element approximate model of an aero-engine dual-rotor system which includes nonlinear models of rolling element bearings and squeeze film dampers and can introduce single or multiple typical rotor faults in. The rolling element bearing is using a kind of nonlinear multi-body dynamic model, and the squeeze film damper is using the fluid-film forces model based on the cavitated short bearing theory. Steady-state vibration response of the dual-rotor system shows main frequency components of the two rotation speed.

The typical rotor faults simulated include rotor unbalance, rotor misalignment, support looseness, and dynamic-static rub-impact. Those faults are introduced to the system by adding nonlinear forces to the generalized external force vector, and fault features are obtained by solving the system equation. The unbalanced fault causes an increase in the amplitude of the rotational frequency. The misalignment fault causes an increase in the amplitude of twice the rotational frequency. The looseness fault brings in fractional frequency components. The rub-impact fault brings in many low and high frequency components. SFDs could weaken the impacts caused by looseness or rub-impact. And the looseness fault and the rub-impact fault could enhance the nonlinear phenomenon of each other when they both exist.

Many parameters could have influences on the characteristics of rotor faults, including the rotor parameters like mass, stiffness, damping and rotation speed, or the fault parameters like the location and the severity of faults. This paper does not discuss all these parameters in detail. But interesting result might be attained by studying fault characteristics in a nonlinear dual-rotor system.

References:

[1] Chiang, H. W. D., Hsu, C. N., Jeng, W., Tu, S. H., & Li, W. C. (2002, January). Turbomachinery dual rotor-bearing system analysis. In ASME Turbo Expo 2002: Power for Land, Sea, and Air (pp. 803-810). American Society of Mechanical Engineers.

[2] Hai, P. M., & Bonello, P. (2008). An impulsive receptance technique for the time domain computation of the vibration of a whole aero-engine model with nonlinear bearings. Journal of Sound Vibration, 318(3), 592-605.

[3] Sinou, J. J. (2009). Non-linear dynamics and contacts of an unbalanced flexible rotor supported on ball bearings. Mechanism & Machine Theory, 44(9), 1713-1732.

[4] Chen, G. (2015). Vibration modelling and verifications for whole aero-engine. Journal of Sound & Vibration, 349, 163-176.

[5] Liu, M., & Gorman, D. G. (1995). Formulation of rayleigh damping and its extensions. Computers & Structures, 57(57), 277-285.

[6] Gupta, P. K. (1975). Transient ball motion and skid in ball bearings. Journal of Tribology, 97(2), 261.

[7] Sawalhi, N., & Randall, R. B. (2008). Simulating gear and bearing interactions in the presence of faults: part i. the combined gear bearing dynamic model and the simulation of localised bearing faults. Mechanical Systems & Signal Processing, 22(8), 1924-1951.

[8] Mohan, S., & Hahn, E. J. (1974). Design of squeeze film damper supports for rigid

rotors. Journal of Engineering for Industry, 96(3), 976.
[9] Burrow, C. R., Sahinkaya, M. N., Kucuk, N. C., & Taylor, D. (1987). Dynamic performance of squeeze-film bearings.
[10] Inayat-Hussain, J. I., Kanki, H., & Mureithi, N. W. (2003). On the bifurcations of a rigid rotor response in squeeze-film dampers. Journal of Fluids & Structures, 17(3), 433-459.
[11] Dewell, D. L., & Mitchell, L. D. (1984). Detection of a misaligned disk coupling using spectrum analysis. Journal of Vibration & Acoustics, 106(1), 9.
[12] Ma, H. (2011). Analysis of dynamic characteristics for a rotor system with pedestal looseness. Shock & Vibration, 18(1-2), 13-27.
[13] Muszynska, A., & Goldman, P. (1995). Chaotic responses of unbalanced rotor/bearing/stator systems with looseness or rubs. Chaos Solitons & Fractals, 5(9), 1683-1704.

NEW POSSIBILITIES OF REDUNDANT DATA TRANSMISSION FOR INTELLIGENT SENSOR NETWORKS

Wojciech STASZEWSKI, Adam JABLONSKI and Tomasz BARSZCZ

AGH University of Science and Technology
Department of Mechanical Engineering
Aleja Adama Mickiewicza 30
30-059 Kraków, Polska
Telephone: (48) 509-805-483
ajab@agh.edu.pl

Abstract: Condition monitoring systems (CMS) that are currently available offer many types of tools, such as stationary monitoring systems, portable on-site instrumentation, and finally wireless, autonomous systems. Nowadays, the technology related to electronic systems is developing rapidly resulting in the overall improvement within division of data processing (clock speed and number of processing cores), data transfer (speed and power efficiency of communication), and energy consumption. Unfortunately, this development is not being transferred into Condition Monitoring Systems, probably because the complexity of installation and cost of such systems is holding its manufacturers from introducing significant changes in its architecture. A potential point where all modern technologies will culminate is in a technology initiative that is already underway - this is the "Industry 4.0". It leverages existing and emerging technologies to improve the efficiency, effectiveness, and service that need to be provided in order to be competitive in the future. Recently developed MEMS technology, ongoing trend to create smaller, more energy efficient electronics, followed by rapidly growing tendency for non-traditional devices being connected to the Internet (IoT) made it possible to design novel, small-size measurement units, which are capable of working autonomously as an individual device or in a set of connected devices, as a distributed system. Such a system might utilize ordinary smart phone device as a part in data transmission chain, as end-user device that allows monitoring current state of machine or as a gateway that can transmit measurements further, into the cloud system. Introducing redundant data transmission for such a device allows creating autonomous CMS, which can react to changes in machines operation and changes of the condition of the device itself. The redundant data transmission allows developing different scenarios including priorities for: extending operation time on battery source, maximizing measurement rate or data transmission rate, prolonging the range of sensors network. Ultimately, these new possibilities might lead to self-learning distributed system and intelligent sensors network.

Key words: Distributed condition monitoring system; diagnostics; IoT; micro data acquisition system; portable machine health management tools; wireless vibration sensors

Introduction: The accuracy of parts that are produced by machining systems is strongly related to the quality their subsystems (e.g., feed-drive systems and spindle units) and therefore if performance degradation due to wear in is not identified, monitored and controlled it may lead into serious production failures. That is why choosing the appropriate maintenance actions is very important as it may delay the possible deterioration and minimizes the machine downtime. Moreover, measuring and monitoring the condition of a machine tool has become increasingly important due to the introduction of agile production, increased accuracy requirements for products and customers' requirements for quality assurance. Condition Based Maintenance (CBM) and Condition Monitoring Systems (CMS) practices, such as vibration monitoring, are becoming attractive, but still challenging. CBM and CMS is usually being used to plan the maintenance action based on the condition of the machines and to prevent failures by solving the problems in advance (before they occurs) as well as controlling the accuracy of the machining operations. By increasing the knowledge in this area it becomes possible to can save money through fewer breakdowns, reduction of inventory cost, reduction of downtime needed for repairs, and an increase in the robustness of the manufacturing processes. The process of monitoring the state of machinery needs the contribution from an early fault and malfunction detection, in order to succeed in prevention of machine failure. Keeping the machine working point at the conditions that are the most optimal leads to machine downtime reduction, which results in economic loss reduction [1][2]. Nowadays, the purpose of monitoring the conditions of an engineering system, especially machinery has become a concern that requires great affection. A Condition Monitoring System (CMS) that bases on vibration monitoring may play a very important role in pushing out the maximum potential of all kinds of machinery by minimizing the downtime. In this paper, the authors try to compare the possible architectures and types of possible communication protocols that can be used for each architecture in order to assess their strengths and weaknesses, which will lead into identification of the threats and opportunities that may have an influence on modern Condition Monitoring Systems. Figure 1 presents scheme of a typical embedded device used for data acquisition for condition monitoring system.

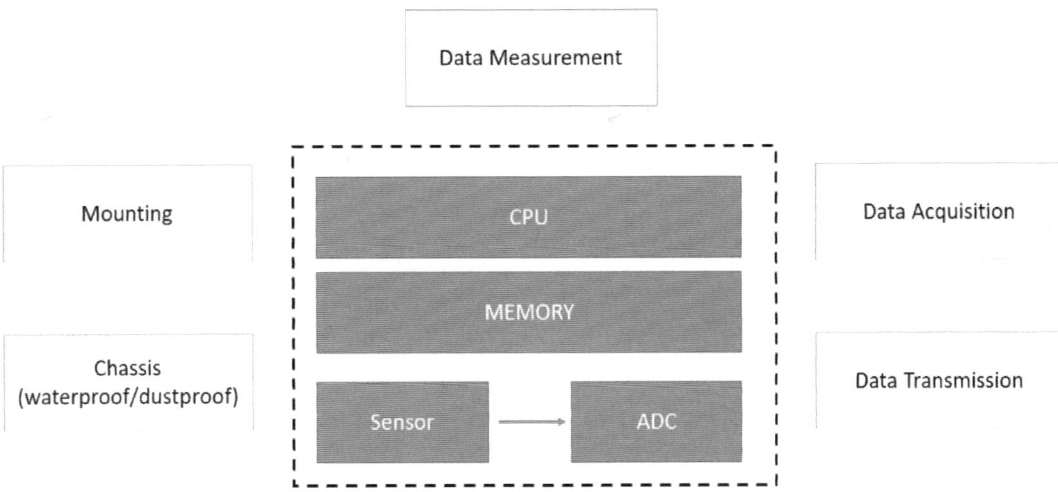

Figure 1: Scheme of an embedded device for modern condition monitoring system.

Condition Monitoring System: Condition monitoring (CM) can be shortly described as a process of monitoring a parameter of condition in machinery, like vibration, in order to identify a significant change in the range of this parameter which usually may indicate a progression of ongoing fault. As this process is a major component of well-known predictive maintenance, the use of CM allows scheduling maintenance or undertaking other actions to prevent further failure as well as avoid its consequences. Condition Monitoring has a benefit in compare to other techniques (breakdown and preventive (or scheduled) maintenance) – deterioration of the crucial parameters that would shorten a typical lifespan of a component can be addressed before they could develop into a major failure [3], causing an unplanned downtime. Condition monitoring techniques are usually used on rotating machinery and other equipment like pumps, electric motors, internal combustion engines, presses, while periodic inspection using non-destructive testing techniques are used for stationary plant equipment such as steam boilers, piping and heat exchangers. The advantages that have been offered by the application of predictive maintenance techniques have in a past few years led to the increase in the development of a number of methods for condition monitoring/fault-diagnostic systems [4]. After detection of possible failure (significant change of monitored parameter), mostly the diagnostic staff is responsible for taking the necessary actions in order to verify the condition of the machinery and later on act accordingly to prevent from machine failure. Those activities may require further data measurement in order to gather more data that might be necessary in order to make appropriate decision regarding machines condition, but at the point where the failure starts to develop it is necessary to notify the engineers that the condition of machinery is changing. Figure 2 presents embedded device that allows End-User to communicate with the device directly, without the use of a gateway.

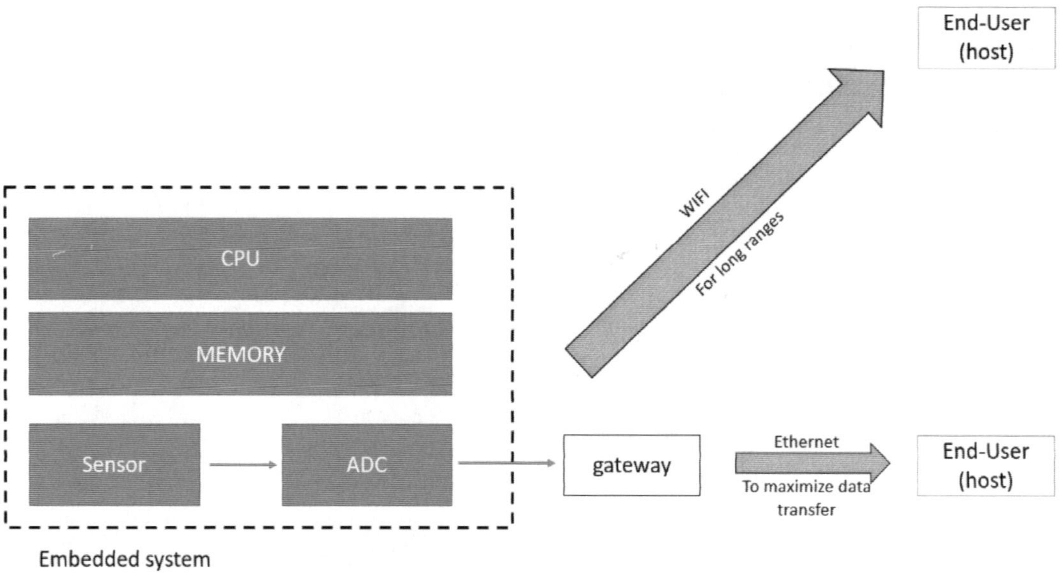

Figure 2: Scheme of an embedded device including direct communication with End-User.

Embedded device

An embedded device is a type of device that is highly specialized for one or very few specific purposes, and is embedded or included within another object or as element of a larger system that serves a greater purpose than the embedded device itself. Data Acquisition Unit (DAQ) is a hardware that acts as the interface between a computer (end-user or host) and signals from the considered machine. Its primary function as a device is to digitize incoming signals so that a computer can interpret them. Many DAQ devices include other functions for automating measurement systems and processes. Figure 3 presents embedded device that allows End-User to communicate with the device with the use of a communication gateway.

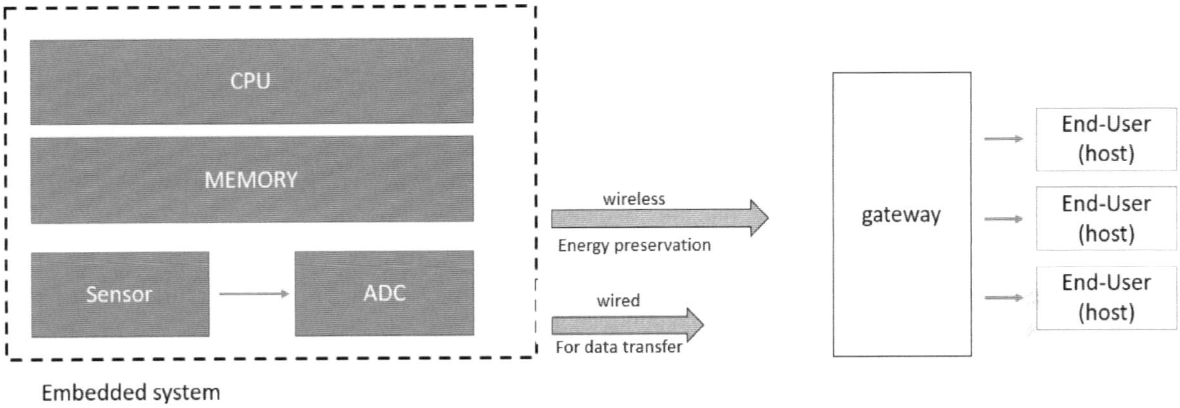

Figure 3: Scheme of an embedded device including communication with End-User using gateway device.

Embedded device and communication protocols

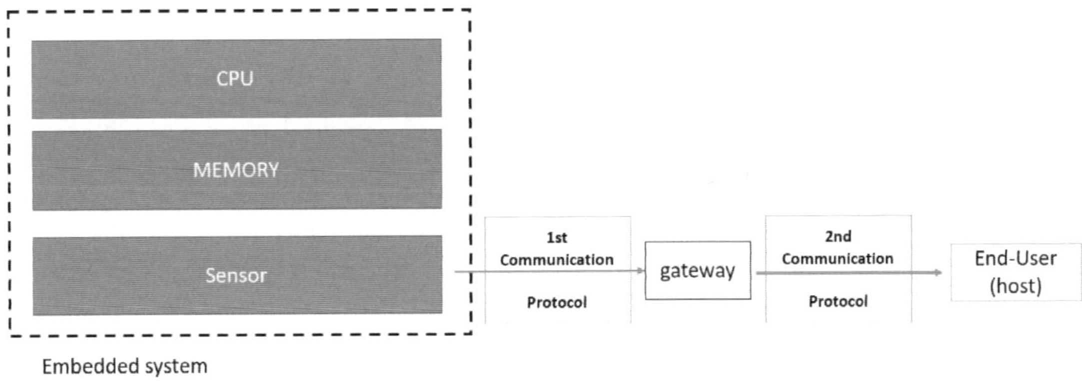

Figure 4: Scheme of a typical communication chain in condition monitoring system.

Figure 4 presents communication chain in condition monitoring system, where different communication protocols are used to connect DAQ with the gateway and the gateway with End-User.

Bluetooth - is a wireless technology standard for exchanging data over short distances (using short-wavelength UHF radio waves in the ISM band from 2.4 to 2.485 GHz) from fixed and mobile devices, and building personal area networks (PANs). Invented by telecom vendor Ericsson in 1994, it was originally conceived as a wireless alternative to RS-232 data cables. It can connect up to seven devices, overcoming problems that older technologies had when attempting to connect to each other [5].

Bluetooth operates at frequencies between 2402 and 2480 MHz, or 2400 and 2483.5 MHz including guard bands 2 MHz wide at the bottom end and 3.5 MHz wide at the top. This is in the globally unlicensed (but not unregulated) Industrial, Scientific and Medical (ISM) 2.4 GHz short-range radio frequency band. Bluetooth uses a radio technology called frequency-hopping spread spectrum. Bluetooth divides transmitted data into packets, and transmits each packet on one of 79 designated Bluetooth channels. Each channel has a bandwidth of 1 MHz. It usually performs 800 hops per second, with Adaptive Frequency-Hopping (AFH) enabled. Bluetooth low energy uses 2 MHz spacing, which accommodates 40 channels.

WiFi – is a technology for wireless local area networking with devices based on the IEEE 802.11 standards. 802.11 is the "radio frequency" needed to transmit Wi-Fi, it was defined by Vic Hayes who created the IEEE 802.11 committee. Devices that can use Wi-Fi technology include personal computers, video-game consoles, smart phones, digital cameras, tablet computers, digital audio players and modern printers. Wi-Fi compatible devices can connect to the Internet via a WLAN network and a wireless access point. Such an access point (or hotspot) has a range of about 20 meters indoors and a greater range outdoors. Hotspot coverage can be as small as a single room with walls that block radio waves, or as large as many square kilometers achieved by using multiple overlapping access points [5].

Depiction of a device sending information wirelessly to another device, both connected to the local network, in order to print a document Wi-Fi most commonly uses the 2.4 gigahertz (12 cm) UHF and 5 gigahertz (6 cm) SHF ISM radio bands. Having no physical connections, it is more vulnerable to attack than wired connections, such as Ethernet.

Zigbee

It is an IEEE 802.15.4-based specification for a suite of high-level communication protocols used to create personal area networks with small, low-power digital radios, such as for home automation, medical device data collection, and other low-power low-bandwidth needs, designed for small scale projects which need wireless connection.

The technology defined by the ZigBee specification is intended to be simpler and less expensive than other wireless personal area networks (WPANs), such as Bluetooth or Wi-Fi. Applications include wireless light switches, electrical meters with in-home-displays, traffic management systems, and other consumer and industrial equipment that require short-range low-rate wireless data transfer.

Its low power consumption limits transmission distances to 10–100 meters line-of-sight, depending on power output and environmental characteristics. ZigBee devices can transmit data over long distances by passing data through a mesh network of intermediate devices to reach more distant ones. ZigBee is typically used in low data rate applications that require long battery life and secure networking (ZigBee networks are secured by 128 bit symmetric encryption keys.) ZigBee has a defined rate of 250 kbit/s, best suited for intermittent data transmissions from a sensor or input device.

Modbus

Communication interface is built around messages. The format of these Modbus messages is independent of the type of physical interface used. On plain old RS232 are the same messages used as on Modbus/TCP over Ethernet. This gives the Modbus interface definition a very long lifetime. The same protocol can be used regardless of the connection type. Because of this, Modbus gives the possibility to easily upgrade the hardware structure of an industrial network, without the need for large changes in the software. A device can also communicate with several Modbus nodes at once, even if they are connected with different interface types, without the need to use a different protocol for every connection. Serial Modbus connections can use two basic transmission modes, ASCII or RTU, remote terminal unit. The transmission mode in serial communications defines the way the Modbus messages are coded. With Modbus/ASCII, the messages are in a readable ASCII format. The Modbus/RTU format uses binary coding which makes the message unreadable when monitoring, but reduces the size of each message which allows for more data exchange in the same time span. All nodes on one Modbus network segment must use the same serial transmission mode. A device configured to use Modbus/ASCII cannot understand messages in Modbus/RTU and vice versa.

Table 1: Communication protocols

	1st Communication protocol			2nd Communication Protocol			
	I^2C	SPI	UART	Bluetooth	WiFi	ZigBee	Modbus
Data rate	1 Mb/s	20 Mb/s	460 kb/s	1 Mb/s	54 Mb/s	250 kb/s	1 Gb/s
Channels	27	n/a	2	8	2000	65000+	254
Range	2m	10m	1m	10m	100m	100m	1000m
Power consumption	6.1 mW	4.5 mW	6.0 mW	7.0 mW	100.0 mW	75.0 mW	600.0 mW
Advantages	Simple addressing, flow control	Full duplex, push-pull mechanism	Simple	Simple	Range, Data transfer	Fixed data rate	Long range
Disadvantages	Half-duplex interface, processing overheads	No flow control, Hardware complexity	Communication between two devices	Range, Number of nodes	Energy consumption	Middle range	Hardware and software complexity

Table 1 presents comparison of properties of available connection protocols that can be used in order to connect DAQ to the gateway and gateway to the End-User. Further analysis will not be conducted since it is not the intention of the authors. The intention is to present the separation between characteristics of protocols in different parts of CMS.

Machine-to-Machine:
Machine-to-Machine SIM (or M2M SIM) – refers to the technologies that enable devices and sensors to communicate with each other – and with other Internet-enabled devices and systems. Even though M2M refers to 'non-human' communication, devices still should have an interchangeable protocol which is used for receiving and sending data. Connectivity is achieved by either inserting or embedding a Machine-to-Machine SIM in the device that is later configured with the home server where all the usage data is collected and can be analyzed for company's beneficial purposes.

Host – universal system with redundant communication protocols:
A network host is a computer or other device connected to a computer network. A network host may offer information resources, services, and applications to users or other nodes on the network. A network host is a network node that is assigned a network layer host address.
Computers participating in networks that use the Internet Protocol Suite may also be called IP hosts. Specifically, computers participating in the Internet are called Internet hosts, sometimes Internet nodes. Internet hosts and other IP hosts have one or more IP addresses assigned to their network interfaces. The addresses are configured either manually by an administrator, automatically at start-up by means of the Dynamic Host Configuration Protocol (DHCP), or by stateless address auto-configuration methods.
Every network host is a physical network node (i.e. a network device), but not every physical network node is a host. Network devices such as modems, hubs and network switches are not assigned host addresses (except sometimes for administrative purposes), and are consequently not considered to be network hosts. Devices such as network printers and hardware routers have IP addresses, but since they are not general-purpose computers, they are sometimes not considered to be hosts.
Network hosts that participate in applications that use the client-server model of computing are classified as server or client systems. Network hosts may also function as nodes in peer-to-peer applications, in which all nodes share and consume resources in an equipotent manner.

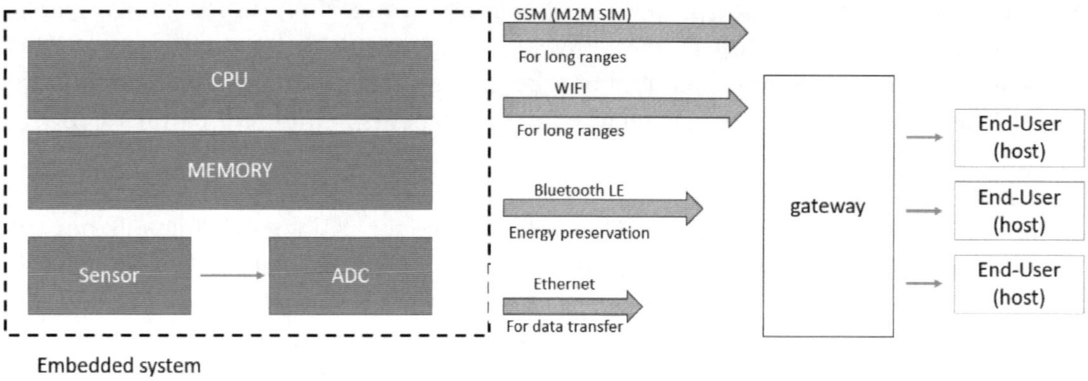

Figure 5: Scheme of proposed communication for condition monitoring system

The host PC serves as the control center of the CMS system. It can be used to:

- control the DAQ,
- provide a graphical user interface (GUI) for the user to operate the system (with use of web gui based on http protocol)
- store the acquired information on mass storage devices
- provide the tools to display and/or analyze the acquired data in real or non-real time

The host computer is also used to define and implement the algorithms performed by the CMS system.

Conclusion: Nowadays, that technology related to electronics systems is developing rapidly – this trend could be observed by looking into the direction of mobile devices development. The Internet of Things (IoT) is the rapidly growing trend of non-traditional devices being connected to the Internet. Industry analysts from International Data Corporation (IDC) suggest that the IoT devices and their ecosystem is expected to create $8.9 (a five-year annual growth rate of 16,8%) trillion market in 2021 with an estimated 20 billion devices [6]. A point where all modern technologies are being culminated is already underway in the industry - this is the Industry 4.0. It leverages existing and emerging technologies to improve the efficiency, effectiveness and service that must be provided in order to be competitive in the future. Presented approach allows creating modern – Industry 4.0 ready – devices that could be integrated into a complex Condition Monitoring System. By now, everyone uses modern devices for ex.: smart phones, wearable accessories like smart watches or smart bands, wireless headphones and speakers or smart ecosystems like smart home automation. Each of those devices provides different ways of communication, usually redundant ones in order to allow their users the most possible functionality. The authors want to point out that by now such a trend is not visible in the industry, but within following years it might be the direction of changes in different branches of industry.

References:

[1] Blanke, M., Kinnaert, M., Lunze, J., Staroswiecki, M., Diagnosis and fault tolerant control. Springer, Berlin, 2006, 2nd edition.

[2] Feldman M., Seibold S., Damage Diagnosis of Rotors: Application of Hilbert Transform and Multihypothesis Testing, Journal of Vibration and Control, 5, 421.1999

[3] Milind A. Siddhpura, Arti M. Siddhpura, S.K. Bhave, Vibration as a parameter for monitoring the health of precision machine tools, ICDM-08, Coimbatore, India, February 01-02, 2008.

[4] Wang W., Tse P.W., Lee J., Remote machine maintenance system through Internet and mobile communication, International Journal of Advanced Manufacturing technology 31:783-789, 2007, doi:10.1007/s00170-005-0236-1.

[5] Khan J.A., Qureshi H.K., Iqbal A., Energy management in wireless sensor networks: A survey, Computers and Electrical Engineering, vol. 41, pp. 159–176, 2015.

[6] Valenta J., Cellular IoT Is the Fastest-Growing Opportunity for CEE Telcos, IDC Media Center, June 2017

Bibliography:

[1] A. Muszyńska, Rotordynamics, Taylor & Francis Group, LLC, Minden (2005), USA.

[2] M. Strączkiewicz, A. Jabłoński, T. Barszcz, Detection and classification of alarm threshold violation in condition monitoring systems working in highly varying operational conditions, Journal of Physics, Conference Series (2015), vol, 628, s. 012087-012087-B.

OPTOMIZED TESTING CAMPAIGN AND CREATING ONEROUS BASELINE DATA FOR A CENTRIFUGAL COMPRESSOR - A PROPOSAL TO MANUFACTURER

Mantosh Bhattacharya
Petrofac International UAE
Petrofac Tower 1
Telephone 00971501469853
mantosh.b@petrofac.com

Abstract This paper proposes an optimized testing agenda for centrifugal compressor with a compact manufacturing and testing schedule. As API mandates to conduct mechanical run tests (MRT) of all compressors of similar deign and geometry, it takes a good amount of time to conduct these tests on OEM test bed particularly if numbers of compressors are large. This paper proposes to conduct only one Full load Full speed (FLFS) test as mechanical run test at vendor works among the full lot of same model and design. The intention of the paper is to classify the criticality of compressors and then deliberate on the extent / type of tests taking account of OEM test bed capability and schedule of delivery of machines and then optimize the test agenda .To supplement the FLFS test, the paper proposes to undertake extensive design audit activities in terms of rotor-dynamics, aerodynamics taking account of case histories of past failures which can raise the confidence level of the end user. Assurance of dimensional repeatability in terms of metrology backed by latest methodology of fault identification of multi-layered manufacturing process with PQM, which can avoid multiple tests of similar machines, may avoid multiple third party inspection at various stages thus saving time.
In No load MRT site conditions such as influence of process parameters such as gas pressure / density, foundation dynamics are not replicated. The MRT do not identify the region of incipient surge, effect of aerodynamic cross coupling instability as well. The paper proposes to introduce various instruments and sensors to detect the above instability region for the proposed extensive test .With the rotor-dynamic data taken from the proposed test set up, it shall be easy to further enhance the base line data after site performance test .The same can be used for pre-alarm configuration based on zone wise amplitudes in vibration spectrum which can be very useful for reliability engineer engaged in diagnostic and prognostics. For critical machines located in hostile environment, anomaly detection may be carried out with shape identification of plot / spectra.

Keywords - MRT , Full Load test , optimizarion of schedule, rotor-dynamic data

Introduction - As mandated by API 617(standard for Axial and Centrifugal Compressors and Expander-compressors for Petroleum, Chemical and Gas Industry Services), all centrifugal compressors have to undergo a four hour mechanical run test at full speed.

API 617 is not explicit about the pressure, type of gas used as flow medium hence all the tests are conducted either in a partial vacuum or at very low suction pressure using inert gases. During the test , shaft displacement measurements are taken by proximity probes - eddy current sensors mounted orthogonally at same probe tracks. During the test, criteria of direct (overall) shaft vibration are met - √ (12000/N) or 1.0 mil peak to peak whichever is less (while N is the Rotor Speed in RPM). API 617 also requires that the discrete nonsynchronous vibration shall not be more than 20% of maximum allowed value for overall vibration. The standard also requires the frequency range of shaft vibration measurement shall be in the range of 0.25 to the 8 times of the Maximum Continuous Speed. A brief refresher is added as below paragraph before we weigh the real value addition this type test renders .

Conventional Mechanical Run Test -

It is a normal belief that MRT (Mechanical run test) results can surface all possible defects in rotor bearing system as there is no damping of rotor vibration by process gas at pressure. During the test conducted in very low pressure gas mixture, pressure and temperature of lubricating oil for bearings are varied to see the rotor response. The bearing metal temperature, lube oil temperature and pressure, seal oil temperature and pressure, bearing housing vibration, shaft vibration, and power consumption are the parameters that are measured in the mechanical running test. As per API 617, the job bearings, job vibration probes and job seals are used for the mechanical running test.

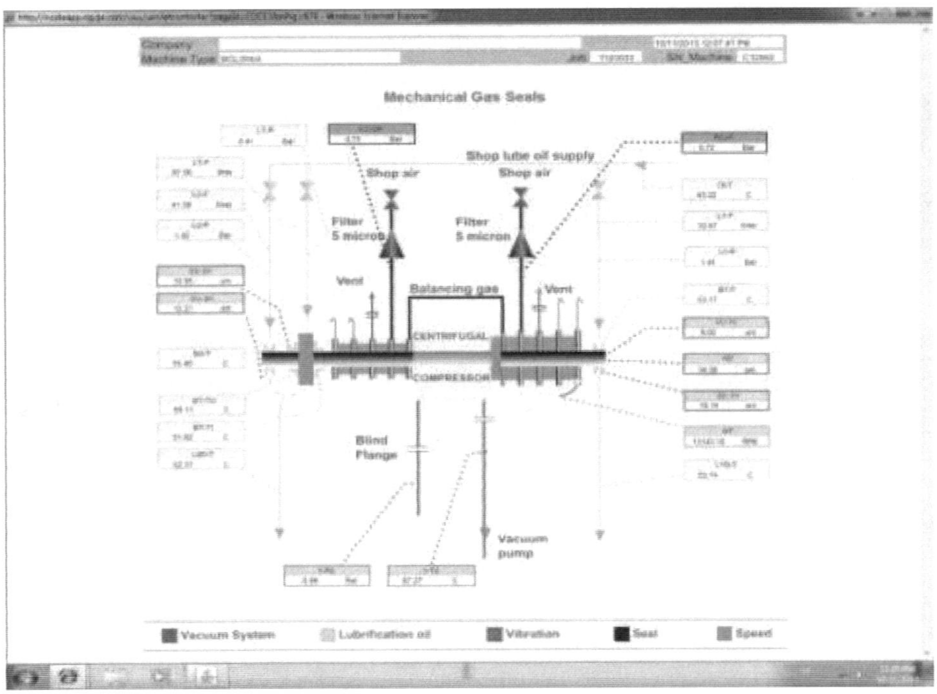

Fig 1 - Typical screen shot of monitoring screen during compressor MRT

In MRT, shop driver and shop base frames are used with contract / moment simulated couplings Shop verification of unbalance response test on main rotor at the end of four-hour MRT is carried out by some of OEM. This is to validate the calculations carried out to predict 1st critical speed. Unfiltered and filtered displacement amplitudes and the corresponding phase angles are recorded and plotted vs. speed during coast down. Bode plots during coast down are also generated during MRT .It is a common practice that OEM (compressor designers & manufacturers) carry out factory internal test prior to customer (or end user) witnessed test. Normally, compressor performance test (ASME PTC10 type 2) is conducted either before, or after the mechanical running test. If a compressor is first subjected to mechanical running test but fails the performance test, the time, resources, and efforts expended in arranging the mechanical running test would be deemed wasted. From this viewpoint, some users prefer to run performance test first, followed by mechanical running test. The requirements of performance test are also included in API Standard 617.On completion of MRT , the loop is left for cooling and then bearings strip-down , visual check and assembled compressor gas leakage test is carried out . The compressor end seals (contract seals) are kept in the installed condition during assembled compressor gas leakage test.

Fig 2 -Compressor MRT - Project specific base frame not used

The compressor is subjected to soap water bubble test. The soap bubble test is done based on the requirement of the ASME Section V - Article 10 - Leak testing or OEM test procedure. The specific leak test solution is used for the test. The applicable standards for centrifugal compressor testing are: -

- For Mechanical performance - API Standard 617; Centrifugal Compressor for Petroleum, Chemical and Gas Service Industries

- For Thermodynamic performance - ASME PTC 10; Performance Test Code on Compressors and Exhausters OR BS- ISO 5389 Turbo compressors Performance Test Code. After performance test, Reynolds Number correction is done in accordance to ICAAMC only after mutual agreement with end user and OEM.

Component checks carried out before MRT (Spin test of compressor) by OEM

1. Centrifugal Compressor Individual Impeller Over speed Test (Spin Test). In this test, the impeller is subjected to the thorough dimensional inspection before the test. The inside diameter (bore), outside diameter, throat are measured and documented. The compressor impeller over-speeding test is performed before rotor assembly. The impeller mounted in the machine in a vacuum chamber (to avoid windage) and then is rotated at the speed of 115% of the MCS for 1 minute. The impeller is subjected to the strong centrifugal force during this 1 minute after over-speed test; a second dimensional inspection is performed and checked against first dimensional inspection. The change in the dimension is inevitable, but the change shall be within the acceptance range. API 617 does not provide acceptance range and refers that to the manufacturer specification. The inspector reviews before and after dimensional inspection reports and check that with manufacturer acceptance criteria and make sure the change in dimension fall within acceptance range after impeller over speed test . In the fourth step, the impeller is subjected to the dye penetration or magnetic particle test to make sure no crack developed during impeller over speeding test

2. High speed balancing / Low speed balancing of assembled rotor using influence coefficient methods In addition to rotor balancing , Rotor stability can be assessed by perturbation tests at balancing machine in particular cases where mass of rotor is comparatively low and OEM has a less fleet record of similar design.

3. Dry gas seal test is carried out at Seal supplier works according to approved procedure.

4. Rotor assembly dimensional check with end float and gap check between rotoric and statoric part is also carried out and verified with API 617 RCV analysis considering highest Amplification factor at 1st critical speed and recorded. The alarm and trip setting of high radial and axial vibration is done accordingly.

In addition to above, multiple checks are carried out by OEM at different stages of manufacturing as per internal test notes or engineering practice. A tight manufacturing tolerance is maintained in such a way that summation of all dimensional errors during manufacturing still lies in acceptance band.

Complete Unit Test is included in API Standard 617 under the section ‘Optional Tests’. Frequently, it is called as string test. Some users stipulate full-speed, full-load / full pressure mechanical running test for compressors while some ask for no-load full speed test of unit. Complete Unit Test with no-load full speed test is like a spin test of complete unit assembled.This type of test cannot detect - torsional lateral coupling, super-synchronous vibration in pinion of speed increaser gear box. Shorter plant commissioning schedule of critical machines in the field is one of the deciding factors to enforce complete unit test of one among the multiples of similar trains. In such situation, end users may want to pay more money upfront for complete unit test than having to spend much more in rectifications at site, if something were to go wrong. Known rotor-dynamic issues or concerns arising from stability analysis of the compressor are validated in full load test. Compressor with special design features such as with multiple side loads, fitted with honeycomb seals, whole pattern seals shunt holes, or swirl brakes may require more exhaustive testing than conventional performance test and mechanical running test. Complete unit test results can be set as bench mark for multiple criteria. If a complete unit test is specified for compressors installed in identical, multiple production trains, it is be carried out on the compressor(s) in only one train.

The question is - Do we achieve a desired outcome of repeated MRT of all centrifugal compressors from a same model number in a testing campaign or a no load complete unit test ?

Based on review of numerous MRT results of such compressors, it is found that critical speed measured are within 3-5% variation with adequate separation margin , 1x compensated shaft displacement at resonant speed do not defer more than 20% and 99.5% of machines meet the criteria of 25.4 um peak-peak vibration. So, the MRT results of all compressors of same model and configuration can be considered as "monotonously" asymptotic. Statistically, the failure history of compressors in MRT is very rare although some of those machines failed to meet acceptance criteria in FLFS test or at site with actual process condition. What we measure in MRT such as bearing pad temperatures , oil consumptions , effect of unbalance (calculated) , slow roll run out data for 1x compensation , is measured in FLFS (Full load Full speed) / FSFP (Full load Full pressure) test with more onerous level .We can measure / witness Coupling enclosure windage effects , oil foaming issue, effect of possible extrusion of bundle O-ring inside the casing, sub-synchronous vibration due to stability in FLFS / FSFP test which cannot be measured / witnessed in MRT.

There are many cases of sub synchronous vibration were encountered in compressors due to aerodynamic instability in a particular impeller stage, flow related instability due to incipient surge, stall in diffuser / impeller stall, SSV due to pad instability and effect of cross coupling.

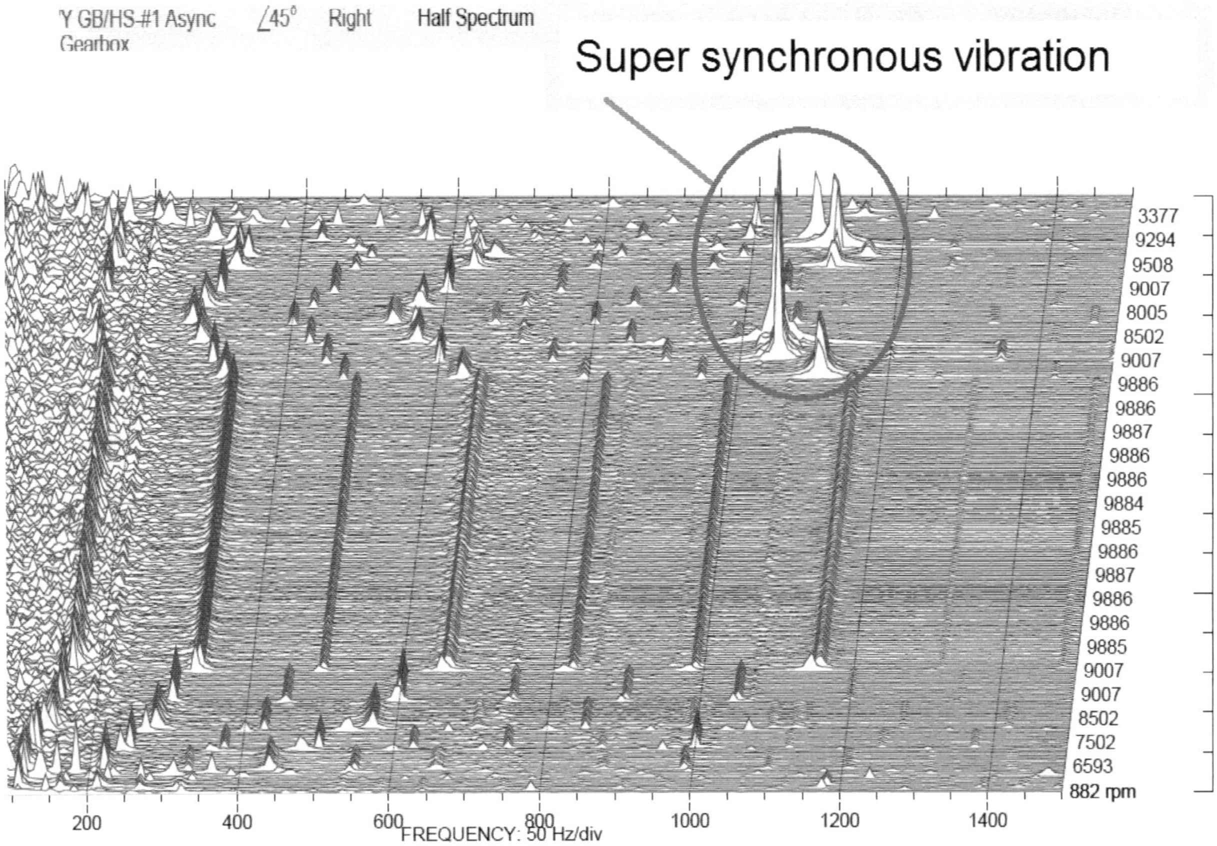

Fig 3- Waterfall showing pinion super-synch vibration at particular speed -load range

During some cases high sub synchronous and super synchronous vibration of pinion of speed increaser gear box were also encountered with a turbo compressor unit.

All above incidents were detected either in full speed, full load test at OEM works or at site. Since full load full speed tests were not conducted at OEM works for some compressors these issues surfaced at site and took lot of time to rectify.

Consider there is a fleet of 9 compressors for three sites of same model but have slightly different operating conditions. Time line for the API 617 MRT and PTC 10 tests if conducted as per conventional methods, shall be around 4- 5 months. Time line for preparation of full load test / full load test at test bed shall be around 2-3 months with a higher level of confidence to end user in machine offered by OEM. So, if OEM conducts a full load full speed test, it can give a good vibration base line data and higher level of integrity assurance.

To optimize the testing campaign and creation of onerous base line data a series of road map is proposed for an OEM to consider -

Road Map of integrity assurance at design phase

It is prudent to have extensive in house design audits by OEM pertaining to mitigate rotor-dynamic and aerodynamic instabilities and prepare reports. The analysis and report should consider -

- Finite element based rotor-dynamic modelling (using linear and non-linear parameters / coefficient of bearings and seals wherever applicable),
- Considering pivot stiffness using best suited software for different designs
- Evaluation of loss of interference ,
- Using API 617 level 2 stability analysis over calculating rotor stability ratios rather using statistical approach such as Futon -Sood and Donald -Kirk charts ,
- Ensuring AF(amplification factor) at 1st critical speed below 8 (using soft bearings) ,
- Exploring use of rub tolerant inter-stage seals ,
- Squeeze film damper bearings for lightweight compressor rotor, use of Helmholtz resonator / Low solidity vanes and proper evaluation of such design ,
- Impeller dynamic modelling and SAFE analysis.
- Modified Seenoo and Nishida criterion of critical angle to be incorporated along with OEM design data to avoid impeller and diffuser stalls.

A quick first hand estimation calculation for stress estimation on impeller can be done as-

$\sigma = \{(u \text{ squared } \times \rho\} / C$ where u is impeller peripheral velocity at MCOS (Maximum continuous operating speed), ρ is density of impeller material, σ is actual stress being applied on impeller and C is a constant in order of 160000 based on flow coefficient. MCOS is maximum continuous operating speed.

The calculated stress must be less than 75% of yield strength of material selected for impeller when used in sweet gas service and 65% of yield strength of material selected for impeller used in sour gas service as defined in NACE standard. Impeller dynamic analysis must be carried out to identify the potential resonance of impeller for compressor handling high density gas such as CO_2 or having vaned stator / diffuser using SAFE diagram. The combined design experience OEM and on site experience of End-users can be utilized to undertake a joint review of design report.

Road map of integrity assurance at manufacturing phase -

There should be a pre-inspection meeting where it should be communicated and agreed that end user / purchaser quality control / rotating equipment specialist shall visit major sub supplier premises to ensure the ISO norms and imposed standard by compressor OEM are followed during various manufacturing stages.

Component manufacturing checks and reports should be shared with end users and any repairs must be notified for acceptance. End user shall exercise his experience and prudence to accept / reject the repair proposal. In case of any repair comes under rotating components, API RP 687 can serve as a good reference to compare OEM proposed repair procedure.

Every component out of forging place shall have a heat number / unique identification number so that there is no mix up with same components of different origin / purpose. Some of advanced metrology systems are described below which are used in precision manufacturing and can be adopted by OEM subject the manufacturing infra-structure and system.

The in-process measurement control systems can be used in cylindrical grinding machines, for example. OEM uses them to measure small, medium or large work pieces during the production process in the machine. The control units operate with maximum precision, reliability and repetitive accuracy. They continuously check the quality of your components, thereby ensuring that the production results are of consistently high quality.

To ensure process reliability, in-process measurement control systems automatically compensate for any tolerance variations, which can arise, for example, when grinding discs become worn or as a result of temperature fluctuations. In such cases, the in-process control systems monitor the measured values in real time, thus controlling the grinding process continuously until the nominal dimension has been reached. This process helps to minimize part rejection and necessary rework.

Precision, accuracy, repeatability, efficiency, and industry knowledge are all critical when we manufacture or install large or assemble complex parts. That same care and attention should also be applied when engaging dimensional metrology service. With 3D metrology as proposed, the manufacturer can also monitor strategic control points on assembly jigs and automated equipment, quickly reverse engineer some parts, and generate repetitive CAD-to-Part analysis and tool certifications with the highest degree of accuracy available. With latest 3D metrology tools such as high-speed, high-precision portable laser trackers, coordinate measurement machines , OEM can perform contact and non-contact measurement in one operation.. The profile projector is used for measuring two-dimensional contours of precision specimens and other work pieces produced. The part to be measured is magnified by an optical system and projected on a screen. The reading on the screen gives the dimension of the part. The following is the photo of a profile projector.

Fig 4 - Typical Coordinate measurement machine

A coordinate measurement machine (CMM) provides instant measurement results without complicated setup and operating procedures. It combines surface plate, micrometer and Vernier type inspection methods into one easy to use machine. CMM can check the dimensional and geometric accuracy by using probe supported on three mutually perpendicular (X, Y & Z) axes. Each axis has built-in reference standard circuit boards.

Impellers which are most sensitive components in compressor are machined with an aim to achieve high dimensional accuracy and acquire specific performance. However, manufacturing uncertainties result in dimensional deviations causing incompatible operational performance and assembly errors. Process capability limitations of the manufacturer can cause an increase in part rejections, resulting in high production cost. To ensure that a turbo-machine is robust to some of manufacturing uncertainties, impeller can be parameterized and evaluated using a commercial computational fluid dynamics (CFD) solver. A sensitivity analysis can be performed initially to identify the critical geometric parameters which influence the performance mainly. Sensitivity analysis can be followed by the uncertainty propagation and quantification using Monte Carlo simulation. Finally, a robust design optimization can be carried out using a stochastic optimization algorithm leading to a robust impeller design. In engineering, optimization problems are often of this type, when we do not have a mathematical model of the system (which can be too complex) but still want to optimize its behavior by adjusting certain parameters.

Road map of integrity assurance at testing phase-

It is advisable to conduct high speed balancing of rotor and then validate the readings with the coupling hub attached. Over-speed test should conducted with an extra probe mounted in such a way that rotor mode shape plot can be drawn and validated with analysis before it goes to test bed . The balancing machine should have similar type of bearing principal stiffness and damping which is used in contract machine (Compressor).

A pre-test meeting should be held at OEM test bed so that end user / purchaser are clear about objective and confident that calibrated measurement devices being used. The skid foundation may use choking / clamping devices at various places to enact the effect of epoxy grout under base frame at site condition.

For full load full speed test of compressor, it is imperative to measure all the transient vibration and record the plots and analyze them with help of a certified vibration analyst. Any sub-synchronous vibration component / small hump in spectrum should not be discarded as noisy signal. The plots should be shared with witnessing team at test bed itself. All concerns (genuine and benign) should be addressed with technical professionalism .Dry gas seal primary vent flow and pressure must be monitored continuously as it is one of the tell-tale signs of possible transient rotor bow.

To check the rotor susceptibility to aero dynamic instability such às incipient surge, impeller / diffuser stall, OEM can use various acoustic pulsation detection device / pressure monitor probes / sensors at stage drains as port. The sensors should have high resonance frequency so that the measurement is correct .Dynamic Pressure Monitor with a signal booster is a specialized monitor that monitors the pressure pulsations with a high accuracy even at high temperature.

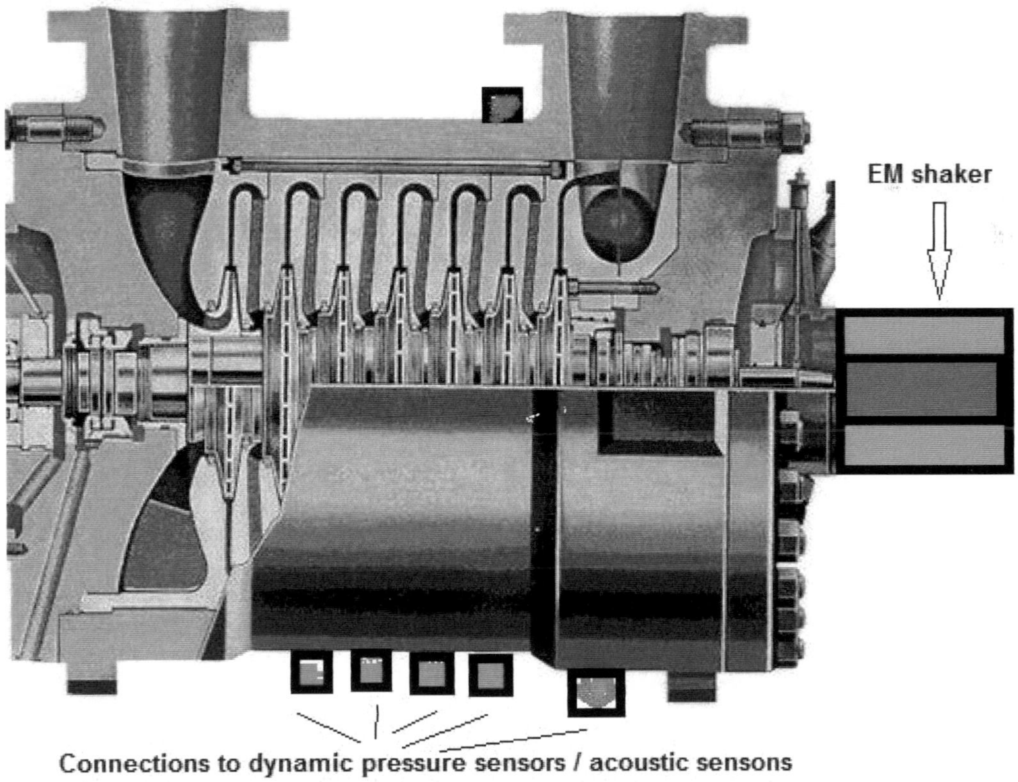

Fig 5 - Extra sensors and Electromagnetic shaker installed for FLFS extended test

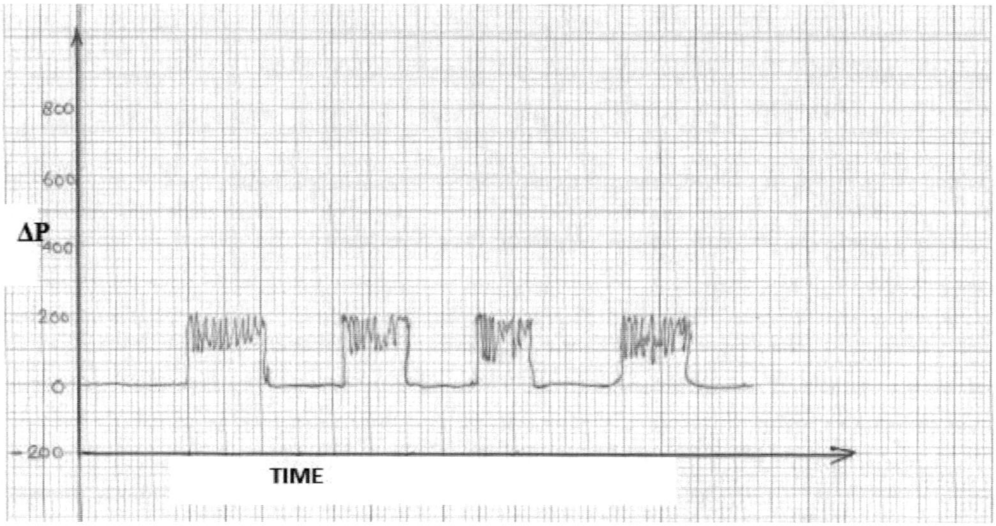

Fig 6 –Pressure Fluctuation ΔP across time shaft

The Dynamic Pressure monitor accepts signals via its associated transducer I/O module from dynamic pressure sensors and dynamic pressure charge amplifier. The monitor further conditions these signals into the appropriate measurement units, compares them to user-programmable alarm set points, and generates appropriate alarm signals for communication to the host control system. The measurement can be peak-to-peak dynamic pressure or RMS dynamic pressure. The results can be further cross checked with particular region of vibration spectrum. In some cases, test results had led to modification in return channel / diffuser of compressors at OEM works which could have resulted into enormous delay to mitigate at site.

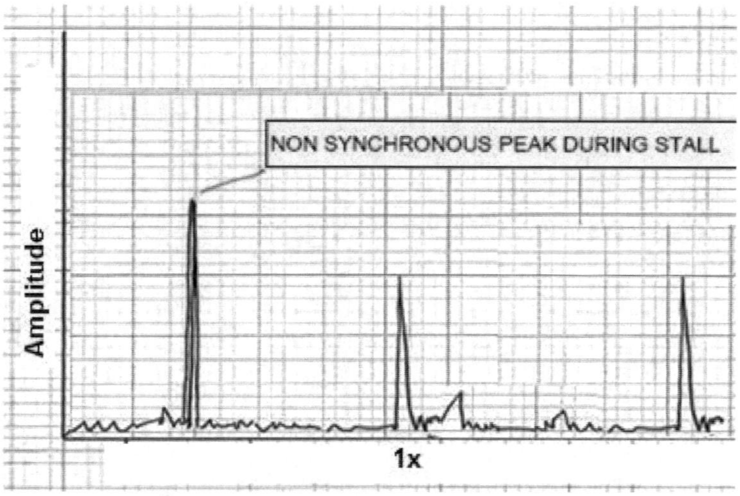

Fig 7 Vibration spectrum during stall

To validate the robustness of rotor bearing design for any cross coupling forces, a magnetic shaker can be used to generate perturbation from a sweep of speed above slow roll to operating speed during full load full speed test. During the test, the casing vibration also can be measured so that with proper processing of actual shaft vibration in peak- peak is carried out. It is to be noted that lateral vibration analysis report always predict absolute shaft displacement in peak -peak.

After a short shutdown during its FLFS test, the compressor should be subjected to a hot restart. With this test, it can be found if high vibration at critical speed can reach alarm / trip level during start up and does not allow a quick start of machine at site. The high vibration also can be an indication of possible rotor to stator rub. A bode plot and orbit plot (direct and 1x filtered) can be used to check above aspect. The measured value shall be used to calculate RCV (remaining clearance value) and to determine if the threat of rotor to stator rub exists.

Validation of train robustness against torsional instability during FLFS complete unit test - If the system has a VFD as driver and / or a speed increasing gear box suitable device such as strain gauge telemetry , torsiograph shall be used to fix the speed band which can be avoided if out of compressor operating speed map . If they fall in speed map, then a quick and proven design solution should be exercised after joint deliberation with end user and third party specialist

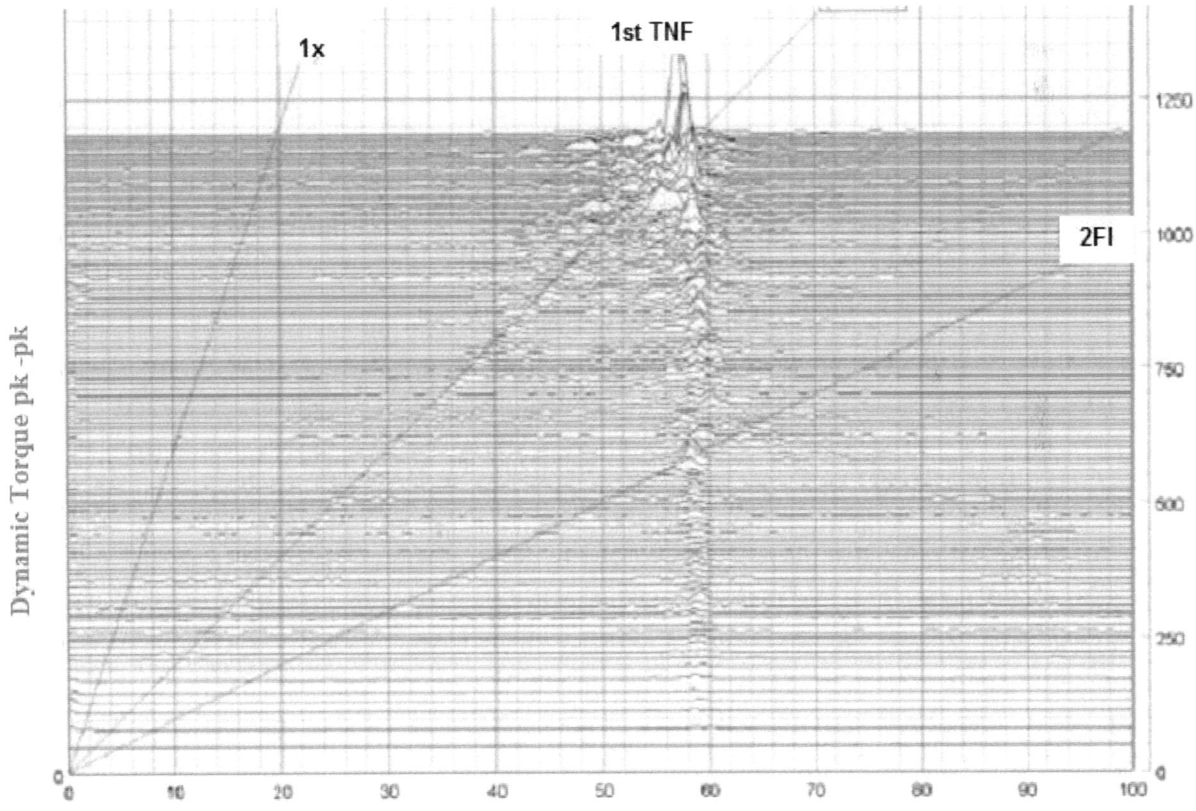

Fig 8 - Waterfall of dynamic torques measured with strain gauge telemetry

Road map of integrity assurance during start up and regular operation phase-

Piping alignment, flange management and proper pipe hanger setting is "a must do "to avoid casing distortion which leads to misalignment and high 1x vibration at site . Axial rubs can result from a mis-match in the thermal growth rates between the rotor and casing. During a cold startup, the rotor may expand faster than the more massive casing. A related problem can occur when a slide key on the machine casing hangs up and prevents free movement of the casing during startup or shutdown. The constrained machine casing can deform, displace internal parts, and cause an internal rub.

For all above such reasons, before starting the machine a joint review by end user, site support engineer from OEM should happen and outcomes must be recorded and shared with central engineering team of OEM. For flexible couplings after assembly balancing, the set shall be etched marked for unbalance location to mitigate 1x vibration and canning effect. No safety critical device should be bypassed. Trip multiplier / alarm annunciation should not be used in centrifugal compressor to start the machine unless agreed and requirement is proven at test bed. Suction strainer (temporary) should have high and high- high differential pressure alarm and trip which should be half of value of design pressure of strainer. If clean up run with nitrogen is suggested by OEM then correctly sized orifice and suitable speed band shall be advised by OEM

Before machine is put on forward flow, it is imperative to configure a pre- alarm level for all vibration probes based on test bed data, which can be further modified as the machine sets in after some months of operation. The pre alarm data shall provide an alert to reliability / condition monitoring engineer. The pre alarm can be further refined for individual forcing frequencies.

While the machine is put on forward flow service, compressor performance map is to be revisited (a repeat ASME PTC 10 test is not required) and fine tune to set capacity control and load sharing management. In this work, as built compressor map from PTC 10 performance test conducted at OEM works should be kept as base line data. Testing conducted during commissioning will establish a baseline of actual performance as it gathers the data from those instruments which shall remain in loop. Periodic field tests are often conducted to verify the overall performance and signal changes that may predict mechanical damage, internal fouling, or other deteriorating conditions. If periodic tests by operating personnel are not possible then online performance system can be installed based on compressor configuration, criticality of compressor in terms of process upsets.

It has now been realized by most of end users, that merely having vibration monitoring system does ensure the high system availability. Hence in addition, diagnostic tools and software are installed in a dedicated computer which is manned by a reliability / condition monitoring system.

Machinery condition based monitoring (CBM), fault and failure classification and remaining useful life are challenges that are gaining much interest by industrial communities. Study reports from various sectors of academia and industry shows the expected improvements of using automatic or semi-automatic systems for condition monitoring, where a significant improvement on maintenance costs and total productivity can be realized .The operator must first determine whether an anomaly has occurred, then separate faults from failures and isolate the problems to be studied and/or modelled.

It is to be encouraged that designer of machine fault diagnostic software works with OEM in tandem to employ the data mining system , shape (bode plots , spectrum , orbit , polar) identification (where the data with allowed deviations can be encrypted) and support vector machines (SVM) to have anomaly detection of system to enhance decision making process . The Shape and Directivity Index (SDI) of journal filtered orbits is an additional diagnostic parameter whose evaluation can be combined with the full spectrum analysis. The ellipticities of the filtered orbit, as well as the amplitude and the inclination angle of the major axis of the orbit, are parameters whose analysis can provide important diagnostic information by shape identification.

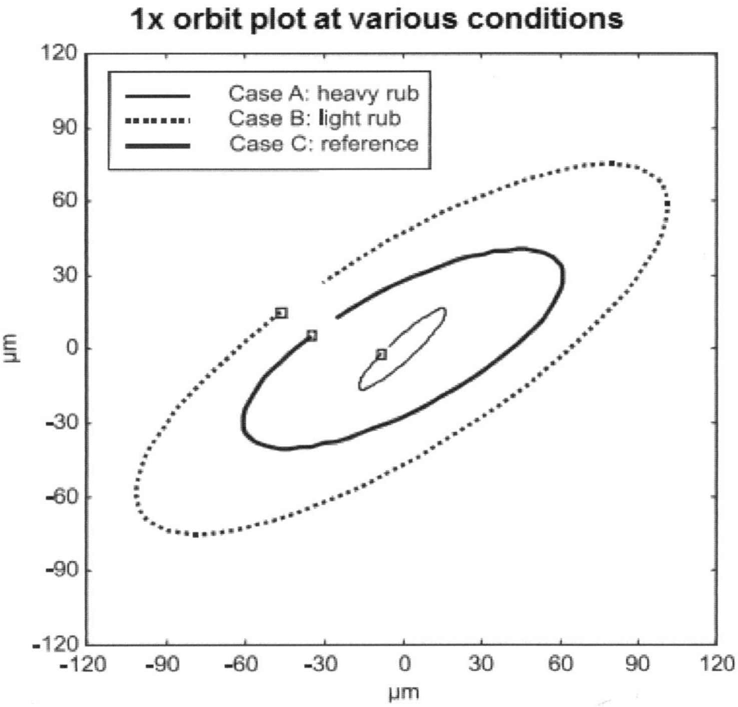

Fig 9 Using shape identification (SDI) for fault detection

Optimization of testing campaign for process gas compressors on various duties.

If above proposals on various road maps are concluded relevant and plausible, then we can deliberate to optimize the testing campaign on case to case basis with high level of prudence described as below with fig 10.

One MRT of lot and one PTC 10 Type 2 of other compressor from same lot can be carried out as test campaign optimization if -

1. Total population is less than 4. Optimizing the testing campaign does not benefit the schedule if number of compressors is less shown as A in figure 10.

2. OEM does not have a facility or time line of project prevents to do so.

3. They are equipped in toothed , rub tolerant / soft labyrinths as inter-stage seals , having complete unit stand by and having high log decrement >0.25

4. Compressors are medium pressure duty with a less than pressure ratio of 3

The type 2 test can further validate the rotor critical speed as measured in MRT of 1st compressor as most of compressors operate above 1st critical speed but lower than 2nd critical speed. During the PTC10 type 2 test sub- synchronous vibration (Precursor of rotating stall) readings can be taken and a similitude prediction can be made for actual gas for a good estimate.This type of test carries hidden risks at site when compressor is put under full speed full load service, so if the above points 1 and 2 are not applicable, it is advisable to opt for FLFS test (shown as B in figure 10)

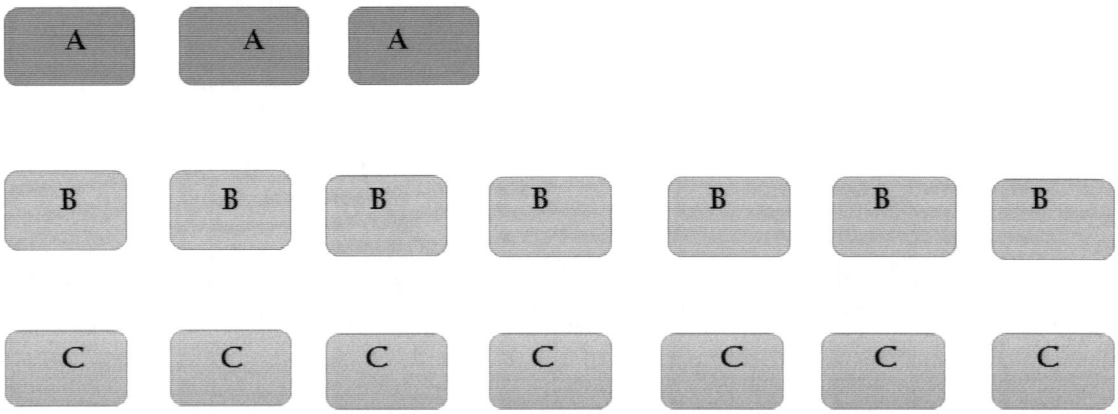

Figure10 - Testing campaign based on criticality

For compressors having pressure ratio higher than 4 or suction pressure higher than 50 bar or discharge pressure higher than 100 bar or employing swirl brakes / hole pattern seals / honey comb seals it is important to have a full load full speed test . Same understanding is applicable for compressors having side loads / side stream .Case histories have proven the fact that if FLFS test is not conducted, some compressors had faced the issues of sub synchronous vibration due to aerodynamic / cross coupling related instability which took a lot of time to rectify at site. If the compressors are high pressure injection compressors, it is important to include validation of cross coupling instability using an electromagnetic shaker. The FLFS test of one compressor can represent the complete rotor-dynamic behavior of all the compressors of that lot. (shown as C in figure 10).Controlled surge test shall be carried out at site backed by OEM calculation of surge tolerance of impeller and diffuser and compressor operating window has enough margins so that a conservative surge control line can be drawn.

Before mandating the FLFS complete unit test we must weigh the ability of the OEM centralized shop to conduct such test. Logistics of transporting gear, turbines / heavy duty electric motors and VFD panel to the test shop must be analyzed before placement of offer. It should be understood does the project schedule allow for additional time duration for transportation, assembly of string and dismantling and packaging?

Once the above facts are jointly understood and risk and rewards are weighed , the complete unit FLFS test at OEM works can be shifted to site activity as "site machinery performance test ".

The above proposals does not absolve OEM from performing their internal test to build up the confidence level of fleet and such tests can be offered for remote witness of end user specialist. The duration of test need not to have a four hour steady state run but all relevant transient data can be captured and analyzed. For fast track project, a modular designed unit may be considered by OEM which may save time in installation and that time can be utilized for extensive site testing.

References

[1]Vibrational Diagnostics of Rotating Machinery - Agnes Muszynska IJRM 1995

[2]Sleeve Bearing Diagnostics Using Proximity Probes - www.vibration.org/Presentation

[3]Transient Speed Vibration Analysis -Insights into Machinery Behavior -Stanley R. Bognatz Presented at: Vibration Institute Piedmont Chapter Meeting 2007

[4]Experience in empirical criteria for rotating stall in rotating stall in radial vaned diffuser- Fulton & Blair -Proceedings of 24th Turbo-machinery symposium TAMU

[5]Recent Experiences in Full Load Full Pressure Shop Testing of a High Pressure Gas Injection Centrifugal Compressor," Proceedings of the 23rd Turbo-machinery Symposium, TAMU

[6]ASME PTC 10 modified test for mechanical test assessment for centrifugal compressors - M. Miranda & Noronha - Proceedings of the 26th Turbo-machinery Symposium, TAMU

[6]Back to Basics – Rubbing or Not? - Richard Thomas, Vibration Institute Piedmont Chapter 2015 SETPOINT™

[7]A Novel method to detect surge - Mantosh Bhattacharya Gas Machinery Council paper 2013.

[8] Reliable Rotor Dynamic Design of High-Pressure Compressors Based on Test Rig Data- By Wagner- Journal of Engineering for Gas Turbines and Power | Volume 123 | Issue 4 |

[9] Start-up Issue with Centrifugal Compressors after Short Shutdown and Mitigation to Enhance Reliability and Availability- Mantosh Bhattacharya Proceedings of MFPT 2016 Dayton Ohio

[10]Layered manufacturing - a Tutorial by Sara Mcmains

[11]Diagnostic Significance of Orbit Shape Analysis and its Application to Improve Machine Fault Detection - N. Bachschmid, P. Pennacchi and A. Vania -Politechnic of Milan

[12]Data Modeling for fault detection - HM Jainsch Huntsville Alabama

[13]Fault detection and prediction to application of rotating machinery - Thesis desertion by Gary.R.Halligan 2009 Missouri University of science and technology

[14]Condition monitoring fault diagnosis and prognostics of industrial equipment by Enrico Zio -Polytechnic of Milan

[15]Machine Fault Signature analysis - by P Jayaswal - IJRM 2008

[16]Predicting time to failure of industrial machine based on temporal data mining - Thesis desertion by Jean Nakamura 2007 University of Washington

[17]Early detection of rotating stall phenomenon in centrifugal compressor by means of ASME PTC 1 type 2 test - L. Baldassarre et al -Proceedings of the 41st Turbomachinery Symposium TAMU

[18]Rotating component modal analysis and resonance-by F Kushner -Proceedings of the 23rd Turbomachinery Symposium TAMU

[19]How To Justify Equipment Improvements Using Life Cycle Cost and Reliability Principles - Paul Barringer

Bibilography

[1]API 617 Standard for Axial and Centrifugal Compressors and Expander-compressors for Petroleum, Chemical and Gas Industry Services 8th edition

[2]Forsthoffer's Best Practice Handbook for Rotating Machinery -By William E. Forsthoffer

[3]Forsthoffer's Rotating Equipment Handbooks: v. 3: Compressors by William E. Forsthoffer

[4]Principles for alarm system design - Norwegian Petroleum Directorate.

[5]Applying Condition Monitoring to Various Machinery- Application Techniques ALLEN BRADLEY

[6]Vibration Spectrum Analysis: A Practical Approach - by Steve Goldman

Ed Spence[a], David Siegel[b]

a The Machine Instrumentation Group LLC
b Predictronics Corporation

Society for Machinery Failure Prevention Technology
Norfolk, VA May 2018

Abstract

Emerging technologies are expanding the implementation of Condition Monitoring to markets and applications not previously practical or economically feasible. This trend is particularly observable as Industrial OEMs develop CBM programs for their own hardware. Data driven analytic approaches to CBM can result in the development of additional monitoring tools and expanded suite of condition indicators using both existing and new instrumentation, augmenting traditional approaches developed for rotating/reciprocating equipment. Transferring learning from the lab to the field introduces challenges and obstacles to deployment that can be overcome by leveraging the convergence of IIoT technologies.

The paper will highlight a pump condition monitoring case study, which used data driven analytics to develop new health indicators from vibration signals.

Pump Vibration Analytics Case Study and the Need for More Deployable Instrumentation

Introduction

Condition Monitoring techniques, systems and services today represent a mature eco-system that has been developing for decades. CBM services have traditionally been provided by service providers and CBM equipment vendors, often in collaboration with plant maintenance engineers. This after-market installation and service of condition monitoring instrumentation means that plant maintenance, system vendors or service providers capture the value of CBM by measuring and analyzing machine data once installed in the plant.

An observable trend to expand CBM coverage to new equipment involves the industrial component OEMs themselves, who are developing machine health programs of their own. Implementing a program for a new class of machines can be a multi-faceted challenge for the OEM requiring mastery of new analysis skills, the capture and application of tribal knowledge, and deployment of new sensor technologies.

This is particularly true for specialized or complex machines that are not currently or easily monitored with traditional tools and techniques. In many cases, end users of complex machinery are the first movers in this direction, as they encounter problems in the field which are opaque to the OEM buffered by their sales channels. Field experience and recommendations from plant maintenance or other service providers may be adapted by the OEM as the foundation for a program to pre-instrument their equipment.

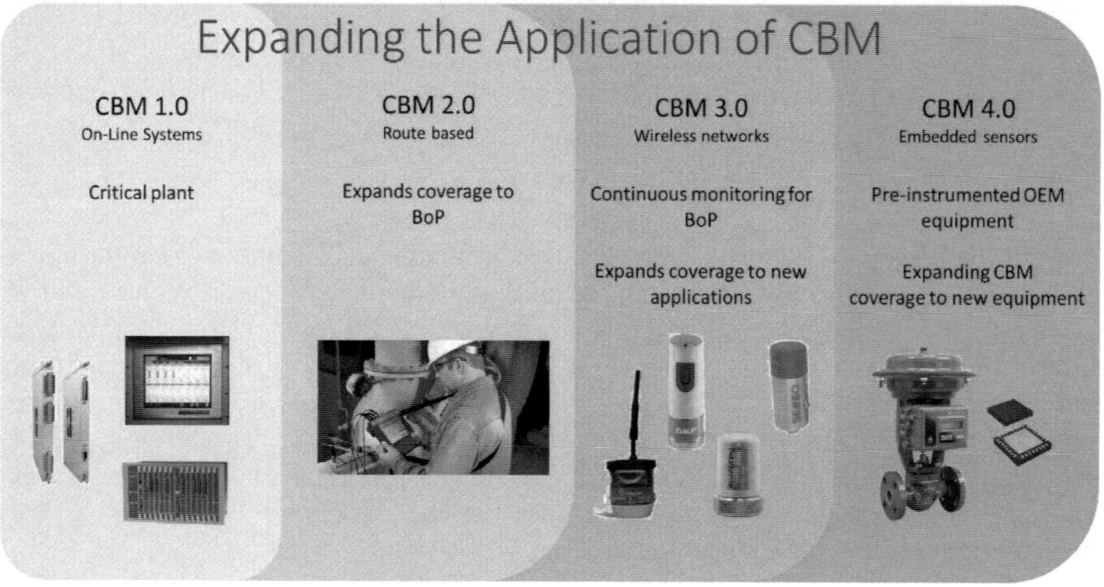

Figure 1 The emergence of CBM as applied by the Machine OEMs may be considered a fourth wave of market expansion, extending CBM coverage to previously unmonitored hardware.

In this universe of specialized equipment, the machine behavior or operating environment often resists traditional vibration analysis approaches. Automation equipment or machines with complex mechanics, mobile platforms or hardware that doesn't rotate at fixed frequencies present challenges to tradition tools. Extending the art of CBM to new equipment with effective monitoring and diagnostics may require development of new health indicators and deployment of new instrumentation. In particular, data engineering and advanced pattern recognition algorithms are joining traditional signal processing approaches to help identify patterns within available sensor data. Once the various signals associated with machine behavior and operation are modeled, higher accuracy machine health indicators can be developed, providing earlier identification of new faults, and leading to more detailed diagnostics.

Help Wanted: A CBM Program Development Process for New Types of Equipment

The process for development of a program starts with knowledge of the common problems. FMEA results, maintenance experience and other 'tribal knowledge', coupled with existing operational and monitoring data can all contribute to better modeling of machine behavior. With this foundation, machine characterization based on available data can proceed. Data engineering techniques and algorithms are selected and applied to extract unique features from the data. Once the features are extracted from the measured data, informed by domain experts, new health indicators can be defined and tested. With effective health indicators in place, more accurate diagnostics are also possible. The duration of this stage is often limited by the availability of useful data.

It may turn out that additional data would further enhance the ability of this process to deliver highly effective indicators. In these cases, new instrumentation will have to be deployed to collect additional field data. This requires the cooperation of supportive end users, who can benefit from the program by early identification of problems that lead to down time, and a front row seat to the prospect of extending CBM coverage to equipment that wasn't previously monitored or can now be monitored more effectively.

Thankfully, new CBM systems developed with IIoT capabilities minimize the time and effort to deploy new instrumentation. New solid-state sensor technologies, low cost wireless joins data engineering analytics to enable new CBM instrumentation solutions tailored toward the machine. A pilot study can determine in a relatively short time whether progress on new health indicators can be accomplished with available data, or whether new instrumentation is needed.

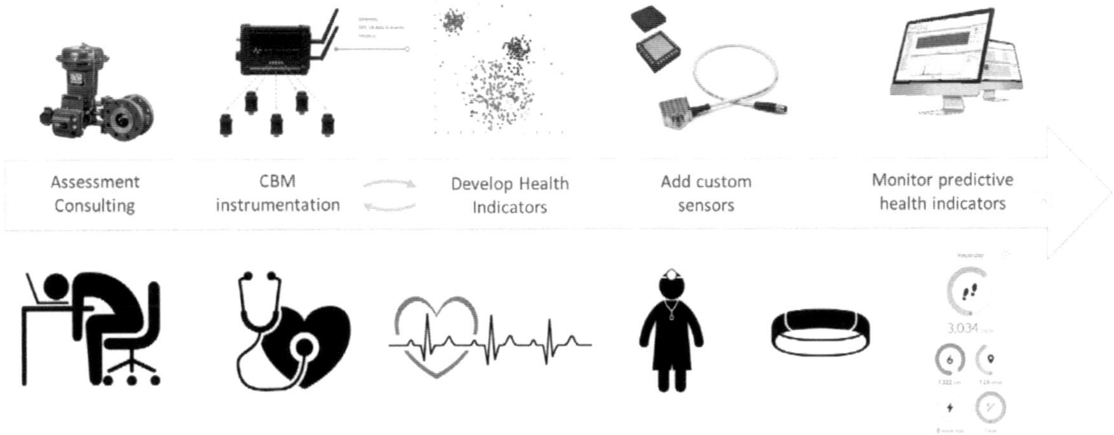

Figure 2 A process for development of CBM instrumentation on new equipment may start with analysis of existing data. A short-term pilot study should help determine whether new instrumentation (sensors) are needed or not.

New Instrumentation Options:

Acquiring the new field data necessary for the development of effective health indicators may be accomplished by deploying a wireless sensor system that enables relatively easy access to the data. More and more wireless CBM systems are appearing on the market. Equipped with low power accelerometers, many systems also have the ability to deploy additional sensors measuring temperature, rotation speed, pressure, etc. For OEM field testing, it's also necessary that the system have the means to transport the data to a location within the test site, so that it can be easily extracted by the OEM for analysis. Many wireless systems today provide data via OPC/UA, Modbus, Ethernet and other standard protocols to port the data to a PLC, SCADA system or even a local server or laptop set up for the purpose of collecting data for later analysis.

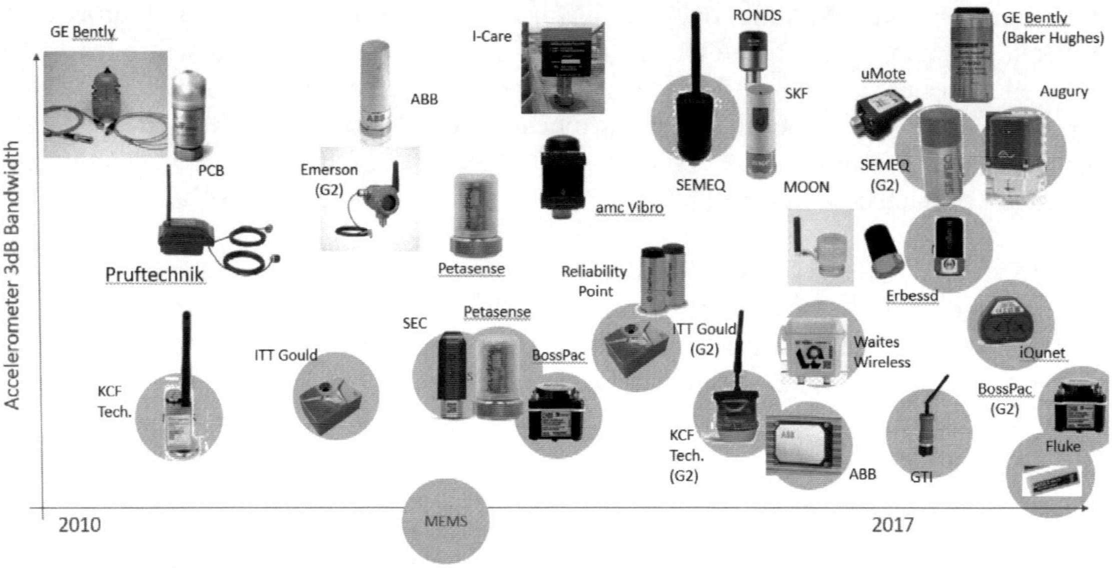

Figure 3 An example of the convergence of IoT related technologies is the explosion of wireless options for CBM instrumentation. Many of these products were developed by vendors with an industrial or CBM pedigree, with interface options enabling relatively easy integration within current plant OT.

Another relatively recent and potentially game changing development is the emergence of MEMS (solid state) accelerometers. Measurement of machine vibration with a solid state sensor enables several characteristics useful to the machine vendor. For starters, high integration levels are now more achievable, offering the potential for 'smart' sensors with embedded pre-processing of the data. Small form factors, low power operation and standard electronic interface options also contribute to the design flexibility of a new sensor solution. In particular, enabling implementation of standard digital interfaces in small form factors is particularly useful for the placement of the sensors, as well as data transport and management of the output data. If the problem is understood well enough, some sensor-based processing can be performed as well, generating useful information (indicators) rather than simply data. Adding this all up, fully integrated and highly capable CBM sensor solutions can be defined - in collaboration with the OEM - specifically for the machine.

Case study: Development of a Health Indicators for a Water Processing Pump

For this application, the end-customer wanted to develop a predictive health monitoring system for their pumps in which they wanted to have an early indication and diagnosis of any potential degradation that was occurring in the pump. In terms of the signal measurements, accelerometers were installed for measurements at several locations, along with a tachometer signal (to measure the speed of the pump) and a pressure signal. Much of the analysis focus was spent on developing analysis methods for the vibration signal, however the pressure and tachometer signal provided important context information. For this study, there was data from seeded faults conducted in a laboratory setting and data from two pumps in the field. The laboratory data was used to develop some of the initial analysis methods. However, the emphasis of this case study will be on the field study since there was an actual failure in the field and it provided an interesting data set for evaluating the pump health monitoring analysis methods.

Signal Analysis and Feature Extraction Methods:

The pump vibration might offer important insight in both the temporal and frequency domain. With this mind, a short-time Fourier transform of the vibration signal was calculated and the spectrogram was one of the vibration analysis tools used in this study. This spectrogram provided a wealth of information for visually understanding the vibration signal and for extracting information to infer the health condition of the pump. An example spectrogram from a pump in the field is shown in Figure 4, in which the left plot contains the time signal and angle-frequency spectrogram from the pump at the start of the data collection (assumed to be healthy) and the plot on the right is from a data record that is one day prior to a failure event occurring on the pump.

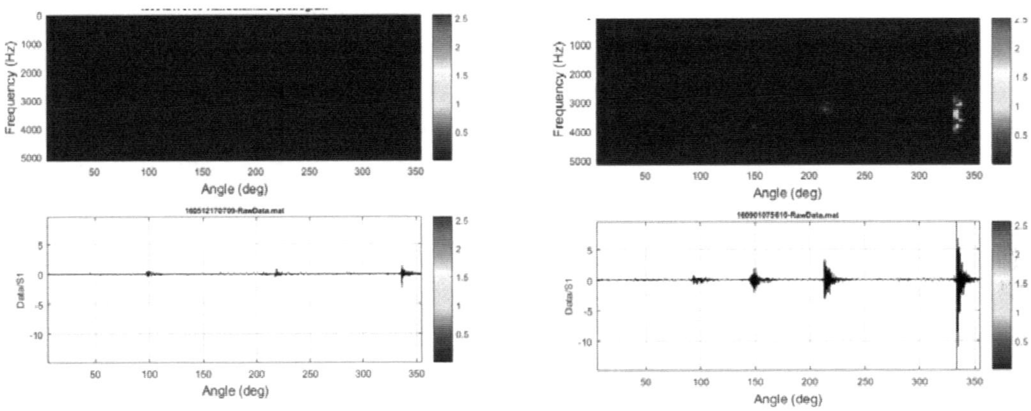

Figure 44. Angle-frequency spectrogram (left plot – healthy system, right plot – one day prior to a failure event).

Various metrics were considered from the time domain, the frequency domain, and the spectrogram in this study. This set of condition indicators were reviewed and ranked using the laboratory data, and this learned information was then applied to the field data.
An example feature plot is shown on Figure 5, in which the magnitude of a frequency band is plotted over time. This band appears to provide a very clear increase and early indication of the failure event (approximately 1 day prior to the failure event). The health index method aims to combine multiple features to provide a more robust assessment of the pump and to have better coverage for the various failure modes that can occur.

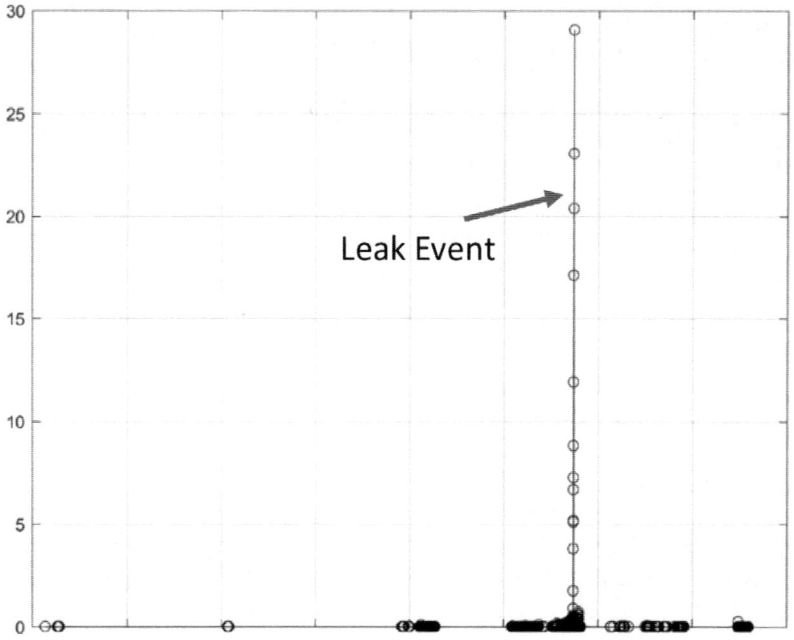

Figure 5. Example feature trend which shows a large increase prior to a failure event.

Health Indication of Pump Condition:

With this notation that a multivariate assessment of the pump might provide a more robust assessment of the pump condition and provide more coverage for the various failure modes that occur, using a baseline-based health index method was a reasonable choice for this application. What was needed was an assessment that could learn the nominal relationship between the vibration features and the operation conditions and provide a predicted vibration feature response that could be compared to the actual vibration feature value. A residual based health index method was considered, and this health index could be trended and monitored over time. It also should be noted that the health index can be scaled by an appropriate statistical threshold, in which the convention is that any health value above one is considered anomalous.

An example health index is shown for one of the two monitored pumps in the field in Figure 6. As one can observe, there is a clear increase in the health trend before one of their maintenance events (failure event). For this failure event, the health index is above the threshold rather consistently one day prior to the failure occurring. Potentially, this event could have been prevented if this system was deployed (the data analysis methods were developed with this historical data).

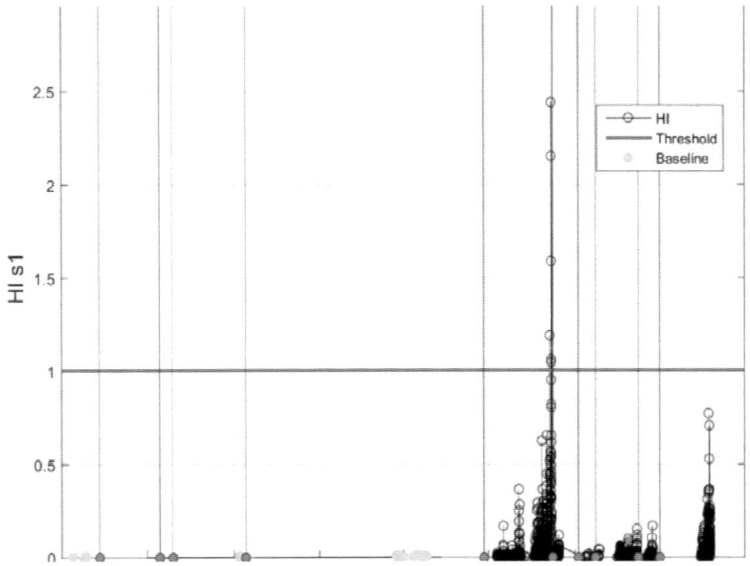

Figure 6. Health indicator result for one of the pumps in the field.

Diagnosis Method:

After reviewing the historical data from the pumps in the field, various types of anomalous vibration patterns were observed. This included the failure vibration pattern, a precursor to the failure pattern, and accelerometer related failures. In addition to using a health index method that might providing a warning that the pump vibration is anomalous, it was also important to provide more context and a potential diagnosis of what problem could be occurring. This would help provide the maintenance technicians and operators with more actionable information based on the assessment and diagnosis of the pumps health. In order to facilitate this diagnosis information, the diagnostic approach considered the frequency patterns for each fault type. This frequency domain pattern recognition method was quite effective and had an accuracy at nearly 90%. With this developed method, one could monitor the pump, assess its health condition, find early indications of potential problems, and diagnosis the problem that was occurring.

Discussion and Future Work:

Further analysis of the available data will be used to further improve and refine the health indicators, particularly if additional measurands are also available for analysis from operational and control sensors. In this case, pressure, tachometer, temperature and other sensors may provide contextual information that may increase diagnostic value of the health indicators as well as increasing the warning time of an emerging fault. Additional field data can also be acquired to increase the resolution of the known fault signals, and to identify additional fault patterns.

Optimizing the instrumentation may be considered as well, applying lessons learned to reducing the amount of instrumentation or to improve the performance of the sensors to deliver higher quality data, which may in turn may improve the resolution of the analysis. Some of the fault patterns were related to damage occurring to the sensors themselves, so finding improvements in the instrumentation will help alleviate these sensor failures in the field. In addition, if there are lower cost sensor accelerometer measurements that could be used, this would allow for more measurement points and an improvement in the fault isolation and diagnosis accuracy. Solid state MEMS accelerometers enable a higher level of integration with digital communications options that provide flexibility to connect the sensors to the existing OT, and might also be considered in the future work.

With a clearer understanding of how effectively the indicators flag and diagnose machine faults, growing confidence may lead to at least some interpretation of the signals closer to the edge of the network, transforming the transmission of data to that of information. Pre-processing in the sensor (or gateway) from raw data to condition indicators may reduce the amount of transmitted data, easing the burden of OT-side data management and integration with enterprise data bases. This can also have benefits for a wireless sensor (longer battery life) or a control network where the channel capacity is limited (less load on the channel bandwidth).

Application of data engineering techniques to characterize the machine, coupled with the deployment of IIoT technologies provides the hardware OEM with both the process and the tools to implement their own CBM strategy, delivering higher uptime by shipping 'smart', pre-instrumented equipment that incorporates domain specific intelligence. This paper highlights the emerging trend in instrumentation technology, the OEM's desire to incorporate this technology into a future service business, and how the analytics can be used to convert sensor data into actionable health information. Although this concept was demonstrated with a pump condition monitoring case study, the same methodology is being considered by various OEM's and can be readily applied to other engineering assets.

ROTOR DESIGN CONSIDERATIONS TO PREVENT IMPELLER AND PREMATURE BEARING FAILURES IN CENTRIFUGAL FANS

Robert J. Sayer, PE
The Vibration Institute
2625 Butterfield Road, Suite 128N
Oak Brook, IL 60523

Abstract: Centrifugal Fans are subjected to blade-pass pulsation and mass imbalance forces as part of normal operation. Fan impellers have several n-nodal diameter modes of natural frequency that can be sensitive to blade-pass pulsation forces. Excitation of these modes can lead to catastrophic failure. The principal flexural mode of a fan rotor is sensitive to mass imbalance force, and if excited, can result in amplified stresses in the shaft and amplified force transmission to the bearings. In the case of SWSI fan rotors, where the 1-nodal diameter mode of the impeller couples with the flexural mode of the shaft, excitation of the rotor mode can lead to catastrophic failure of the impeller.

Key Words: Blade-pass pressure pulsation, centrifugal fan, critical speed, fan rotor, fan wheel, finite element analysis (FEA), natural frequency, mode shapes, nodal diameter modes, rotor dynamics.

Introduction: Fans are used for a wide variety of industrial and commercial applications. There are two primary types of fans, centrifugal (sometimes referred to as radial) and axial. This paper concentrates on the structural dynamics of centrifugal fans. A centrifugal fan system consists of a rotor (fan wheel and shaft), stationary components (fan housing, scroll, inlet cone, dampers, pedestals and foundations) and interface components (bearings).

There are two basic types of centrifugal fans. The first type is called a double-wide, double-inlet (DWDI) fan. A picture of a large DWDI fan rotor is shown in Figure 1. It consists of a shaft and a fan wheel. The fan wheel is typically mounted at the mid-span of the shaft, an equal distance from each bearing. Because the fan wheel is located at mid-span, its stiffness does not contribute significantly to the rotor dynamic properties. However, its mass is a significant factor in the rotor dynamic properties.

The second type of centrifugal fan is the single-wide, single-inlet (SWSI) fan. The location of the fan wheel on the shaft can vary for this type of fan. It can be cantilevered outside of the bearings, or located between the bearings as shown in Figure 2. In either case, the center-of-gravity of the fan wheel is not located at mid-span of the shaft, a condition which complicates the rotor dynamics thereof.

Figure 1: Photograph of a DWDI Fan Rotor Hanging in Slings

Figure 2: Photograph of a SWSI Fan Rotor Hanging in Slings

Air enters the DWDI fan wheel from both sides, flowing radially from the center of the fan to the trailing edge of the blades. Air enters the SWSI fan wheel from only one side. Dynamic forces produced by centrifugal fans as part of normal operation include imbalance forces and blade-pass pulsation forces. Fan rotors must be designed to operate subjected to these dynamic forces.

The natural frequencies of the fan wheel can be very sensitive to dynamic pressure pulsations produced either as a part of normal operation or by unusual aerodynamic circumstances. Catastrophic failures have been attributed to the excitation of natural frequencies of fan wheels by dynamic pressure pulsations (Figure 3).

Figure 3: Examples of Fan Wheel Failures Attributed to BPPF Pulsations

Blade Pass Pressure Pulsations: All fans produce dynamic pressure pulsations as the blade tip passes by the cutoff point in the housing scroll. Figure 4 shows a cross section of a housing scroll. The outer diameter of the fan is indicated as a dotted line. The fan in this illustration would be rotating in a counterclockwise direction. Air would exit the fan wheel radially into the housing and flow along the scroll until it exits the housing at the upper left quadrant.

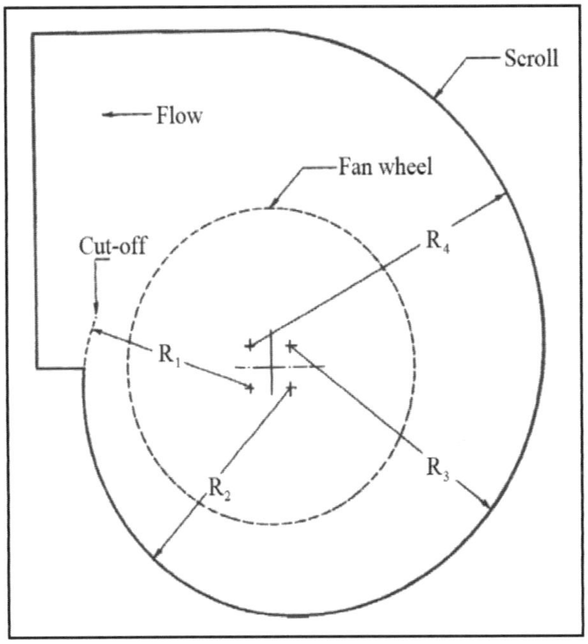

Figure 4: Cross Section of Housing Scroll

The distance between the outer diameter of the fan and the scroll gradually increases from the cut-off point to the discharge. When a blade rotates past the cutoff point, it develops a force that produces a dynamic pressure pulsation. The blade pass pulsation frequency (BPPF) is equal to the rotational speed of the fan multiplied by the number of blades. For example, a 10-bladed fan rotating at 1200 rpm would produce dynamic pressure pulsations at 12,000 cpm or 200 Hz.

Blade pass pulsations can be detected by spectral analysis of dynamic pressure data. Testing for blade pass pressure pulsations is easily done by installing a dynamic pressure transducer into the sidewall of the housing or duct. The response of the housing or duct to blade pass pulsations can also be detected with an accelerometer placed on the surface of the housing or duct.

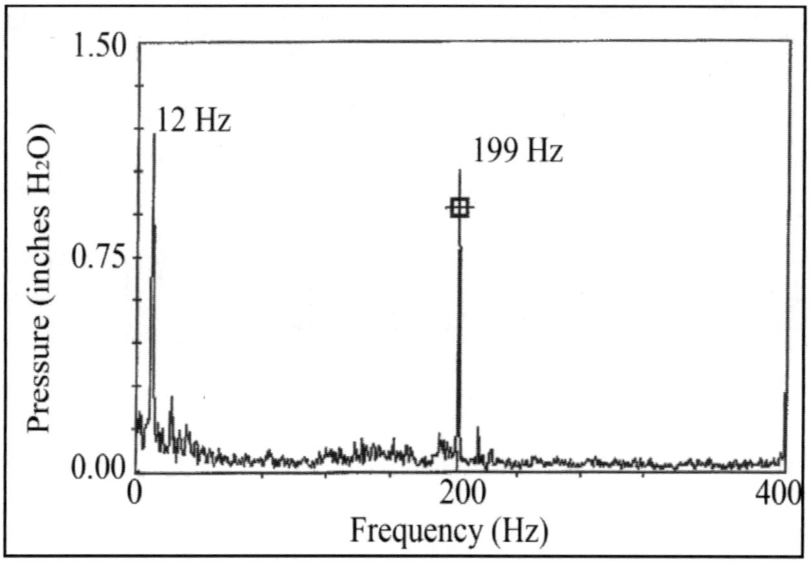

Figure 5: Frequency Spectrum of Pressure Pulsations

Figure 5 contains a spectrum of data acquired from a large 10-bladed induced-draft (ID) fan operating at 1195 rpm (19.9 Hz) using a piezoelectric dynamic pressure transducer. A spectral peak is clearly evident at the blade-pass pulsation frequency (199.0 Hz). The magnitude of the dynamic pressure developed from the BPPF was around 1.0 inches H_2O. The magnitude of this dynamic pressure is very small and is not typically large enough to damage a fan wheel. However, if the frequency of the dynamic pressure is near a natural frequency of a fan wheel, resonant amplification of stresses can cause failure. The mode of the natural frequency must be sympathetic to the BPPF force in order for resonance to be a problem.

An understanding of the modes of natural frequency of a fan wheel is imperative in order to assess the potential severity of resonant excitation by dynamic BPPF pressure. A finite element analysis (FEA) of a ten-bladed DWDI centrifugal fan was performed to illustrate the various modes of natural frequency of a fan wheel. The fan wheel used for this example was a type that is typically employed for large induced draft (ID) fans, similar to that shown in Figure 1.

n-Nodal Diameter Modes: All centrifugal fans have a set of modes of natural frequency that are called the n-nodal diameter modes. The general form of the mode shapes for these natural frequencies are such that the maximum deformations occur at the outer diameter of the wheel, at the trailing edge of a blade. The n-th degree of the mode represents the number of maximum deformations (nodal) points in either the positive or negative directions.

The n-nodal diameter modes can be very sensitive to dynamic pressure pulsations if the frequency of the dynamic pressures is close to the natural frequency of the wheel. Excitation of these modes can lead to the catastrophic failure of the fan wheel. These modes are especially sensitive when the n-th order of the mode is an integer factor of the number of blades in the fan wheel. For instance, the two-nodal diameter and five-nodal diameter modes would be very sensitive to blade pass pulsations produced by a ten-bladed fan. Likewise, the two-nodal diameter and four-nodal diameter modes would be very sensitive to blade pass pulsations produced by an eight-bladed fan.

Typically, only a few of the n-nodal diameter modes are present in most fans within the frequency range bounded by the blade pass pulsation frequency (BPPF). The natural frequency associated with these modes will increase with the n-th order of the mode. For example, the natural frequency of the four-nodal diameter mode will be greater than the natural frequency of the three-nodal diameter mode, which will be greater than the natural frequency of the two-nodal diameter mode.

With the exception of the nodal circle and wheel wobble modes, the participation of the shaft in any of the other n-nodal diameter modes is negligible. Since the deformation of the fan wheel is independent of the shaft, the transmission of vibration to the shaft and bearings is minimal when these natural frequencies are excited. Therefore, the excitation of these modes is difficult to detect during operation. If excited, catastrophic failure can occur with very little warning.

2-NODAL DIAMETER MODE: The two-nodal diameter mode is often referred to as the "chip" mode or "butterfly" mode since the deformed shape resembles a potato chip or butterfly. Figure 6 is an end view of the 2-nodal diameter mode for the example fan. The natural frequency was calculated at 128.5 Hz by the FEA. The mode shape is made up of a combination of axial and radial response. The radial and axial deformations of the side plates have two maximum positive and two maximum negative locations. The axial deformations of the side plates are in-phase while the radial deformations of the side plates are out-of-phase with each other.

The radial deformations dominate in wide fan wheels while the axial deformations dominate in narrow fan wheels. The 2-nodal diameter mode can be sensitive to dynamic forces applied either in the axial or radial direction and, thus, would be susceptible to excitation by BPPF forces if the natural frequency and blade-pass frequency were to be near each other. For most DWDI fans, the separation margin between the natural frequency of the 2-nodal diameter mode and the BPPF is sufficient to preclude the resonant excitation of the mode. For the 10-bladed fan, the rotational speed would have to be reduced to around 780 rpm for the resonant excitation of this natural frequency, which was calculated at 128.5 Hz, to become an issue (BPPF = 780x10/60 = 130 Hz). Excitation of this mode is more common in variable-speed fans that operate at lower speeds.

If the natural frequency for this mode is excited, the deformation at any location in the wheel, either axial or radial, is countered by an equal and opposite deformation at another location. Because of these deformation characteristics, the inertia of this mode does not transfer to the shaft and the shaft does not participate in this mode. The 2-nodal diameter mode of a fan wheel can be excited, but not detectable by transducers at bearings. Strain gage testing is required to confirm excitation of this mode.

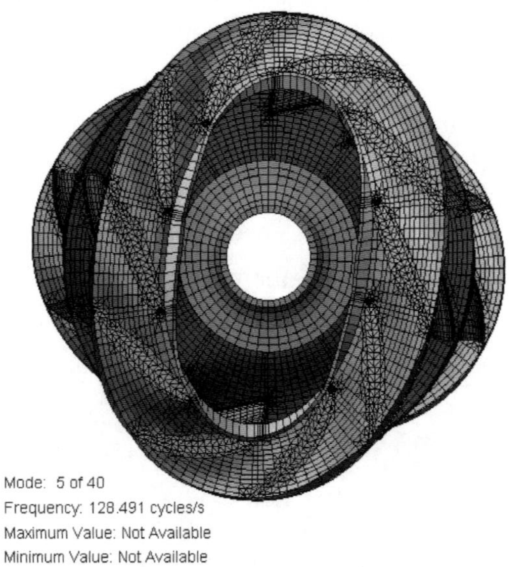

Figure 6: 2-Nodal Dia Mode

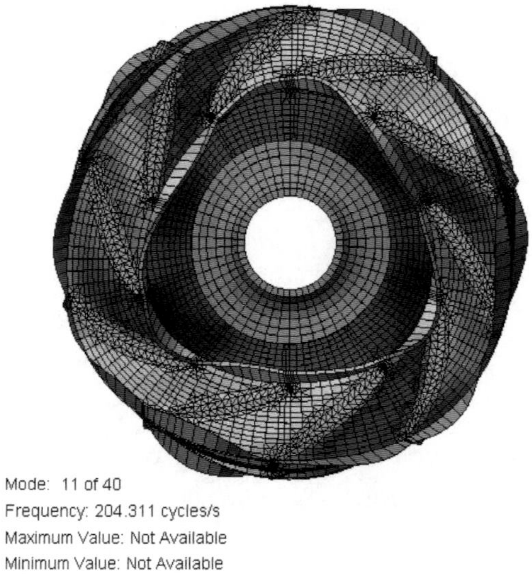

Figure 7: 3-Nodal Dia Mode

3-NODAL DIAMETER MODE: Figure 7 shows the end view of the FEA result for the 3-nodal diameter mode. This mode has both axial and radial response. The radial and axial deformations of the side plates have three maximum positive and three maximum negative locations. The axial deformations of the side plates are in-phase while the radial deformations of the side plates are out-of-phase with each other.

The natural frequency for this mode was calculated to be 204.3 Hz by the FEA . Note that this natural frequency is greater than the natural frequency of the 2-nodal diameter (128.5 Hz) mode. This is typical for all fan wheels because the complexity of the shape increases as the order of the mode increases. It is more difficult to deform the fan wheel into a 3-lobed than a 2-lobed pattern.

The 3-nodal diameter mode can be sensitive to dynamic forces applied either in the axial or radial direction. It would be susceptible to excitation by BPPF forces if the natural frequency and blade-pass frequency were to be near each other. The 3-nodal diameter mode is more sensitive in 9-bladed and 12-bladed fans since the n-th order of the mode is an integer multiple of the number of blades (9/3 = 3 and 12/3 = 4).

For the 10-bladed fan, the BPPF would be around 200 Hz if the rotational speed of the fan were 1,200 rpm (BPPF = 1,200x10/60 = 200). In this case, the natural frequency of this fan would be in close proximity to the blade pass frequency.

Like the 2-nodal diameter mode, the deformation at any location in the wheel, either axial or radial, is countered by an equal and opposite motion at another location. Since the inertia of the wheel does not transfer to the shaft, the shaft does not participate in this mode. Even though the excitation of this natural frequency results in significant vibration within the fan wheel, it typically is not detectable by vibration probes mounted to the bearings.

4-NODAL DIAMETER MODE: Figure 8 shows the end view for the 4-nodal diameter mode. This mode also consists of a combination of both axial and radial response, each having four maximum positive and four maximum negative locations. The axial deformations of the side plates are in-phase while the radial deformations of the side plates are out-of-phase with each other.

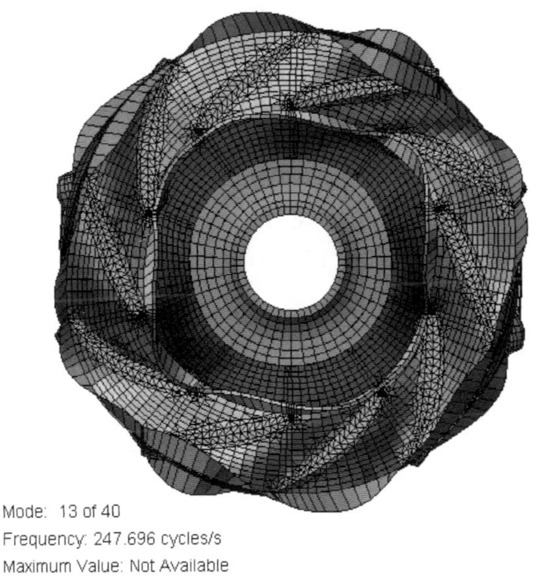

Figure 8: 4-Nodal Dia Mode

The natural frequency for this mode was estimated by the FEA as 247.6 Hz. This natural frequency is greater than the natural frequency of both the 2-nodal diameter (128.5 Hz) and 3-nodal diameter (204.3 Hz) modes.

The 4-nodal diameter mode can be sensitive to dynamic forces applied either in the axial or radial direction. The 4-nodal diameter mode is more sensitive to BPPF in 8-bladed and 12-bladed fans since the n-th order of the mode is an integer multiple of the number of blades (8/4 = 2 and 12/4 = 3).

For the 10-bladed fan, the BPPF would be around 248 Hz if the rotational speed of the fan were 1,488 rpm (BPPF = 1,488x10/60 = 248). In this case, the natural frequency of this fan would be in close proximity to the blade pass frequency.

Once again, the motion at any location in the wheel, either axial or radial, is countered by an equal and opposite motion at another location. The inertia of this mode does not transfer to the shaft. Because of this, the shaft does not participate in this mode. Since the shaft does not participate, even though the excitation of this natural frequency results in significant vibration within the fan wheel, it typically is not detectable by vibration probes mounted to the bearings.

5-NODAL DIAMETER MODE: Figure 9 shows the end view of the FEA result for the 5-nodal diameter mode. This mode has both axial and radial response. The radial and axial deformations of the side plates have five maximum positive and five maximum negative locations. The axial deformations of the side plates are in-phase while the radial deformations of the side plates are out-of-phase with each other.

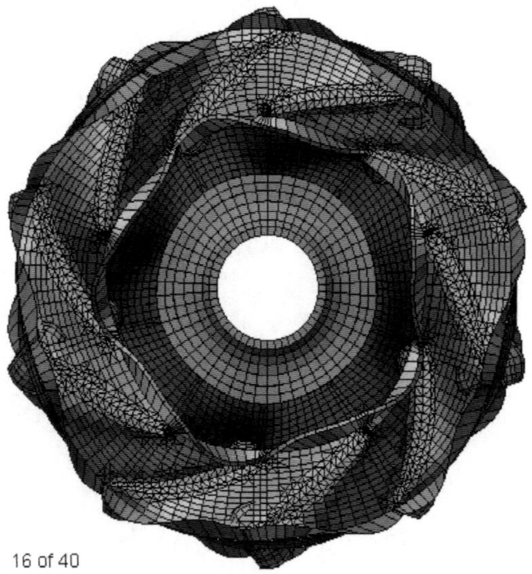

Figure 9: 5-Nodal Dia Mode

The natural frequency of this mode was calculated to be 264.3 Hz by the FEA. Note that this natural frequency is greater than the natural frequency of the 2-nodal diameter (128.5 Hz), 3-nodal diameter (204.3 Hz), and 4-nodal diameter (247.6 Hz) modes. The 5-nodal diameter mode can be sensitive to dynamic forces applied either in the axial or radial direction. The 5-nodal diameter mode is more sensitive in 10-bladed fans since the n-th order of the mode is an integer multiple of the number of blades (10/5 = 2).

For the 10-bladed fan, the BPPF would be around 264 Hz if the rotational speed of the fan were 1,584 rpm (BPPF = 1,584x10/60 = 264). In this case, the natural frequency of this fan would be in close proximity to the blade pass frequency.

Once again, the motion at any location in the wheel, either axial or radial, is countered by an equal and opposite motion at another location, the inertia of this mode does not transfer to the shaft. Because of this, the shaft does not participate in this mode. It typically is not detectable by vibration probes mounted to the bearings.

Sensitivity of n-Nodal Modes: The results of an impact test can be used to illustrate the sensitivity of the n-Nodal diameter modes of a centrifugal fan wheel. Figure 10 is the time waveform and transfer function for an impact test performed on a large induced-draft (ID) fan wheel. This test was performed with the rotor hung from flexible nylon straps, approximating a nearly free-free support condition (Figure 1).

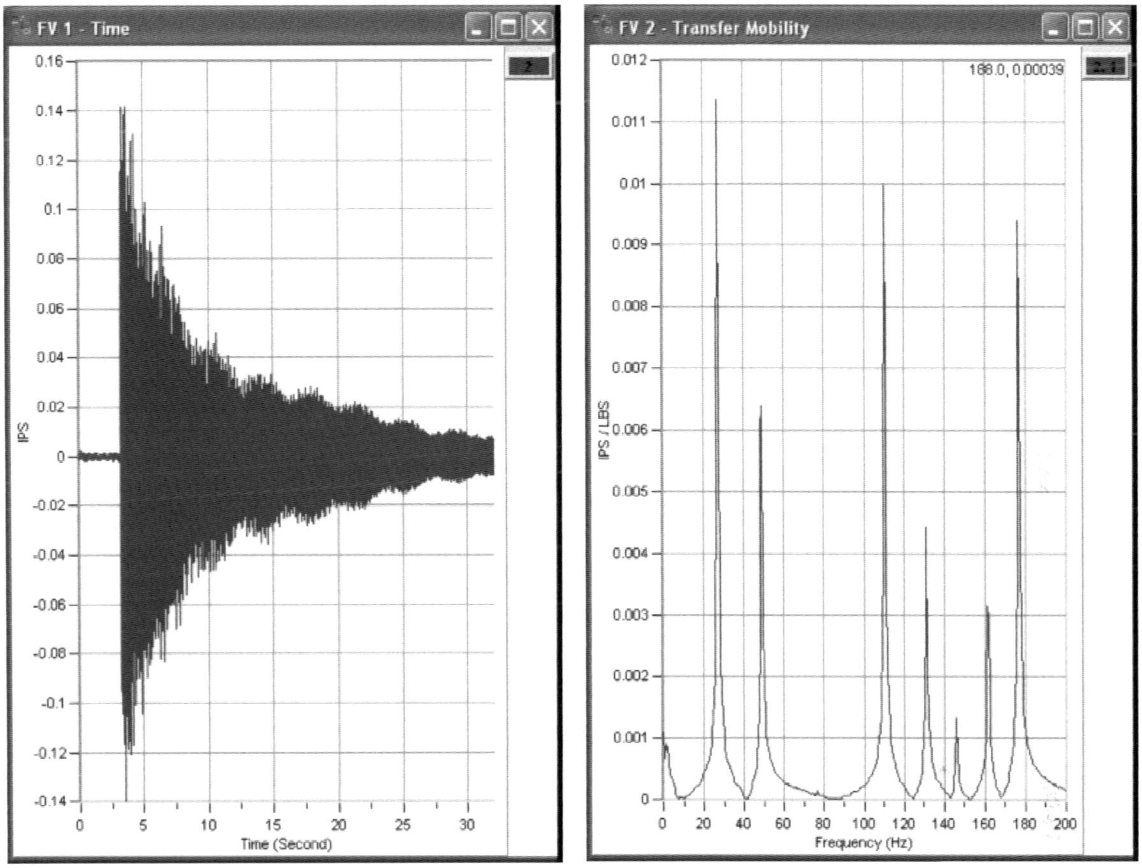

Figure 10: Waveform & Transfer Function (Natural Frequency Test of Fan Wheel)

The time waveform of the "ring-down" response shows that the fan wheel continues to vibrate over 30 seconds after application of the impact force. This is indicative of very lightly damped modes of natural frequencies. In fact, the damping factor for these modes are typically around 0.10% of critical damping ($\zeta \sim 0.001$). This low amount of damping would result in an amplification factor of around 500:1. This means that a cyclic operating stress of only 20 psi, caused by pulsation in a non-resonant fan wheel, would increase to 10,000 psi if the BPPF and the n-Nodal mode frequency would coincide.

The transfer function curve shows the many n-nodal diameter modes of the fan wheel. The narrow and steep transfer function curves also indicate lowly damped sensitive modes.

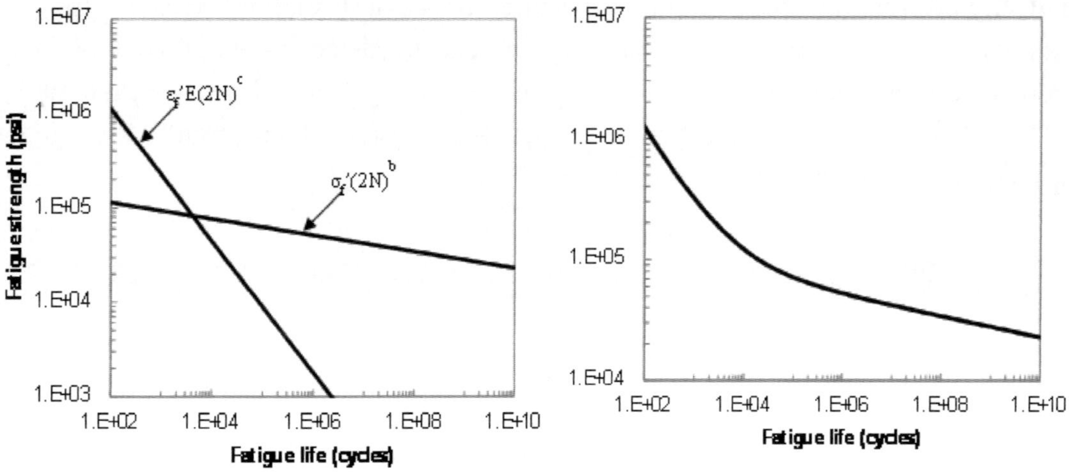

Figure 11: Notch-Strain Fatigue Life Curve

The *notch strain-life method* can be used to demonstrate the sensitivity of the n-nodal diameter modes. This method is described by a low-cycle fatigue component and a high-cycle fatigue component. Per Equation 1, the addition of these two components provides an estimate for the number of fatigue cycles to failure (2N) when subjected to a given cyclic stress ($\Delta\sigma/2$) and constant mean stress (σ_m).

$$(\Delta\sigma/2) = (\sigma_f` - \sigma_m)(2N)^b + (\varepsilon_f`)E(2N)^c \qquad \textbf{Equation 1}$$

Where σ_f` is the fatigue strength coefficient, b is the fatigue strength exponent, $\varepsilon_f$`is the fatigue ductility coefficient, and c is the fatigue ductility exponent. References [1][2][3] are available that provide values for these coefficients and exponents for various materials.

The left side of Figure 11 contains the log-log plot of each term. When added together, these terms provide the fatigue life curve on the right side of Figure 11.

Fan wheels are subjected to numerous cycles of stress reversal in a very short time. For instance, consider a 10-bladed fan operating at 1,180 rpm. The fan wheel is subjected to 196 stress reversal per second. The accumulation of stress reversals in a single day would equal nearly 1.7 million.

For cases of high-cycle fatigue, the fatigue life equation is dominated by the first term of Equation 1 and can be simplified to Equation 2.

$$(\Delta\sigma/2) = (\sigma_f` - \sigma_m)(2N)^b \qquad \textbf{Equation 2}$$

Sensitivity Example: Consider two double-wide, double-inlet (DWDI) fan wheels of identical design, operating at 1,180 rpm (19.67 Hz). Both fan wheels have 10 blades and are subjected to pressure pulsations at blade-pass pulsation frequency (BPPF = 10 x 19.67 = 196.7 Hz). One fan experienced a fatigue failure shortly after initial start-up while the second did not experience any failures and had been operating on another unit at the same power plant for some time.

Figure 12 shows the fan rotor after the initial catastrophic failure of the first wheel. All of the blades and the side plates are missing from the wheel. This failure is consistent with the excitation of an n-nodal diameter mode of natural frequency, but since the second wheel had not failed, the failure investigation concentrated on other possible root-causes (i.e. foundation inadequacy, process parameters, acoustic amplification of pulsations, turbulence).

Figure 12: Fan Rotor after Wheel Failure

Modal testing and finite element analysis of the fan wheel provided an estimate of 196.0 Hz for the "stress stiffened" natural frequency associated with the 4-nodal diameter mode. Figure 8 shows the deformation shape for this mode. If excited, it results in distortion within the wheel that leads to large stresses in the welded connections between the blades and side plate. Failure of the side plates then leads to excess stresses in the welded connections between the blades and center plate, resulting in the failure pattern displayed in Figure 3 and Figure 12. If this were indeed the root-cause of the catastrophic failure, why didn't the second fan wheel fail?

The mean tensile stress developed by centrifugal forces, at the intersection of the blade tips and the side plate, was 40,000 psi for both fans. For the material used to manufacture the fans, (σ_f` = 170,000, b = -0.087) the high-cycle fatigue life equation was:

$$(\Delta\sigma/2) = (\sigma_f` - 40{,}000)(2N)^b = 130{,}000(2N)^{-0.087} \qquad \textbf{Equation 3}$$

The damping factor (ζ) for this mode was 0.003, or 0.30% of critical. This very low level of damping resulted in a very large resonant amplification factor. The ratio of the blade pass pulsation frequency to the natural frequency for the fan wheel that failed was r = 196.7/196.0 = 1.0035.

The amplification factor for resonance is given by Equation 4.

$$AF = = 1/[(1- r^2)^2 + (2\zeta r)^2]^{1/2}$$ **Equation 4**

Without resonance amplification, the cyclic stresses that would have been developed in the fan wheel due to BPPF were estimated at +/- 240 psi. Substituting the frequency ratio (r = 1.0025) for the fan that failed into Equation 4 yields an amplification factor of 108.4. This means that the cyclic stresses caused by the blade pass pulsations could have been as much as 26,000 psi. Substituting this cyclic stress into Equation 3 yields an estimated fatigue life of 100E06 cycles.

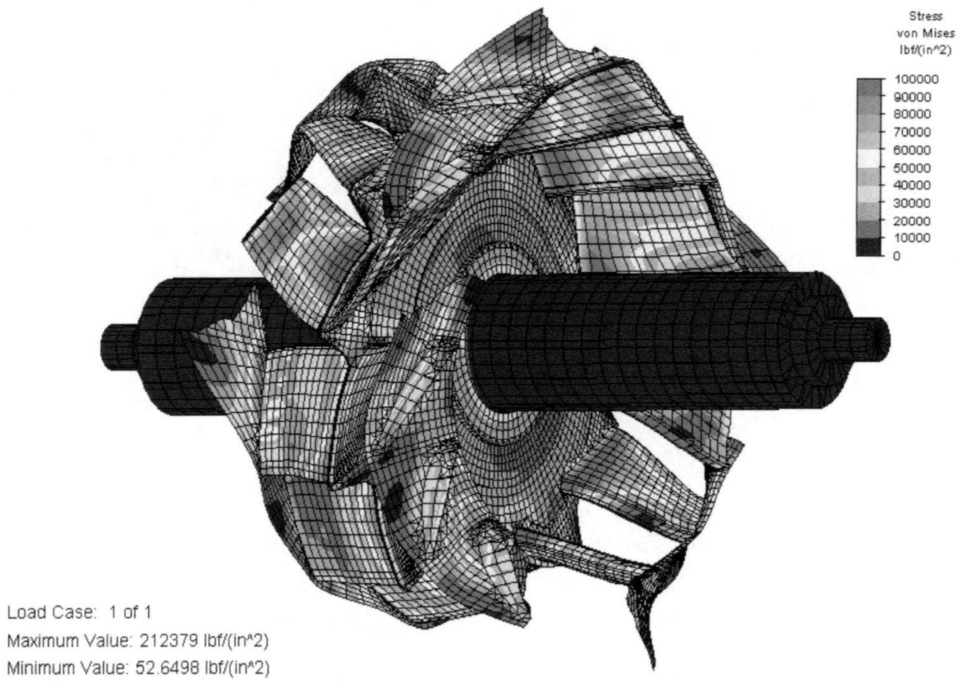

Figure 13: Final Failure Scenario after Crack Propagates Through Side Plates

The fan was subjected to 199.7 cycles of stress reversals per second, or 17.25E06 cycles per day. This means that the predicted fatigue life to crack initiation would be around 140 hours, which was consistent with the time of the actual catastrophic failure. Figure 14 is a photograph of a fan wheel that was removed from service before fatigue cracking propagated to final failure. In this case, fatigue cracks initiated in the side plate at the trailing edge of almost every blade tip.

Figure 13 is a finite element representation of the final failure scenario. Fatigue cracks that initiated at several locations on the side plates, have propagated completely through the side plates. The pieces of blades and side plates become masses cantilevered off of the center plate. Centrifugal forces acting on each cantilevered section result in high stresses at the intersection point of the blades and center plate. The final catastrophic failure at this point is not fatigue, but overload due to the large centrifugal stress, in this case exceeding the ultimate tensile strength of the material.

Now consider that the natural frequency of the second wheel was lower than that of the first wheel by only 1.0% (f_n ~ 194.0 Hz). It is not unusual for the natural frequencies associated with the n-nodal diameter modes to differ slightly from one wheel of identical design to another. The frequency ratio for the second fan wheel increased to 1.014. Because of the steep gradient of the amplification curve, the resonant amplification factor decreased to 35.1 and resulting dynamic cyclic stress was around 8,420 psi. This seemingly small reduction in natural frequency resulted in a significant reduction in stress. The predicted fatigue life for this lower cyclic stress was 50E12 cycles, which was more than the expected useful life of the fan.

This example illustrates the sensitivity of the n-nodal diameter modes and the importance of designing a fan wheel such that these natural frequencies are not close to the blade-pass pulsation frequency.

There are other sources of pulsation that a fan can produce that are not considered as normal operating conditions, such as rotating stall pulsations and vortex shedding from dampers. These occur at frequencies other than the BPPF. It is difficult to design a fan wheel not to be resonant at these other frequencies since they are not expected to occur under normal operating conditions and their frequencies vary from fan-to-fan.

Figure 14: Fatigue Crack Initiation in Fan Wheel

Rotor Dynamics of SWSI Fan Rotor: For a DWDI fan, the wheel "wobble" mode and the shaft flexural modes will be independent of and orthogonal to each other. The modal characteristics of a SWSI fan rotor are more complex. For the SWSI fan rotor, the shaft flexural and wheel "wobble" modes will combine with each other and form two orthogonal flexural modes. Textbook formulae that treat the wheel as a lumped mass, ignoring its flexibility, cannot provide an accurate estimate of the actual natural frequency. This is illustrated in the following example.

Consider the engineering formula given in Equation 5 for the SWSI fan rotor shown in Figure 15. For this example, the fan wheel weight (W) equals 1,450 lbs., the span between bearings (L) equals 39.5 inches, the overhang distance (c) equals 14.7 inches, and the diameter of the shaft is 5 7/16 inches. The shaft is made of carbon steel (E = 29,600,000 psi).

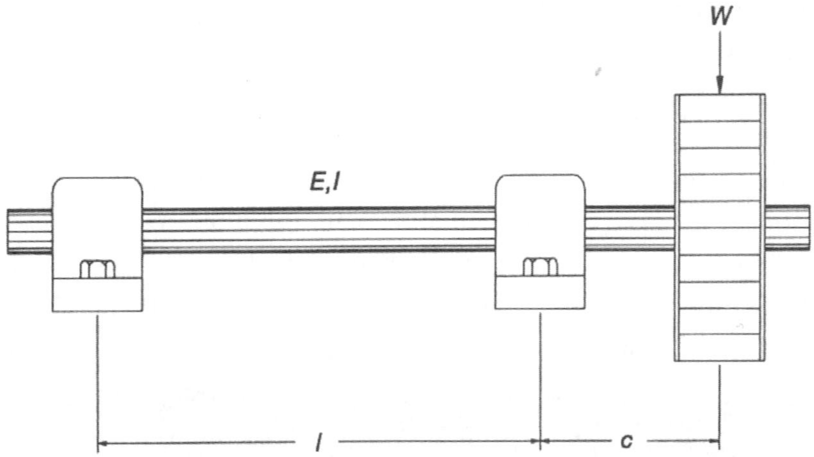

Figure 15: SWSI Fan Rotor Dimensions

Natural Frequency = $f_n = (1/2\pi)[k/M]^{1/2}$ **Equation 5**

Where: k = flexural stiffness of the shaft = $3EI/(L + c)c^2$

M = mass of the fan wheel

Substituting the known values into Equation 5, provides a calculated natural frequency of 46.4 Hz for the flexural mode of the rotor. Note that the mass of the shaft is neglected in this equation. Will the exclusion of the mass of the shaft result in an overestimation of the natural frequency? The answer is yes, but it is not the only reason that the actual natural frequency will be overestimated using the textbook formula. This equation not only neglects the mass of the shaft, assuming it to be negligible compared to that of the fan wheel, but also treats the fan wheel as a rigid component.

In reality, centrifugal fan wheels are very flexible structural components. An assumption that the fan wheel is rigid compared to the shaft is an oversimplification that will lead to erroneous results.

A single-wide, single-inlet (SWSI) fan wheel has blades on only one side of the back plate, and only one inlet side plate, sometimes referred to as the inlet shroud. The blades and the side plate are supported by and cantilevered off of the back plate (see photo in Figure 16). Since the back plate is a flexible structure, it results in another spring in the spring-mass system. Therefore, it cannot be treated as a single degree-of-freedom system, for which Equation 5 is only valid.

Figure 16: SWSI Fan Wheel

Figure 17 shows two separate spring-mass systems. Equation 5 is analogous to the lower system in the figure. There is only one spring, and it is associated with the flexural stiffness of the shaft. The mass of the fan wheel (M1) is lumped along with the mass of the shaft (M2).

The actual condition is better represented by the spring-mass system in the upper system. The lower mass represents the effective mass of the shaft. The lower spring represents the flexural stiffness of the shaft. The upper mass and spring represent the fan wheel.

Since the upper spring-mass system is a two degree-of-freedom system, it will have two natural frequencies. The motion of the masses will be in-phase with each other in the first mode of natural frequency. The motion of the masses will be out-of-phase in the second mode.

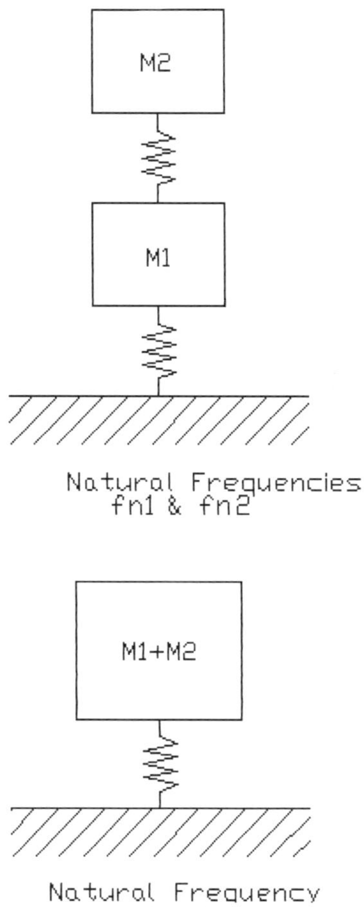

Figure 17: Spring-Mass Models

The natural frequency of the single degree-of-freedom model will be somewhere between the natural frequencies for the modes in the two degree-of-freedom system.

If f'_{n1} is the natural frequency of the single degree-of-freedom system and f_{n1} and f_{n2} are the natural frequencies of the two degree-of-freedom system, then:

$$f_{n1} < f'_{n1} < f_{n2}$$

Figure 18 shows the mode shape for a finite element analysis (FEA) of the example SWSI fan rotor. FEA [4] is a numerical technique that can account for the mass of the shaft and the flexural flexibility of the fan wheel. The mode shape in Figure 18 is for the case where the flexural displacement of the fan wheel is in-phase with the flexural displacement of the cantilevered section of the shaft. The natural frequency estimated by the FEA is 26.0 Hz, which is much less than the natural frequency (46.4 Hz) provided by Equation 5.

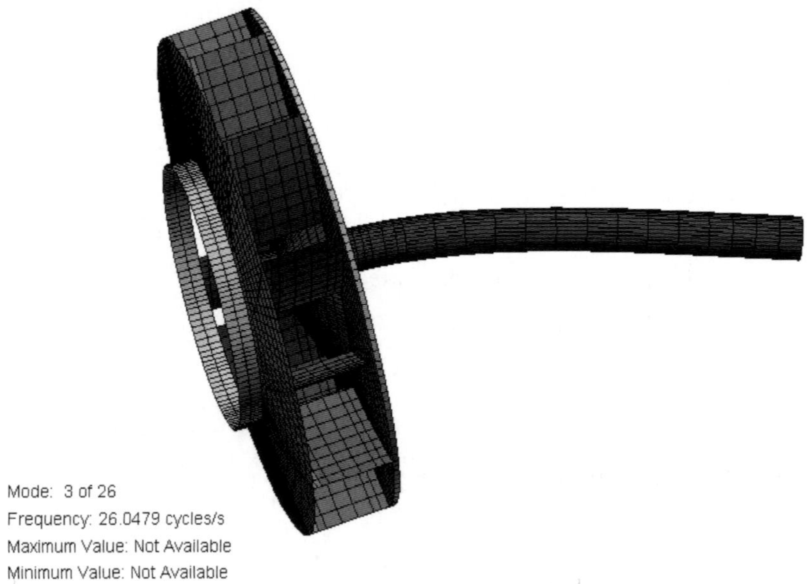

Figure 18: Mode Shape for Natural Frequency = 26.0 Hz

Figure 19 shows the mode shape for another natural frequency. This mode shape is for the case where the flexural displacement of the fan wheel is out-of-phase with the flexural displacement of the cantilevered section of the shaft. The natural frequency estimated by the FEA is 107.3 Hz, which is much greater than the natural frequency (46.4 Hz) provided by Equation 5.

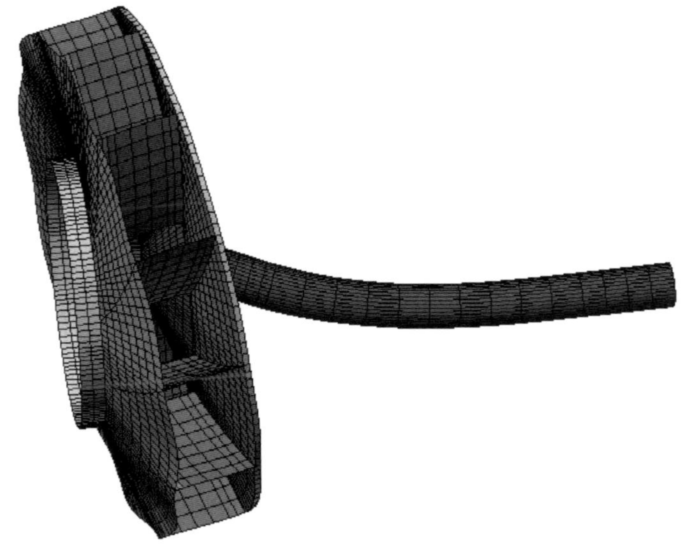

Figure 19: Mode Shape for Natural Frequency = 107.3 Hz

The significance of this analysis is that the textbook formula will always overestimate the natural frequency and provide an unrealistic estimate of the separation margin between the operating speed of the fan and its natural frequency. If the fan in the example above were to operate at 1800 rpm (30 Hz), the textbook formula provides a rather significant separation margin (f_o/f_n = 46.4/30 = 1.55). The more realistic FEA analysis would indicate a much lower separation margin (f_o/f_n = 26.0/30 = 0.87). The textbook formula indicates that the principal "at rest" natural frequency of the rotor is 55% above operating speed, while the actual "at rest" natural frequency is closer to 13% below operating speed.

Stress Stiffening: The operating natural frequencies of rotating parts of machines, such as fan wheels, pump impellers, and turbine blades, can be different from the "at rest" natural frequencies. Impact tests are performed on a rotor when it is in an idle condition. Therefore, the impact test provides the "at rest" natural frequencies of the rotor. During operation, the rotor will become stiffer due to centrifugal stresses and gyroscopic affects. Since natural frequency is a function of stiffness and mass, and the mass of the rotor is unchanged, the natural frequencies will increase as the centrifugal stress stiffening effects increase. Figure 20 displays the relationship between the natural frequency of a centrifugal fan wheel versus its operating speed. Note that since centrifugal stress increases with the square of rotational speed, the rate of increase in natural frequency is more rapid at higher speeds. The amount of increase in the natural frequency is not only dependent upon the rotational speed of the fan, but on the amount of centrifugal stress. Thus, it can vary widely from one fan design to another.

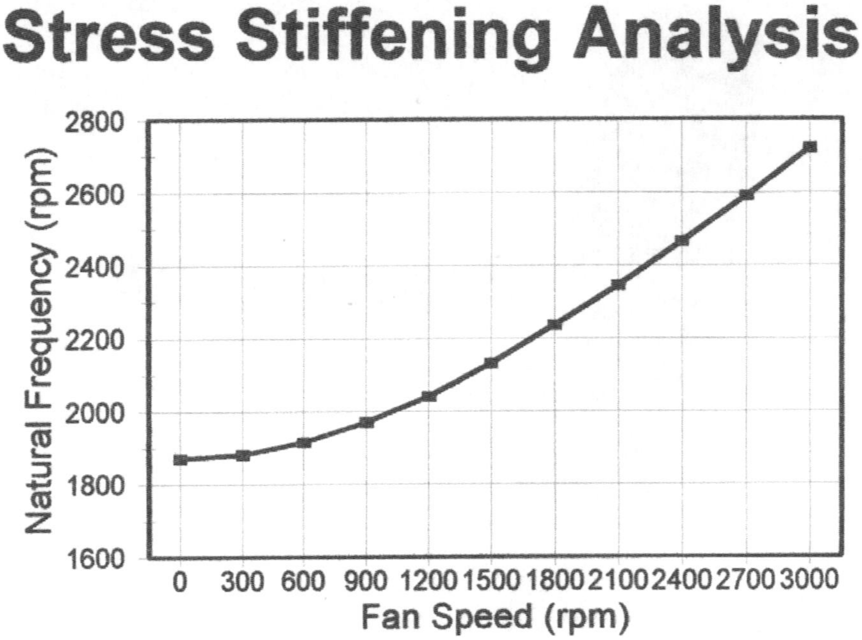

Figure 20: Stress Stiffening Curve

The importance of stress stiffening is that, if an inappropriate single dof model is used to estimate the natural frequency of a SWSI fan rotor, there is a good chance that the actual "at rest" natural frequency will be less than the operating speed of the fan. Since stress stiffening will increase the natural frequency during operation, there is a chance that the natural frequency and operating speed will be too close to each other. It is imperative that the numerical technique used to design the fan rotor provide an accurate assessment of the nature frequency.

References:

[1] J. A. Graham (Editor), *Fatigue Design Handbook*, Advances in Engineering Series (Volume 4), Society of Automotive Engineers, Warrendale, PA, 1968.

[2] R. L.Wetzel (Editor), *Fatigue under Complex Loading*, Advances in Engineering Series (Volume 6), Society of Automotive Engineers, Warrendale, PA, 1977.

[3] H. O. Fuchs and R. I. Stephens, *Metal Fatigue in Engineering*, John Wiley and Sons, New York, 1980.

[4] R. J. Sayer, *Finite Element Analysis; A Numerical Tool for Root-Cause Vibration Analysis*, Proceedings of the 27th Annual Meeting of the Vibration Institute, New Orleans, LA, July 2003.

SLOW ROLLRUN OUT PROBLEM OF ROTOR – GENUINE OR BENIGN

Mantosh Bhattacharya
Petrofac International UAE
Petrofac Tower 1
Telephone 00971501469853
Mantosh.b@petrofac.com

Abstract The issue of Rotor "run-out" which causes high overall vibration (displacement) readings on rotor has always been a point of concern during spin test and site acceptance test of turbo-machines. Rotor run-out is sub-divided to TIR (total indicated run out) commonly known as Mechanical run out-and electrical run out. While API (American Petroleum Institute) standards have mandated a certain limit of Slow roll run out, the proposed paper discusses to find which of contributing factors of high run-out should be considered as genuine or benign during machinery health assessment and monitoring as its main objective .Proposed paper differentiates rotor bow and slow roll run out , effect of run out during rotor balancing and to points out the lacunae of understanding exits for point / stage of such measurement. A short case study is presented to explain the issue and delays caused for repeated repair works due to such misunderstanding .The paper discusses the feasibility and overall technical effectiveness of eddy current probe, capacitance probe and other devices used . The proposed paper suggests the use of various diagnostic tools such as band and envelope spectrums for creating pre-alarm set up , using shape identification algorithm in diagnostic system based on reference data captured by initial slow roll measurement with a detailed set up .

Keywords - Mechanical run out, electrical run out rotor bow and slow roll eddy current probe, capacitance probe, laser triangulation

Introduction Slow roll run out is a combination of electrical and mechanical run out measured as shaft dynamic motion by vibration sensor at slow speed where the dynamic effects associated with rotation are negligible**.** The measured vibration signal of a rotating shaft will contain the run out of the probe target track area so the actual shaft vibration (irrespective to its geometrical cross-sectional shape / profile) is always different than measured shaft displacement as microns / mils pk-pk . A high value of slow roll run out is not damaging to rotor although interferes with actual vibration readings.

Slow roll run out termed as the vectorial sum of mechanical run out (MRO) and electrical run out (ERO) and is measured for a full rotation with an electronic device such as eddy current proximity probe. The signal from an eddy current proximity probe is a function of the gap between the probe tip and the target material. However, it is also a function of the electrical conductivity and magnetic permeability of the target material. These metallurgical variations lead to variations in the electrical conductivity and magnetic permeability of the shaft, affecting the proximity probe signal which is known as electrical run out. The electrical component of run out, in contrast, is not dimensional and instead represents metallurgical variations around the circumference of the shaft.

For a rotating shaft, a physical out-of-roundness result is a change in gap read by shaft displacement sensor. However, a perfectly round shaft with non-uniform electro-magnetic properties will also result in a change in probe output, even though the physical gap is uniform. In some cases two shafts may give identical probe outputs even though they have different physical shapes.

In practice, mechanical run out can be somewhat sinusoidal whereas electrical run out is characterized by a noisy waveform with numerous spikes. The combined run out shall have a waveform with numerous spikes riding on 1x and / or 2x waveform.

With Slow Roll run out, two possible extreme scenarios exist. If the slow-roll run out vector is additive to the vibration vector, then the proximity probe will report larger vibration amplitude than truly exists. This could lead to the situation where the machine goes into an alarm or trip condition prematurely. If, on the other hand, the slow-roll run out vector is subtractive to the vibration vector, then the proximity probe will report smaller vibration amplitude than truly exits. In this scenario, the true vibration may actually exceed an alarm or trip condition, but the monitoring system will not detect this due to the slow-roll vector making apparent vibration less than real vibration.

Scratches in the rotor surface viewed by the probe are another form of mechanical run out. Scratches(indentations) will produce negative-going voltage spikes on the transducer signals as instantaneous gap increases.

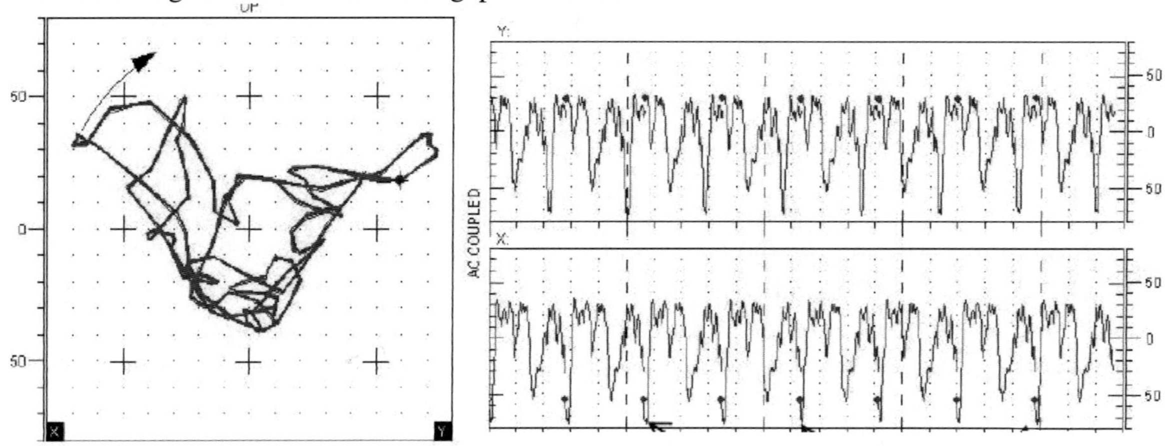

Fig -1 Unfiltered Orbit and slow speed time wave form of rotor with scratch

There are various sources of Mechanical Run out such as machining processes -

- If center-less grinding machines are used , then variations in shaft hardness can result in a non-circular geometry. This is known as Lobing of shaft.
- Improper feed rate, blunt tool and speed of cutting tools. Surface finish is strongly affected by cutting tool feeds and speeds and cause high mechanical run out.
- Defective or worn bearings on Lathe machine,

- Dents caused by mishandling and / or rust patches cause high mechanical run out.

Electrical run out in the probe track area is caused by:

- variations in conductivity and magnetic permeability of the steel
- sub surface defects in the steel
- variations in surface stress during heat-treating
- localized burnt areas due to grinding / residual stress concentration.
- generation of small magnetic fields such as those left by during MPI (magnetic particle inspection) without proper degaussing. As a rule of thumb, residual magnetism of 5 gauss can cause an electrical run out as much as 12.5 microns .

Synchronous and Asynchronous Shaft Run out Components

Some run out components such as shaft out-of-roundness will repeat at certain angular locations of the rotation; these are synchronous error motions. Other shaft run out components such as run out due to out-of-roundness of rolling elements in the bearing are cyclic but do not repeat at the same angular locations and are called asynchronous error motions. Slippage of rotor on V block also may give asynchronous error motions. So it is quite important that we identify and mitigate asynchronous error motions first before correcting synchronous error motions.

Vibration measurement affected by slow roll run out

The out of roundness of a shaft at both the 90° and 270° positions (oval shape) produces a time wave form which has a frequency of twice the rotational speed. In this case, if run out is not considered then it would appear as if we have a large 2x vibration when in fact there is none. An even more misleading situation would be if the run out were in anti-phase (180 degree) to the true vibration. In such a case it may appear as if there is little or no vibration when in fact there is a significant amount of vibration.

The slope of the curve, the linear range, and the DC output by a probe corresponding to a given gap will vary with changes in a target's geometric surface profile, its conductivity and permeability.

If a eddy-current probe and oscillator demodulator calibration for 4140 steel are used without recalibration on a material as such stainless steel or chrome , the curve shifts to the left, producing a higher-output voltage for a given gap than expected . Since the magnetic field of the probe penetrates the surface of the observed material, any repair which results in an interface between two materials (when the shaft is plated or metal sprayed, HVOF coated) may introduce distortion in the output signal measured by an eddy current displacement transducer. Porosity of spray coating on probe track area will also cause serious issues as the transducer shall read this as scratches on shaft probe track .Eddy-Current sensors are normally calibrated to a flat target. If the shaft diameter is less than 8-10 times the eddy-current probe diameter as a minimum we may have inaccurate measurements or possibility of sensor cross talk.

A scratch is the most common problem in rotor probe track which produces spikes on an unfiltered orbit that usually point away from the probes. A single scratch on the rotor will produce a 1X component and its harmonics in the vibration signal, but they will be visible in the spectrum over all speeds. If a shaft has multiple scratches, it is possible for two scratches to affect two probes simultaneously. When this happens, one spike will appear on the orbit that moves away from both probes (but not directly away from either), and two more will appear that move directly away from each probe. The three spikes will be spaced at 90° of rotation from each other. Direct orbit and full spectrum show the effect of a scratch on the shaft in the probe track area. As the shaft rotates, the scratch first passes under the X transducer, producing a sharp, and negative-going voltage spike in the orbit that points away from the X transducer. The scratch then passes under the Y transducer, producing another spike in the orbit that points away from the Y transducer. In the full spectrum, the 1X frequency consists of both the 1X rotor dynamic response and the once-per-turn scratch response. The scratch response persists to slow roll speed, and the sharpness of the scratch produces a rich, harmonic spectrum as shown in figure 2.

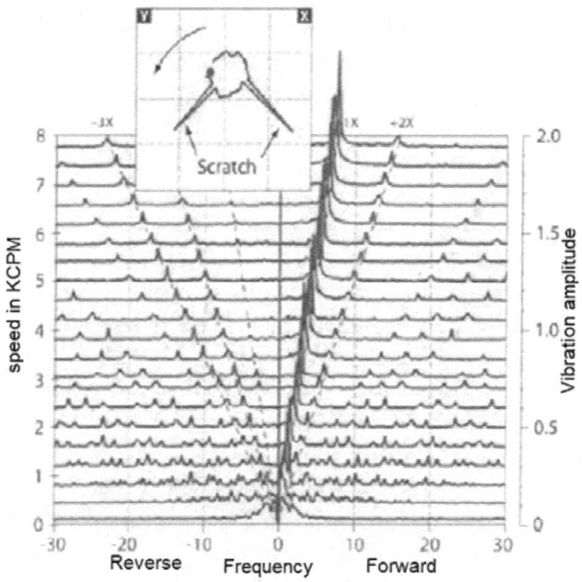

Fig 2 sharp 1x vibration due to scratch on probe sensing area

Effect of run out during rotor balancing -

Run out is the total linear displacement measured on the outside diameter of rotor when it is turned using a dial indicator. The run out is considered as static feature for balancing .Both the run out and mass unbalance are separate and independent quantities. For example, a noncircular part (cam) can be well balanced to run smoothly but it would measure a very large run out. On the other hand, a perfectly round disk (having zero run out) but having a heavy spot might have a serious unbalance .Since we measure the unbalance forces at shaft ends , the displacement probes measure unbalance response along with slow roll run out .

If the rotor is balanced on bearing pedestal and reaction forces on bearing are measured by a velocity transducer, then slow roll run out does not make considerable difference.

If the rotor is being balanced on bearing with X-Y probe then it is imperative to use a mode identification probe to see the displacement at centre span. It is often said that for exact balancing, a slow roll run out must be subtracted from the vibration proximity probe reading. Slow Roll Run out subtraction may be counterproductive or sometimes even dangerous when other dimensional deficiencies exist on the rotor (bow, coupling eccentricity, etc.). Slow roll subtraction could lead an inexperienced balancer to ignore bearing-pedestal vibration up to the point of damaging it while reducing shaft readings. So, when Slow Roll Run out is applied, it must be done with caution and understanding that rotor is free from a bow / bend.

One of plausible method to forego the slow roll run out is to consider Four-Run and Least-Squares Influence Coefficient method (some of computer programme have an optional feature to extract such data).

Measuring Mechanical Run out (MRO)

API 687 (Repair of Special Purpose Rotors) specifications recommend rotation intervals of no more than 20 degrees; but with truthfully , this is a relatively large gap between data points. As the rotor is moved, it may not settle into position repeatedly, leading to significant error. To help counteract this "stick slip" effect, smaller measurement intervals (10 degrees or less) are recommended. When measuring TIR, it is strongly recommended that the same probe series be used to eliminate possible sources of inconsistencies. Two types of devices can be used for measuring MRO.

1 Linear Variable Differential Transformers (LVDT) - LVDT are available with a resolution in the 0.01mil (.0003 mm) range and are particularly well-suited for accurate mechanical run out determination.

2 Dial Indicators-Use of a conventional needle type indicator that can be very difficult to read with the required resolution and for this reason are unsuitable for run out measurements. In addition, they do not allow for automated data acquisition. Dial indicators with a digital display should be used for the necessary resolution.

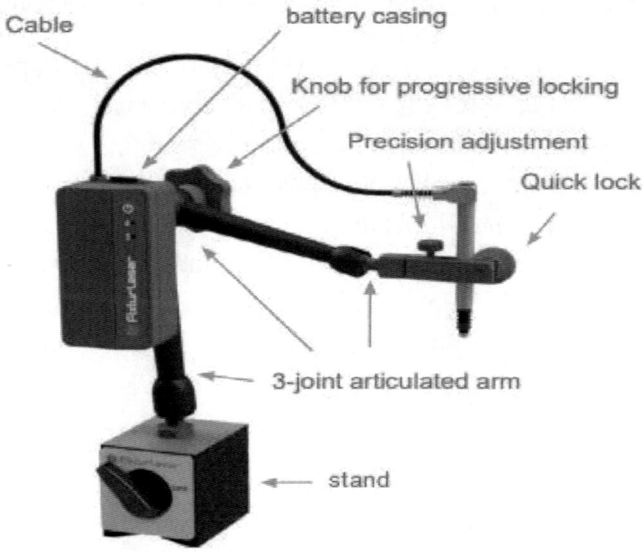

Fig-3 Device to measure mechanical runs out

Measuring Electrical Run out

The bottom probe in the Fig 4 uses an electronic dial indicator. The top one (the cylindrical one) is the electronic run out probe. The device under the shaft, very closely spaced is a large capacitor. The Run out Testing System is a measuring tool is used to check the run out of a rotor's measuring planes used for vibration monitoring. The measuring software takes the measuring data about the total and the mechanical run out to calculate the electrical run out and displays it depending on the angular position of the rotor. That information will then be used to arbitrarily change the actual run out for given limits. Run out is often being tested on lathes. However the rotor bearings of a lathe have limited accuracy if bearings are old and worn.

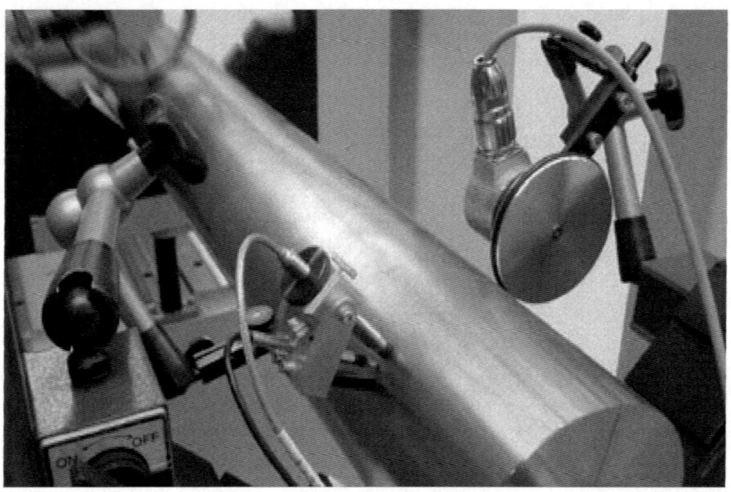

Fig 4 - Set up for Measuring electrical run out- courtesy Hoffmann

It is advised to measure slow roll characteristics of partially machined rotating machine shaft which is supported by hydrodynamic journal bearings .This shall enable to mitigate the issue which ingress in different process in early stage of manufacturing

Removing run out from a vibration signal

Some types of tracking filter enable run out vectors to be automatically removed from the incoming signal, so that the true vibration vector is displaced immediately. This facility can also be programmed into data logging equipment. The nulling is also carried out between critical speeds when producing a Bode plot, to obtain information about phase for use in balancing. It is always preferable to utilize hot shutdown data than to use cold startup data for establishing the correct, actual, operating condition slow roll values. Rotor axial thermal growth at times is significant, which could result in a different observed shaft run out surface cold vs. hot. The figure 5 is the example of removing the slow roll vector from 1x displacement data. The figure shows the high participation of ERO shown in blue as compared to MRO data almost circular in shape.

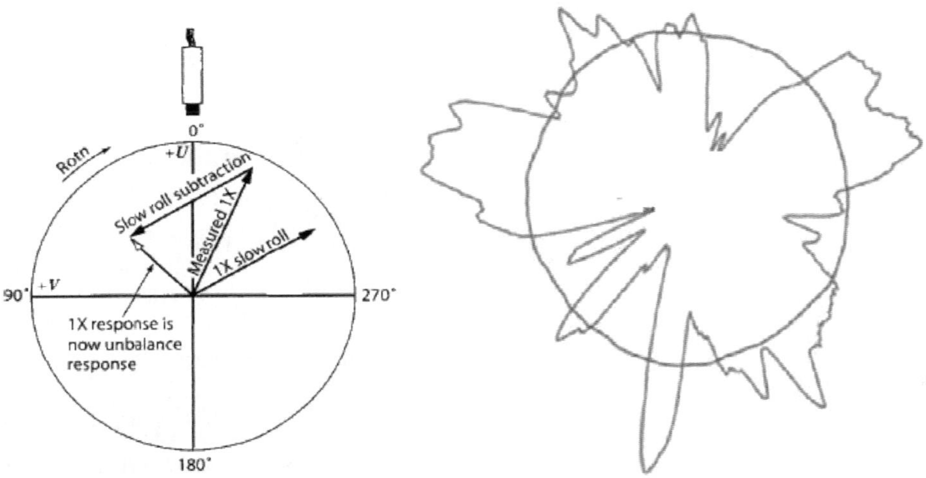

Figs 5 (Left) Slow roll compensation, Right - comparison of MRO and ERO

Slow Roll Waveform Compensation

Slow roll waveform compensation provides the capability to digitally “memorize” an unfiltered (direct) slow roll waveform and then digitally subtract it from dynamic waveforms at different machine speeds and operating conditions.

Each unfiltered time base waveform consists of a sequence of digitally sampled values. Once a suitable slow roll waveform has been selected, the digitally stored slow roll sample values can be subtracted from the corresponding digital sample values in the unfiltered time base waveforms at different machine speeds and operating conditions.

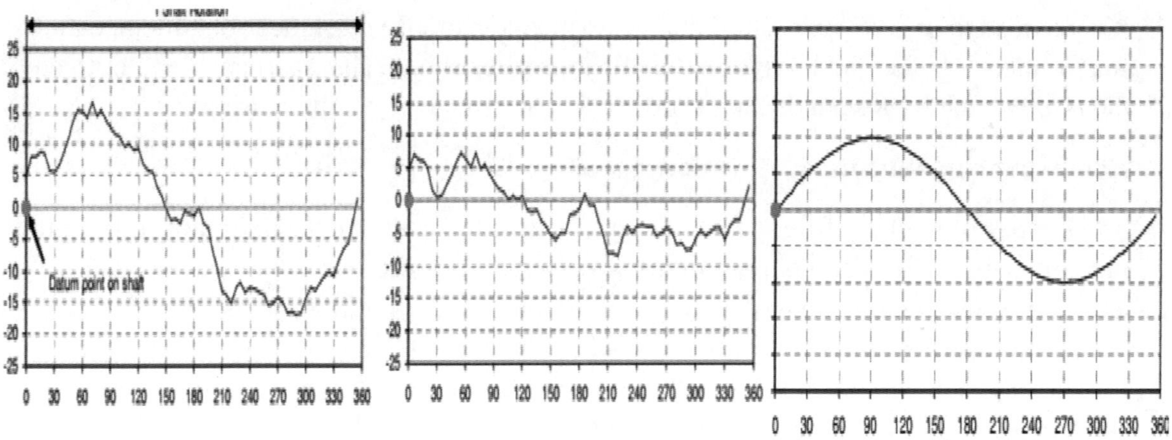

Fig 6 -Left -Unfiltered wave form, middle - Slow roll run out, Right -compensated waveform

Slow roll waveform compensation has the advantage of removing 90% of the total slow roll signal. Slow roll waveform compensation will remove both sub synchronous and super synchronous frequencies, including signal errors due to surface defects, imperfections, scratches, etc.

Waveform compensation can be applied to unfiltered time wave form, orbit plots and spectrum plotted as orders of running speed. This is quite helpful to find actual synchronous amplification factor (SAF) for machinery analysis and diagnosis.

Commentary on API standards -

A detailed study of relevant clauses / mandates as made in various API standards -

API 670 11th edition clause 9.3.3.1 states – shafts shall be machined and finished throughout their length so that TIR is not more than 25/ 40um .This requirement is based on flexible fit and component fit on shaft .
The standard mandates that probe areas shall be properly demagnetized and combined total electrical and mechanical run out shall not exceed 25% of the maximum allowed peak-to-peak vibration amplitude or 6 micrometers (0.25 mil), whichever is less.

API 541 and API 546 define limits for the maximum probe track run out. In simple terms, these run out limits are 25% of the unfiltered vibration limits, or 0.45 mils (thousandths of an inch) for most induction machines and 0.5 mils for most synchronous machines. These correspond to 11.4and 12.7 um respectively.

API 616 5th edition define limits for the maximum probe track run-out as 15 um peak-peak

The surface finish of sensing areas shall have a maximum 0.8um average roughness and area shall not have a residual magnetism more than 2 gauss with a variance not more than 1 gauss at two different places of probe track.

API 616 adds a note as "rotor run out determined by V block placed at journal centres may not be reproduced when the rotor is installed with hydrodynamic bearings.API 670 further mandates that surface areas shall not be plated or metalized whereas API 610 adds a note that use of metalized aluminium of 1mm thickness have been successful. We find this as contradiction to one another.

Difference between slow roll run out and rotor bow

During startup of a turbo machine equipped with hydrodynamic bearing, at the beginning it is difficult to separate slow roll run out and rotor bow up to 70% of its 1st critical speed.

In real world, a rotor shall always have a certain amount of residual unbalance, eccentricity and slow roll run out. Eccentricity is the measurement of Rotor Bow at rotor slow roll which may be caused by any or a combination of: -

- Fixed mechanical bow,
- Gravity bow (sag) and
- Temporary thermal bow.

When the run out is subtracted vectorially, an error is introduced by subtracting rotor bow also. At rotor 1st resonant speed the run out may have a slight rise in peak vibration at that point but effect of rotor bow at resonant speed has a profound effect as this is termed as eccentricity at rotor mid-span. Consider a Jeffcott rotor with all three imperfections listed above as real world example. Due to these imperfections the total response vector δ at vibration sensing area shall be –

$\delta = \delta_o + \delta_r + \delta_x$ where is δ_o response vector due run out, δ_x unbalance and δ_r due to bow .The below pictures of bode plots depicts the difference between rotor run out without bow and rotor bow without run out.

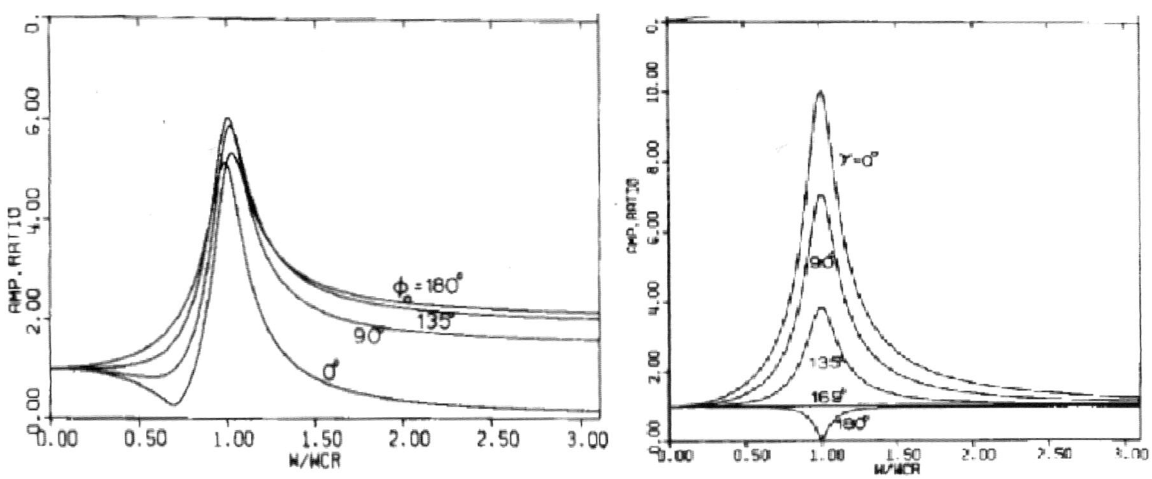

Fig -7-Left - Response with only run out no rotor bow Right - Only Rotor bow no run out

Where γ is angle between bow vector and mass unbalance vector, ϕ is rotor run out vector and timing mark.

Based on the above plots, the test can be done at the time of commissioning and hot restart of machine very carefully with trip limits lowered.

In a full speed range bode plot the slow roll and shaft bend can be differentiated by comparison of phase and amplitude at certain RPM which have fixed relation with 1st resonant speed.

A rotor with a transient bow may continuously change its phase and after repetitive start up the extent shall be considerably decreased where slow roll data remains almost same.

Sample Case study -

One of four Effluent Water Transfer Pumps (API 610 Horizontal, between bearing design) showed high readings at 45 RPM after tripping multiple times even after all possible reasons were eliminated .It was found that probe track area of NDE side had partly affected by corrosion . Before dismantling, slow roll was measured & recorded. High residual magnetism observed in the shaft probe track area at NDE side. NDE side slow roll displacement has reduced from 50 microns pk-pk to 21 microns pk-pk after demagnetization of the track probe location.DE side slow roll displacement maintains around 25 microns pk-pk before and after demagnetization of the track probe location. TIR on both track probes are within 0.02 mm as checked with dial gauge indicator .Bearings were inspected and found to be OK .Rotor mechanical run out checked and found 0.02mm at probe area due to suspected corrosion. (Reduced to 0.005mm after grinding).on both DE & NDE probe track. After that HVOF spray coating was applied to probe track area. Electric run out was found to be higher which was lowered by burnishing. The run out was taken in V-Blocks to meet acceptance criteria mandated by API 610. Once the rotor was fixed and started the pump NDE was showing high slow roll run out. Similar activity was carried out with no tangible benefit. It was suspected that due to porosity in coating, high electric run out was causing high slow roll run out. However any other electrical run-out (other than TIR and magnetic run out) is not possible to be corrected. The observed run-out has no detrimental effect on machine vibration although it affects as cascade effect on 1x vibration and subsequently overall vibration.

EWP	2989 RPM	900 RPM	750 RPM	600 RPM	500 RPM	350 RPM	270 RPM	97 RPM
PUMP DE	µm pk-pk	41	27	27	26	26	25	24
	µm pk-pk	45	28	27	27	27	26	25
PUMP NDE	µm pk-pk	37	36	36	36	36	36	35
	µm pk-pk	36	37	36	36	36	36	35

Table 1 - Slow roll run out readings.

Overall vibration was well below alarm limit (75um). End user reluctantly accepted the pump for operation with an understanding the rotor stator gap is 2 times higher than maximum measured vibration although actual vibration was quite low .The delay in repetitive , unproductive corrective actions was due to lacunae of understanding (which slow roll run out reading should be onerous) .

Considerations for an acceptance criteria

These are important deliberations to be exercised for a rotating machine installed at site during dynamic acceptance testing .The first consideration to take in account is that bearings lubrication, which is by oil ring, will start to work properly at 500-600 RPM, so before that running speed , the oil film that sustain the shaft is not yet fully created. The second consideration is that at low speed the hydraulic sustainment of a pump rotor is not yet archived since the pressure inside the pump casing is not yet built. It is to be noted that when the rotor is coupled with a drive some of other parasitic effect may not give the desired reading as mandated in API standards.

Alternative measurement devices

Other than widely used Eddy-current sensors which employs in a magnetic field to determine the distance to the target; capacitive sensors use changes in capacitance.

Capacitive sensors assume that changes in capacitance between the sensor and the target are a result of a change in distance between them. Another factor that affects capacitance is the dielectric constant (ε) of the material in the gap between the target and sensor. The dielectric constant of air is slightly greater than one. If another material, with a different dielectric constant, enters into the sensor/target gap, the capacitance will increase, and the sensor will erroneously indicate that the target has moved closer to the sensor. Due to the sensitivity to the dielectric constant of the material between the sensor and the target, capacitive displacement sensors must be used in a clean and steady environment when measuring a target position. Eddy-current displacement sensors work in wet, dirty environments and can be mounted further away from the shaft .Capacitive sensors are not affected by material thickness and subsurface irregularities but need a closer view of target material.

Factor	Capacitive	Eddy-Current
Dirty Environments	—	••
Small Targets	••	•
Large Range	•	••
Thin Materials	••	•
Material Versatility	••	•
Multiple Probes	••	•
Probe Mounting Ease	••	•
Resolution	••	•
Bandwidth	•	••
Cost	•	••

Table 2 - Comparison between sensors two dots - best , one dot less compatible

The capacitive sensors measure all conductive materials the same. Once a capacitive sensor is calibrated, it can be used with any conductive target with no degradation in performance. Since the electric field of a capacitive sensor does not penetrate the material, variations within the material do not affect the measurement. Capacitive sensors do not exhibit the electrical run out and can be used with rotating targets of any conductive material without additional error. Capacitive sensors, once calibrated, can be used with any conductive material with no material related errors and they work well with rotating targets. Because of differences in the shape and reactive nature of the sensing fields of capacitive and eddy-current sensors, they have different probe mounting requirements.

The electric fields of capacitive probes are only emitted from the front surface of the probe. The field has a slightly conical shape resulting in a spot size about 30% larger than the sensing area diameter.

Inductive sensors are not as sensitive as the eddy current sensors and it is advantageous. However, there are disadvantages with inductive sensors: (a) There are very few variety of inductive sensors available in the market as it is very costly, (b) The shaft target ring has to be made from laminated silicon steel.

Laser Triangulation Method

Laser triangulation sensors determine the position of a target by measuring the reflected light from the target surface. High resolution lasers are generally utilized in position and displacement monitoring applications but they require stability (much rigid casing), high precision and low temperature drift.

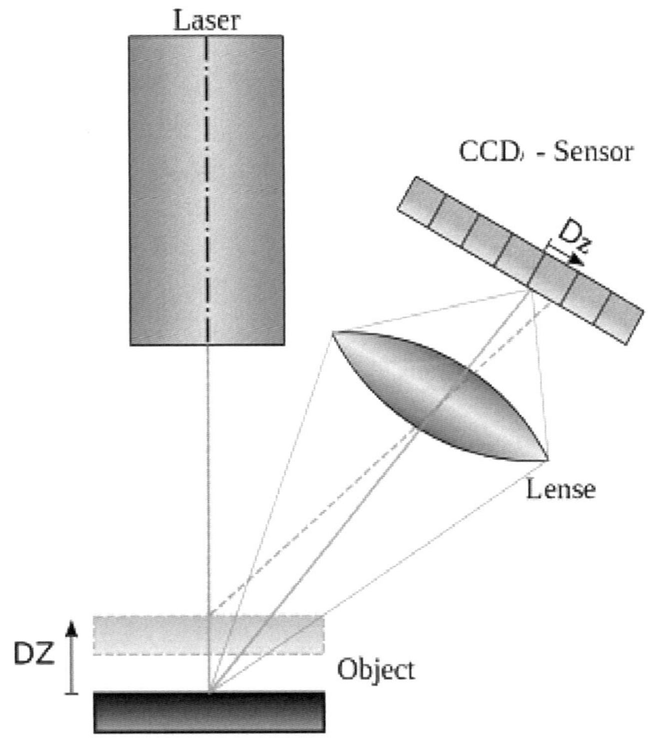

Fig 8 - Laser Triangulation Measuring

Proximity type laser triangulation sensors are also less expensive when compared to other high performance technologies. Laser triangulation sensors include a CMOS/CCD detector and a solid-state laser light source. They operate on the basic principle wherein a laser beam is projected on the target under measurement and a part of this beam is then reflected via focusing optics onto a detector. As and when the target shifts, the laser beam moves on the detector. The detector sends a signal which is utilized to measure the relative distance to the object or target. This data is usually available through a digital (binary) interface, an analog output, or a digital display for processing. Since laser heads have electronic components that are highly sensitive, their operating temperature is somewhat restricted.

Remedial actions to mitigate high slow roll

To remove mechanical run out the shaft must be re-machined; if the shaft is bent it must be straightened prior to re-machining. Once shaft is machined, honing and grinding activities shall be taken up .Electrical run out is present when, around the shaft circumference, there are variations in the permeability of the shaft material to the electrical field set up by the transducer. Residual magnetic fields can be removed by degaussing, if residual magnetism is not the cause of the electrical run out then special treatments such as diamond burnishing, micro peening, or electroplating with non-magnetic material may be necessary. As we can see all these activities to remove electrical run out are time consuming, repetitive and costly.

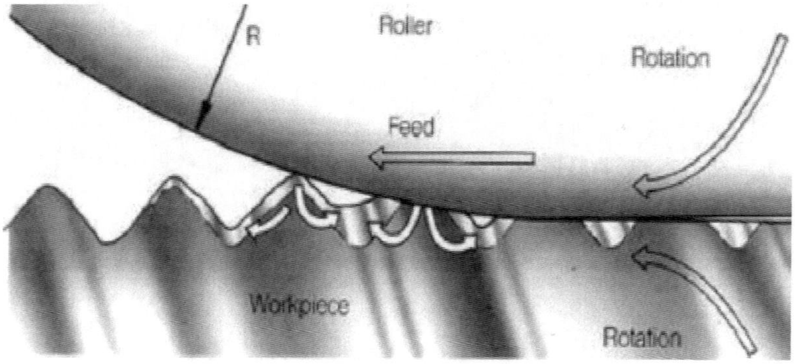

Fig 9- Rotor burnishing

Alternate treatment of probe track area to mitigate slow roll run out

Sleeve - Non- magnetic sleeve properly fit with a radial thickness of 1 mm can be mounted on probe track area. It is to be ensured that interference is not lost due to circumferential stress during high speed operation.

Coatings-Depositing a layer of less run out-prone material onto a shaft can be employed successfully, and there are several technologies for this. The material should not generate a run out signal by their own. The material must also be applied in a thick enough layer to prevent the probe from seeing through to the substrate.

When and how to live with high slow roll run out -

The slow-roll test of rotor is one of the last items to be tested on a machine in factory before machine is assembled in casing . If the machine fails to meet the slow-roll run out requirement, significant rework may be required in order to perform the necessary remedial work. For example, the machine may have to be disassembled and the rotor to be mounted on lathe again. This can cause significant delays in shipping the machine on time, resulting in project delays for the purchaser. Even with diamond burnishing of the rotor shaft, run out levels this low value may be very difficult for machinery manufacturers to achieve. The end result is that this error is costly for both manufacturer and purchaser. It is known that a "true" shaft vibration creates alternating stress and causing shaft fibres being stretched and contracted whereas a high slow roll run out is not a true vibration hence machinery reliability and integrity is not a concern .

Considering the value of highest amplification factor after a slow roll compensation and RCV as 60% instead of 75% as mandated in API we can live with a higher slow roll run out . The numerous data analysed from rotor-dynamic calculation by author shows that in 99% of cases it is achievable .

Use of electrically insulated coupling to mitigate residual magnetism and ERO - On specific demand , some manufacturers can offer an insulated spacer between the coupling ends, The insulation is provided by a fibre disk that, through small lips, also provides piloting between the two halves of the spacer. bushings and washers insulate the bolts. With proper design, the insulated connection can be left intact for the life of the machine.

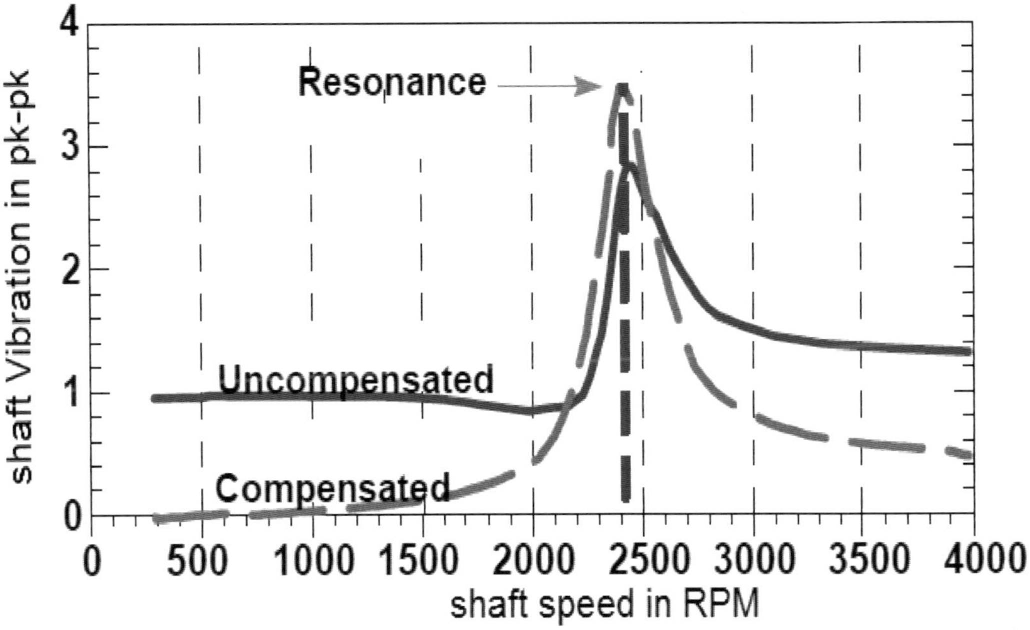

Fig 10 - Effect of slow roll compensation on Bode plot

If the rotor is using antifriction bearings, all vibratory stresses are passed on casing / bearing housing and is a true indication of rotor vibration. With rotor using antifriction bearings, slow roll run out has no meaning . But in case, the rotor is supported by hydrodynamic journal bearing true vibration cannot be extracted at casing / bearing housing due to damping of vibration oil film which enforces us to measure shaft vibration using proximity probes. As it is impractical to mount transducers at mid span on the rotor where displacement would be the highest the transducer(s) are mounted near bearing housing. As slow roll run out is not being measured in antifriction bearing as it has no effect on rotor vibration same can be held true theoretically in perspective of reaction forces on bearings.

If we find that rotor to stator clearances are much higher than alarm / trip limits then we can live with a higher slow roll run out such as 16 um pk-pk at site condition . This value can be further adjusted by OEM for onsite operation of a machinery train.

It is feasible to check if vibration is genuine due to machinery anomaly or benign due to slow roll run out. With consent of OEM expert vibration analyst, we can use the mask or envelope alarm where 1x, 2x and 3x values shall be used as individual alarm set up before triggering a trip. With a high slow roll run out we shall get a 1x and 2 x components higher in spectra which shall increase overall vibration level. The height of other components than unbalance and misalignment response should be kept as baseline as slow roll run out. The alarm should be set based on bearing clearance and rotor to stator clearance so that machinery damage is avoided. With a maximum limit set point, minimum limit set point for 1x and 2x component must be implemented so that possible nulling of actual vibration is also addressed. .

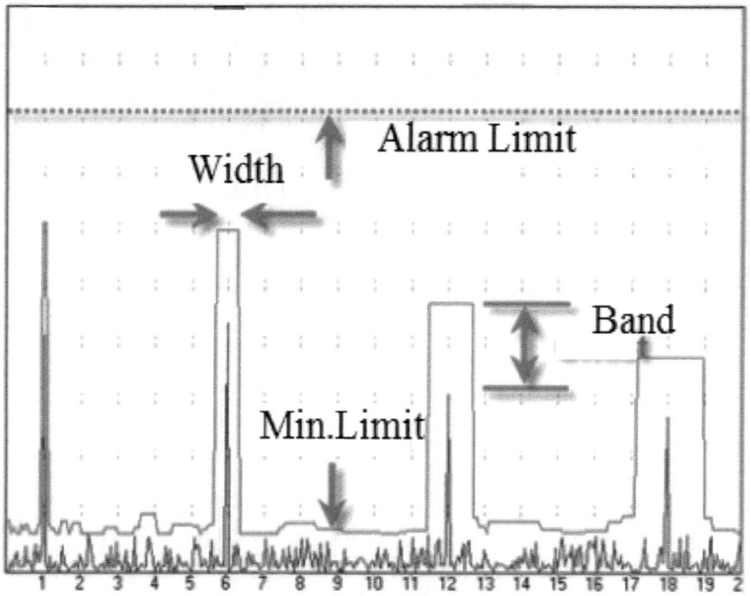

Fig 11 - Masked alarm

Statistical data based mask alarm configuration is further refined way to avoid unnecessary concerns of high vibration.

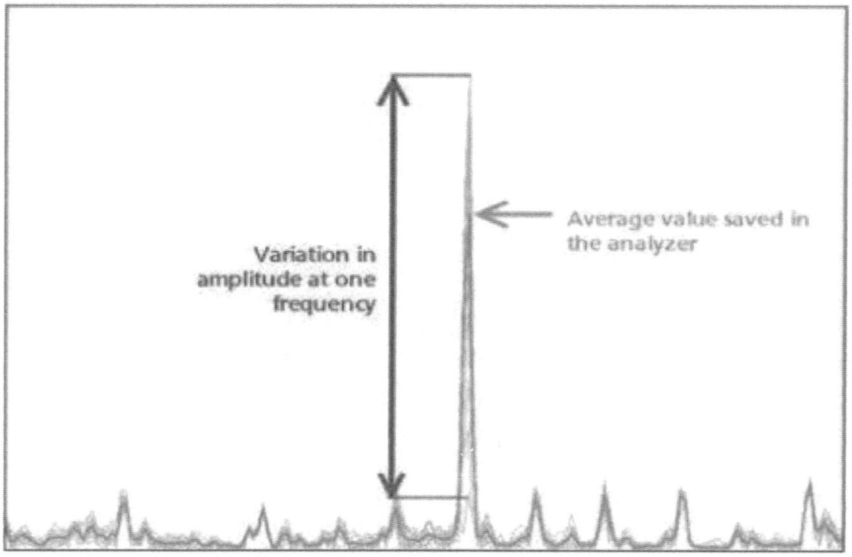

Fig 12 - Statistical data based alarm for a particular frequency

We can explore to create rule based logic to generate alarm to see the true vibration. This can be formulated by using shape identification algorithm in diagnostic system based on reference data captured by initial slow roll measurement and vibration spectrum by a Laser Doppler and eddy current probe as well at OEM works during test. The algorithm shall remember the difference as base data and same shall be applied to see 1 x and 2x vibrations.

By suggesting the above actions with consent of OEM, it should not be construed that OEM should be lenient on requirement of slow roll run out. OEM should ensure that mechanical run out is at minimum extent, free of scratches and required surface finish as mandated in API. The above suggestions are explicitly to avoid spurious trips and concerns due to electrical run out.

In a slow roll run out measurement, capacitive and Eddy current probes were alternately paired with the LVDT probe to observe the motion of the rotor shaft. The probes are set up to measure the mechanical and electrical run out on a lathe at slow speed (7-10 RPM) so that readings are not affected by lathe machine's bearing condition. Under similar conditions, one set of measurements was taken for the Eddy current probe in conjunction with the LVDT probe and one set of measurements is taken for the capacitive probe in conjunction with the same LVDT probe.

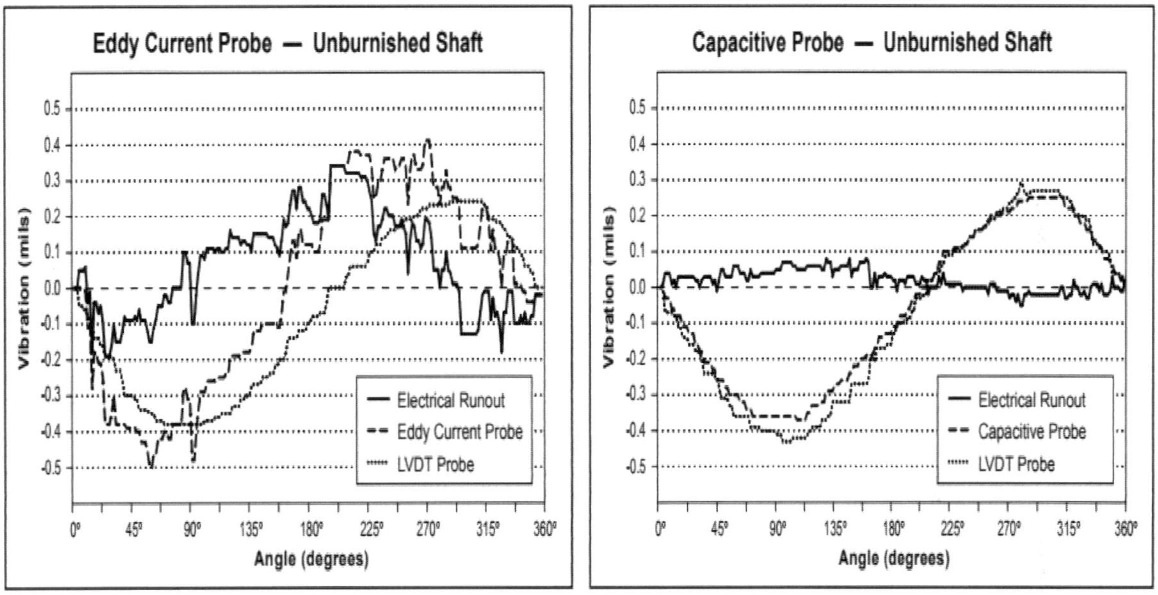

Fig13 - Comparison of readings of ERO by various sensors with LVDT probe

The most remarkable aspect of the capacitive proximity probe measurement is that the probe exhibits a peak-to-peak level of low level of electrical run out (roughly one-quarter of the Eddy current reading). This would pass run out specifications (in particular API-670) without the need for diamond-burnishing. Thus, false readings and their expensive consequences can be avoided.

Prospect of future research

A) Prospect of using capacitive transducer -

Considering the capacitive probe's competitive pricing, it should now be considered a superior replacement for Eddy current probes for almost all noncontact vibration measurement applications. The scope of research lies on reducing its size. The capacitive probe can be interfaced into the same rack via a signal conditioning device to provide the standard 200mV/mil sensitivity required by the existing VMS rack instrumentation

b) Prospect of using Micro electro-mechanical systems (MEMS) - Now a days use of MEMS is highly encouraged for critical equipment installed in remote area .There is a good scope of exploration of using in shaft vibration monitoring (OSVM) in conjunction with on bearing vibration monitoring (OBVM).

c) Prospect of using Laser telemetry - As known to us that shaft centreline plot displays the shaft centre DC position changes within a bearing clearance range. An orbit plot represents the shaft centre AC dynamic motion. If the shaft true centreline is identified in both DE and NDE end and bearing details are kept in data base, it is possible to have real time data in form of orbit plot, shaft displacement and shaft polar and centreline plots. This is same principle on which shaft laser alignment tool works.

References -

[1]Rotor Glitch - An article by GE MCS
[2]Littrell, N. "Understanding and Mitigating Shaft Run out," ORBIT Magazine, Vol. 25 No. 3, 2005, pp. 5-17.
[3]DeBlock, M., McDonnell, J.W., Wood, B. "Predicting Proximity Probe Track Run-out on API Motors and Generators," ORBIT Magazine, Vol. 27 No. 2, 2007, pp. 36-4
[4]GE Bently Nevada System 1 Tips Tricks Vol3 Issue 4 Apr 2012 Collect Slow Roll Data
[5] Comparison of unbalance response of a Jeffcott rotor with shaft bow and/or run-out by Dr.EJ Gunter , Flack et al ASME Paper 81-DET -47
[6]Understanding Vector, Gap, and Waveform Data Compensation- Richard Thomas Machinery Diagnostic Services SETPOINT™ Vibration
[7]Slow roll runout measurement for hydropower - Koncar electrical engineering institute
[8]Development of Electrical Runout Online Measurement System Based on Eddy Current Technique and Laser Triangulation- Wang Yifeng et al -School of Mechanical and Automotive Engineering, Zhejiang University of Science and Technology, Hangzhou, 310023, China
[9]Accu Scan 2-Probe Run-out Inspection System with ERO- Technical sheet by Paris Mountain Consulting, LLC.
[10]Mechanical and Electrical Run-Out Removal on a Precision Rotor-Vibration Research Spindle- G. A. Horattas, M. L. Adams and F. Dimofte Journal of Vibration and Acoustics
[11]Predicting, measuring and correcting for electrical run-out using a burnishing and degaussing process- Tech notes by Thermaspray (PTY) Ltd Olifantsfontein South Africa
[12]Shaft Run-out Measurement With Noncontact Displacement Sensors General Sensing Application Note LA05-0022 - Lion Precision instrument notes
[13]Comparing Capacitive and Eddy-Current Sensors General Sensor Tech Note LT05-0011

[14]Non-contact displacement sensors technology -a review by Chris Jones Power in motion 2008
[15]Method and apparatus for programming and reducing run out in rotating shaft vibration signal detection - US Patent 4135244 by Robert R Davis .
[16]Identification of critical speed of rotating machines using on shaft wireless vibration measurement - ME Elandy , Jyoti Sinha et al University of Manchester UK
[17]Rotor-dynamic measurement techniques - Tech notes by Agilient Technologies
[18]A multi-model approach for anomaly detection and diagnosis using vibration signals - Victor Valencia et al . Conference on PHM 2013
[19]Electrical And Mechanical Run-out As A Problem In Manufacturing Industry - Darshan Jayaram et al - Mysore India
[20]Electrical Run-out using an eddy current sensor for roundness measurement - Byron Knapp et al Minnesota, USA
[21]Electrical & Mechanical Run-out Inspection- Tech notes by Watson grinding system Houston, TX

Bibliography -

[1]Rotating machinery vibration - from analysis to troubleshooting - Maurice Adams
[2]Rotor and Structural Dynamics of Turbo-machinery A Practical Guide for Engineers and Scientists.: By Subbiah, Raj, Littleton, Jeremy Eli
[3]Rotor Systems: Analysis and Identification - Rajiv Tiwati IIT Guwahati India CRC Press, Taylor & Francis 2017.
[4]Practical Machinery Vibration Analysis and Predictive Maintenance-by Cornelius Scheffer and Paresh Girdhar
[5] API 670 Machinery Protection Systems. Fifth edition
[6]API Recommended Practice 687 - Rotor Repair

THE DYNAMIC RESPONSE OF A PLANETARY GEAR TRAIN IN THE PRESENCE OF A SPALLING FAULT

Keming Zhang, Yimin Shao , Diego Galar*
State Key Laboratory of Mechanical Transmission
Shazheng Street 174
Shapingba District
Chongqing, 400044, CHINA
ymshao@cqu.edu.cn
*Division of Operation and Maintenance Engineering
Lulea University of Technology
Lulea 97187, SWEDEN

Abstract: Planetary gear trains (PGTs) are widely used in industrial applications. Failure often occurs under working conditions. Tooth surface spalling is one of the most common defects in a PGT system; it seriously affects the reliability and safety of the mechanical transmission system and may even cause serious incidents. However, research into the faults of planetary gear trains is insufficient, especially with respect to the response characteristics of a PGT in the presence of a spalling defect. The paper designs spalling cases with different localized distributions to demonstrate their influence on the dynamic performance of the planetary gear set. The research provides a theoretical basis for health diagnosis and early fault detection for a PGT system.

Key words: Planetary gear train; spalling defect; dynamic performance;

Introduction:

Planetary gear trains (PGTs) are widely used in industrial applications because they can operate in many different conditions. Tooth spalling is one of the most common failure modes [1]. When a spalling fault appears on a gear tooth, the mesh stiffness will reduce and the dynamic vibration properties of the system will change, leading to the shutdown of the whole system. Establishing a dynamic model of a planetary gear train with a spalling defect and studying the vibration response characteristics of the system would help to monitor the health of a PGT system and detect incipient failures.

Many researchers have studied the vibration properties of PGT systems. Kim [2] established a dynamic model with time-varying pressure angles and contact ratios to investigate their effects on the dynamic behavior of a planetary gear. Kahraman [3-4] used two- and three-dimensional models to examine the dynamic response of both time-invariant and time-varying representations. Kahraman and Vijayakar [5] investigated the effect of internal gear flexibility on the quasi-static behavior of a planetary gear set by using a finite element/semi-analytical nonlinear contact mechanical formulation, while Inalpolat and Kahraman [6,7] studied the modulation sidebands of planetary gear sets. Using a finite element/contact mechanics model, Parker [8] investigated the dynamic response of a planetary gear system; the calculated response showed classical resonances when the harmonic of the mesh frequency coincided with the natural frequency. Sun and Hu [9] studied the nonlinear dynamics of a planetary gear system with multiple clearances, such as backlashes.

Most of the existing models focus on the system under healthy conditions. Studies of faulty planetary gear trains are insufficient, especially with respect to the response characteristics of a PGT in the presence of a localized spalling defect. To overcome this insufficiency, this study establishes a dynamic model of planetary gear sets with spalling defects. It calculates the time-varying mesh stiffness under spalling conditions, solves the responses of the defect system in the presence of a localized sun/planet/ring gear spalling defect, and analyzes the frequency distribution of response signals under these conditions. The research provides a theoretical basis for diagnosis of the PGT system and early fault detection.

Mesh Stiffness Evaluation:

Gear mesh stiffness is one of the most significant internal issues of gear dynamics. Mesh stiffness can change and when it does, the dynamic properties of the gear system will change. Therefore, accurate evaluation of gear mesh stiffness is essential for the dynamic analysis of a PGT system.

This paper uses the potential energy method proposed by Yang and Lin [10] and refined by Tian [11] and Chen et al. [12] to calculate the time varying mesh stiffness (TVMS) of gear pairs. The gear tooth is modeled as a nonuniform cantilever beam. The total energy store in the gear pair is the combination of bending energy U_b , shear energy U_s , axial compressive energy U_a, and Hertzian energy U_z which respectively correspond to bending stiffness K_b , shear stiffness K_s , axial compressive stiffness Ka, and Hertzian stiffness K_h.

Figure 1 shows a gear tooth modeled as a nonuniform cantilever beam. There is a spalling fault on the gear tooth surface. Without loss of generality, the spalling is modeled as a rectangle with dimensions $w \times l_s$ in Figure 2. The depth of the spalling defect is h_s and the width of the gear tooth is L. As shown in Figure 2 and described in more detail below, there are three stages in spalling.

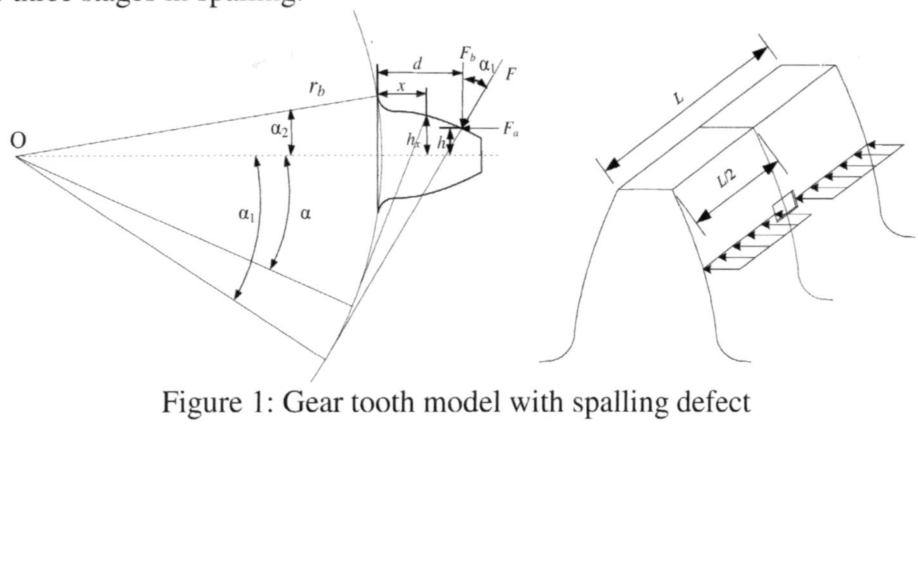

Figure 1: Gear tooth model with spalling defect

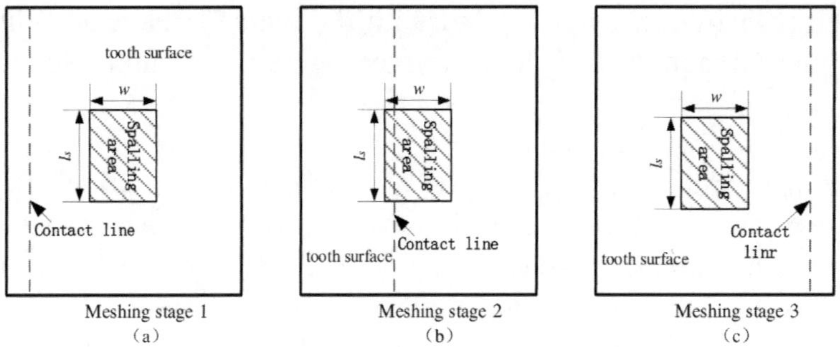

Figure 2: Three stages of spalling

Stage 1: The contact line is below the spalling area in stage 1, as shown in Figure 2(a). In this stage, the width of contact of two meshing teeth is equal to W; the bending stiffness is Kb, shear stiffness is Ks, axial compressive stiffness is Ka, and Hertzian stiffness is Kh. These can respectively be expressed as :

$$k_b = 1/\left\{\int_{-\alpha_2}^{\alpha_1} \frac{3\left[1-\cos\alpha\cos\alpha_1-(\alpha+\alpha_2)\sin\alpha\cos\alpha_1\right]^2(\alpha+\alpha_2)\cos\alpha}{2EL\left[(\alpha+\alpha_2)\cos\alpha-\sin\alpha\right]^3}d\alpha\right\} \tag{1}$$

$$k_{s1} = 1/\left\{\int_{-\alpha_2}^{\alpha_1} \frac{1.2\cos^2\alpha_1(\alpha+\alpha_2)\cos\alpha}{2GL\left[(\alpha+\alpha_2)\cos\alpha-\sin\alpha\right]}d\alpha\right\} \tag{2}$$

$$k_{a1} = 1/\left\{\int_{-\alpha_2}^{\alpha_1} \frac{\sin^2\alpha_1(\alpha+\alpha_2)\cos\alpha}{2EL\left[(\alpha+\alpha_2)\cos\alpha-\sin\alpha\right]}d\alpha\right\} \tag{3}$$

$$k_h = \frac{\pi EL}{4(1-\nu^2)} \tag{4}$$

where L is the tooth width, α_1 is the acting pressure angle of gear tooth, E denotes the modulus of elasticity, and ν denotes the Poisson ratio.

Stage 2: The contact line is engaged in the spalling area in stage 2, as shown in Figure 2(b). In this stage, the spalling area does not engage with the meshing; the length of the contact line changes from L to $L\text{-}l_s$. Correspondingly, the Hertzian stiffness decreases, and can be expressed as:

$$k_h = \frac{\pi E(L\text{-}l_s)}{4(1-\nu^2)} \tag{5}$$

In addition, computation of the TVMS has to take into account the shape of the elementary cross-sectional area and the moment of inertia, as both of these will change. The bending stiffness, shear stiffness, and axial compression stiffness can respectively be expressed as follows:

$$k_b = 1/\left(\int_{-\alpha_2}^{\alpha_{s2}} \frac{3\left[1-\cos\alpha\cos\alpha_1-(\alpha+\alpha_2)\sin\alpha\cos\alpha_1\right]^2(\alpha+\alpha_2)\cos\alpha}{2EL\left[(\alpha+\alpha_2)\cos\alpha-\sin\alpha\right]^3}d\alpha \right.$$
$$\left. +\int_{\alpha_{s2}}^{\alpha_1} \frac{12F^2R_b^3\left[1-\cos\alpha\cos\alpha_1-(\alpha+\alpha_2)\sin\alpha\cos\alpha_1\right]^2(\alpha+\alpha_2)\cos\alpha}{E\left\{8LR_b^3\left[(\alpha+\alpha_2)\cos\alpha-\sin\alpha\right]^3-h_s^3 l_s\right\}}d\alpha\right) \tag{6}$$

$$k_s = 1 \Bigg/ \left(\int_{-\alpha_2}^{\alpha_{s2}} \frac{1.2R_b(\cos\alpha_1)^2(\alpha+\alpha_2)\cos\alpha}{2GL[(\alpha+\alpha_2)\cos\alpha-\sin\alpha]} d\alpha + \int_{\alpha_{s2}}^{\alpha_1} \frac{1.2(\cos\alpha_1)^2(\alpha+\alpha_2)\cos\alpha}{G\{2LR_b[(\alpha+\alpha_2)\cos\alpha-\sin\alpha]-h_s l_s\}} d\alpha \right) \quad (7)$$

$$k_a = 1 \Bigg/ \left(\int_{-\alpha_2}^{\alpha_{s2}} \frac{\sin^2\alpha_1(\alpha+\alpha_2)\cos\alpha}{2EL[(\alpha+\alpha_2)\cos\alpha-\sin\alpha]} d\alpha + \int_{\alpha_{s2}}^{\alpha_1} \frac{R_b\sin^2\alpha_1(\alpha+\alpha_2)\cos\alpha}{E\{2R_bL[(\alpha+\alpha_2)\cos\alpha-\sin\alpha]-h_s l_s\}} d\alpha \right) \quad (8)$$

where α_{s2} denotes the angle of pressure corresponding to the starting point of the spalling defect.

Stage 3: When the meshing line passes the end of the spalling area, stage 3 begins. In this stage, even though the spalling area is not engaged, the spalling will affect the elementary moment of inertia and the cross-sectional area of the gear, so the gear bending stiffness, shear stiffness, and axial compressive stiffness will change because of the spalling fault, and can respectively be expressed as:

$$\begin{aligned} k_b = & \int_{-\alpha_2}^{\alpha_{s2}} \frac{3[1-\cos\alpha\cos\alpha_1-(\alpha+\alpha_2)\sin\alpha\cos\alpha_1]^2(\alpha+\alpha_2)\cos\alpha}{2EL[(\alpha+\alpha_2)\cos\alpha-\sin\alpha]^3} d\alpha \\ & + \int_{\alpha_{s2}}^{\alpha_{s1}} \frac{12F^2R_b^3[1-\cos\alpha\cos\alpha_1-(\alpha+\alpha_2)\sin\alpha\cos\alpha_1]^2(\alpha+\alpha_2)\cos\alpha}{E\{8LR_b^3[(\alpha+\alpha_2)\cos\alpha-\sin\alpha]^3-h_s^3 l_s\}} d\alpha \\ & + \int_{\alpha_{s1}}^{\alpha_1} \frac{3F^2[1-\cos\alpha\cos\alpha_1-(\alpha+\alpha_2)\sin\alpha\cos\alpha_1]^2(\alpha+\alpha_2)\cos\alpha}{2EL[(\alpha+\alpha_2)\cos\alpha-\sin\alpha]^3} d\alpha \end{aligned} \quad (9)$$

$$\begin{aligned} k_s = 1 \Bigg/ \Bigg(& \int_{-\alpha_2}^{\alpha_{s2}} \frac{1.2(\cos\alpha_1)^2(\alpha+\alpha_2)\cos\alpha}{2GL[(\alpha+\alpha_2)\cos\alpha-\sin\alpha]} d\alpha + \int_{\alpha_{s2}}^{\alpha_{s1}} \frac{1.2(\cos\alpha_1)^2(\alpha+\alpha_2)\cos\alpha}{G\{2LR_b[(\alpha+\alpha_2)\cos\alpha-\sin\alpha]-h_s l_s\}} d\alpha \\ & + \int_{\alpha_{s1}}^{\alpha_1} \frac{1.2F^2(\cos\alpha_1)^2(\alpha+\alpha_2)\cos\alpha}{4GL[(\alpha+\alpha_2)\cos\alpha-\sin\alpha]} d\alpha \Bigg) \end{aligned} \quad (10)$$

$$\begin{aligned} k_a = 1 \Bigg/ \Bigg(& \int_{-\alpha_2}^{\alpha_{s2}} \frac{\sin^2\alpha_1(\alpha+\alpha_2)\cos\alpha}{2EL[(\alpha+\alpha_2)\cos\alpha-\sin\alpha]} d\alpha + \int_{\alpha_{s2}}^{\alpha_{s1}} \frac{R_b\sin^2\alpha_1(\alpha+\alpha_2)\cos\alpha}{E\{2R_bL[(\alpha+\alpha_2)\cos\alpha-\sin\alpha]-h_s l_s\}} d\alpha \\ & + \int_{\alpha_{s1}}^{\alpha_1} \frac{\sin^2\alpha_1(\alpha+\alpha_2)\cos\alpha}{2EL[(\alpha+\alpha_2)\cos\alpha-\sin\alpha]} d\alpha \Bigg) \end{aligned} \quad (11)$$

where α_{s1} denotes the angle of pressure corresponding to the terminal point of the spalling defect.

With the sun and planet gear parameters shown in Table.1, in the presence of a localized sun gear spalling defect, the dimensions of the spalling fault, w, L_s, h_s, are 5, 15, and 1.5mm, respectively, and the mesh stiffness can be calculated by Equations (1-11). The single-tooth mesh stiffness and double-tooth mesh stiffness are shown in Figure 3. The simulated results show that a spalling defect leads to a bigger reduction in mesh stiffness.

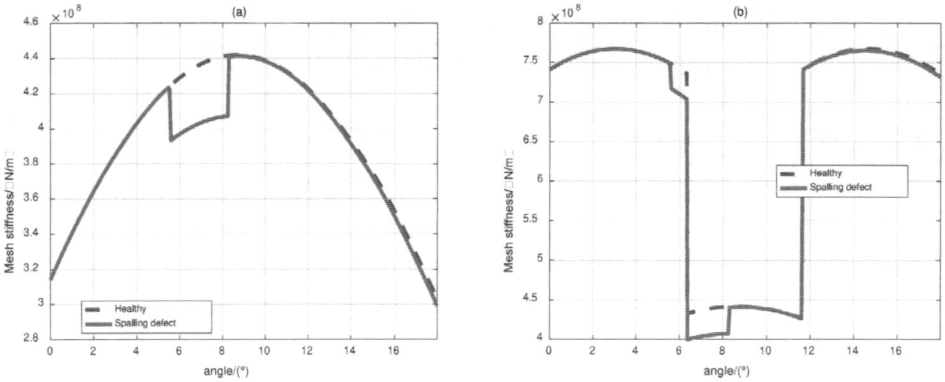

Figure 3: Mesh stiffness of sun(defect)-planet gear pair: (a) single-tooth (b) double-tooth

The internal mesh stiffness with a spalling defect is also calculated this way; because of space constraints, however, we do not discuss it here.

Dynamic Model:

The lumped-parameter model (shown in Figure 2) of the planetary gear system is used in this study to analysis the effect of a spalling fault; in this model, the mass and moment of inertia of each planet gear are identical. The planet gear rotates around the sun gear and ring gear with mesh stiffness k_{spn} and k_{rpn}, respectively, while k_{su}, k_{ru}, and k_{cu} are the support stiffness of sun, ring gear and carrier, respectively, and k_{su}, k_{ru}, and k_{cu} are the torsional stiffness of sun, ring gear and carrier, respectively. Each component contains two degrees of freedom, x_n, y_n (n=r, c, s, p), in a translational direction, and one degree of freedom, θ_m (n=r, c, s, p), in a torsional direction. The central component is a generalized displacement vector expressed as follows:

$$q=[x_c,y_c,\theta_c,x_r,y_r,\theta_r,x_s,y_s,\theta_s,x_1,y_1,\theta_1,\cdots,x_n,y_n,\theta_n]^T \tag{12}$$

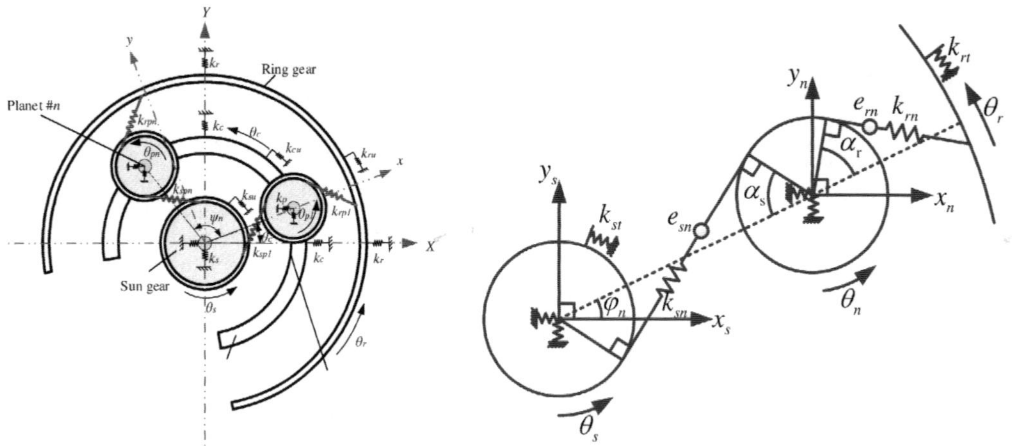

Figure 4: Dynamic model of planetary gear set

Suppose the sun gear and the carrier are the input and output component. The input torque is T_s, and the load is T_c. The ring gear is held stationary. The mass of the sun gear, carrier, ring gear, and planet gear is m_s, m_c, m_r, and m_p, respectively, and the moments of inertia are I_s, I_c, I_r, and I_p, respectively, According to Newton's second law, the motion equation for the carrier, ring, sun gear, and planet gear can be respectively expressed as:

$$\begin{cases} m_c\ddot{x}_c + \sum k_p\delta_{cnx} + \sum c_p\dot{\delta}_{cnx} + k_c x_c + c_c\dot{x}_c = 0 \\ m_c\ddot{y}_c + \sum k_p\delta_{cny} + \sum c_p\dot{\delta}_{cny} + k_c y_c + c_c\dot{y}_c = 0 \\ (I_c / r_c)\ddot{\theta}_c + \sum k_p\delta_{cnu} + \sum c_p\dot{\delta}_{cnu} + k_{ct}\theta_c r_c + c_{ct}\dot{\theta}_c r_c = T_c / r_c \end{cases} \tag{13}$$

$$\begin{cases} m_r\ddot{x}_r - \sum k_{rn}\delta_{rn}\sin\psi_{rn} - \sum c_{rn}\dot{\delta}_{rn}\sin\psi_{rn} + k_r x_r + c_r\dot{x}_r = 0 \\ m_r\ddot{y}_r + \sum k_{rn}\delta_{rn}\cos\psi_{rn} + \sum c_{rn}\dot{\delta}_{rn}\cos\psi_{rn} + k_r y_r + c_r\dot{y}_r = 0 \\ (I_r / r_r)\ddot{\theta}_r + \sum k_{rn}\delta_{rn} + \sum c_{rn}\dot{\delta}_{rn} + k_{rt}\theta_r r_r + c_{rt}\dot{\theta}_r r_r = T_r / r_r \end{cases} \tag{14}$$

$$\begin{cases} m_s\ddot{x}_s - \sum k_{sn}\delta_{sn}\sin\psi_{sn} - \sum c_{sn}\dot{\delta}_{sn}\sin\psi_{sn} + k_s x_s + c_s\dot{x}_s = 0 \\ m_s\ddot{y}_s + \sum k_{sn}\delta_{sn}\cos\psi_{sn} + \sum c_{sn}\dot{\delta}_{sn}\cos\psi_{sn} + k_s y_s + c_s\dot{y}_s = 0 \\ (I_s / r_s)\ddot{\theta}_s + \sum k_{sn}\delta_{sn} + \sum c_{sn}\dot{\delta}_{sn} + k_{st}\theta_s r_s + c_{st}\dot{\theta}_s r_s = T_s / r_s \end{cases} \tag{15}$$

$$\begin{cases} m_n\ddot{x}_n + k_{sn}\delta_{sn}\sin\psi_{sn} + c_{sn}\dot{\delta}_{sn}\sin\psi_{sn} + k_{rn}\delta_{rn}\sin\psi_{rn} + c_{rn}\dot{\delta}_{rn}\sin\psi_{rn} - k_p\delta_{cnx} - c_p\dot{\delta}_{cnx} = 0 \\ m_n\ddot{y}_n - k_{sn}\delta_{sn}\cos\psi_{sn} - c_{sn}\dot{\delta}_{sn}\cos\psi_{sn} - k_{rn}\delta_{rn}\cos\psi_{rn} - c_{rn}\dot{\delta}_{rn}\cos\psi_{rn} - k_p\delta_{cny} - c_p\dot{\delta}_{cny} = 0 \\ (I_n / r_n)\ddot{\theta}_n + k_{sn}\delta_{sn} + c_{sn}\dot{\delta}_{sn} - k_{rn}\delta_{rn} - c_{rn}\dot{\delta}_{rn} = 0 \end{cases} \tag{16}$$

where:

$$\delta_{cnx} = x_c - x_n,\ \delta_{cny} = y_c - y_n,\ \delta_{cnu} = (x_n - x_c)\sin\psi_n + (y_c - y_n)\cos\psi_n \tag{17}$$

$$\delta_{rn} = (x_n - x_r)\sin\psi_{rn} + (y_r - y_n)\cos\psi_{rn} + (\theta_r - \theta_c)r_r - (\theta_n - \theta_c)r_n + e_{rn} \tag{18}$$

$$\delta_{sn} = (x_n - x_s)\sin\psi_{sn} + (y_s - y_n)\cos\psi_{sn} + (\theta_s - \theta_c)r_s + (\theta_n - \theta_c)r_n + e_{sn} \tag{19}$$

The dynamic analysis of a planetary gear train with a spalling fault requires a dynamic model. Depending on the location of the defect in the component, the response will be quite different; therefore, it is necessary to study the conditions of the spalling at different locations.

The spalling fault is localized on the sun gear: At setting t=0, the meshing gear pair between the sun gear and the #1 planet gear is the gear pair with the spalling defect. The tooth on the sun gear with the spalling defect has just engaged in the meshing, and the carrier is regarded as static. In this situation, the torsional displacement of the sun gear relative to the initial position can be expressed as:

$$\theta_{s\_spall} = \mathrm{mod}(\theta_s - \theta_c, 2\pi) \tag{20}$$

where mod(a,b) stands for the remainder after dividing a by b. Because there is a certain meshing phase difference between the sun gear meshing and the planet gear meshing, the torsional displacement interval between the spalling teeth of the sun gear and those of the various planetary gears is not equal to 2π/N. Consequently, the torsional displacement of the sun gear from the time when the spalling tooth begins to engage, $\theta_{s\_\mathrm{spall}\&p_i(\mathrm{in})}$, until it ceases engagement, $\theta_{s\_\mathrm{spall}\&p_i(\mathrm{out})}$, must take the meshing phase difference into consideration. This difference can be expressed as:

$$\theta_{s\_\mathrm{spall}\&p_n(\mathrm{in})} = \mathrm{floor}\left(\frac{z_s(n-1)}{N}\right) * \frac{2\pi}{z_s} + \mathrm{mod}\left(\varphi_n, \frac{2\pi}{z_s}\right) \tag{21}$$

$$\theta_{s\_spall\&p_n(out)} = \text{floor}\left(\frac{z_s(n-1)}{N}\right) * \frac{2\pi}{z_s} + \text{mod}\left(\varphi_n, \frac{2\pi}{z_s}\right) + \frac{2\pi}{z_s} * e_{sp} \tag{22}$$

In Equations (21-22), floor(a) denotes a rounded to the nearest integer which is smaller than or equal to a, and e_{sp} is the contact ratio of the sun and planet gears.

If $\theta_{s\_spall\&p_n(in)} \le \theta_{s\_spall} < \theta_{s\_spall\&p_n(out)}$, the spalling tooth has entered into the meshing, and the meshing stiffness, k_{spn}, changes in the dynamic equation.

The spalling defect is localized on the planet gear: At setting at t=0, the meshing gear pair between the sun gear and the planet gear #1 is the gear pair with spalling. The spalling defect localized on the planet gear #1 has just entered into engaging, and the carrier is regarded as static. In this situation, the torsional displacement of the planet gear relative to the initial position can be expressed as:

$$\theta_{p_n\_spall} = \text{mod}\left(\theta_p - \theta_c, 2\pi\right) \tag{23}$$

The torsional displacement of the planet gear from the time when the spalling tooth begins to engage, $\theta_{s\_spall\&p_i(in)}$, to when it ends engaging, $\theta_{s\_spall\&p_i(out)}$, can be expressed as:

$$\theta_{s\&p_n\_spall(in)} = 0, \theta_{s\&p_n\_spall(out)} = -\frac{2\pi}{z_p} * e_{sp} \tag{24}$$

If $\theta_{s\&p_n\_spall(in)} \le \theta_{p_n\_spall} < \theta_{s\&p_n\_spall(out)}$, the spalling tooth has entered into the meshing, and the meshing stiffness, k_{spn}, changes in the dynamic equation. However, the spalling tooth on the planet gear will also engage with the ring gear. The internal engaging of the sun & planet and the internal engaging of the planet & ring have a certain phase difference, γ_{sr}, which can be obtained following [13]. The torsional displacement of the planet gear with spalling going into and out of the inner meshing, $\theta_{r\&p_n\_spall(in)}$ and $\theta_{r\&p_n\_spall(out)}$, respectively, should take the phase difference into consideration. The difference can be expressed as:

$$\theta_{r\&p_i\_spall(in)} = -\left[\frac{2\pi * \text{ceil}\left(\frac{z_p}{2}\right)}{z_p} + \gamma_{sr}\frac{2\pi}{z_p}\right], \quad \theta_{r\&p_i\_spall(out)} = -\left[\frac{2\pi * \text{ceil}\left(\frac{z_p}{2}\right)}{z_p} + \gamma_{sr}\frac{2\pi}{z_p} + \frac{2\pi}{z_p} * e_{rp}\right] \tag{25}$$

If $\theta_{r\&p_n\_spall(in)} \le \theta_{p_n\_spall} < \theta_{r\&p_n\_spall(out)}$, the spalling tooth has entered into the meshing, and the meshing stiffness, k_{rpn}, changes in the dynamic equation.

The spalling fault is localized at the ring gear: At setting at t=0, the meshing pair between the ring gear and the #1 planet gear is the gear pair with spalling. The spalling defect on the ring gear just entered into the meshing, and the carrier is regarded as static. In this situation, the torsional displacement of the ring gear relative to the initial position can be expressed as:

$$\theta_{r\_spall} = \text{mod}\left(\theta_r - \theta_c, 2\pi\right) \tag{26}$$

The torsional displacement of the sun gear with a spalling tooth from the beginning of engagement, $\theta_{r\_\text{spall}\&p_i(\text{in})}$, to the ending of engagement, $\theta_{r\_\text{spall}\&p_i(\text{out})}$, can be expressed as:

$$\theta_{r\_\text{spall}\&p_n(\text{in})} = -\left[\text{floor}\left(\frac{z_r(n-1)}{N}\right)*\frac{2\pi}{z_r}+\text{mod}\left(\varphi_n,\frac{2\pi}{z_r}\right)\right] \tag{27}$$

$$\theta_{r\_\text{spall}\&p_n(\text{out})} = -\left[\text{floor}\left(\frac{z_r(n-1)}{N}\right)*\frac{2\pi}{z_r}+\text{mod}\left(\varphi_n,\frac{2\pi}{z_r}\right)+\frac{2\pi}{z_r}*e_{rp}\right] \tag{28}$$

If $\theta_{r\_\text{spall}\&p_i(\text{in})} \le \theta_{r\_spall} < \theta_{r\_\text{spall}\&p_i(\text{out})}$, the spalling tooth has entered into the meshing, and the meshing stiffness, k_{rpn}, changes in the dynamic equation.

Numerical Simulation:

The position of the fault has a significant effect on the PGT system. In order to investigate the effect of the spalling fault and the location of the defect on the vibration properties of the system, we propose the following. A tooth surface spalling defect whose dimensions w, L_s, and h_s are 5, 15, and 1.5mm, respectively, is seeded into the sun gear, planet gear, and ring gear. The sun gear is the input component; a constant torque of 600Nm is applied to the sun gear, and its shaft speed is set at 1200rpm. The vibration properties of the PGT system with the parameters given in Table 1 are simulated.

Table 1 Parameters of the planetary gear train

	Sun	Ring	Planet	Carrier
The number of teeth	31	73	21	
Modulus (mm)	4	4	4	
pressure angle (°)	20	20	20	
mass(kg)	4.45	7.28	0.79	13.70
Moment of inertia(kg · m^2)	4.77×10^{-2}	3.72×10^{-1}	7.07×10^{-4}	2.96×10^{-1}
Support stiffness (N/m)	2×10^9	2×10^9	2×10^9	2×10^9
Torsional stiffness(Nm/rad)	0	2×10^9	0	0
Tooth width(mm)	30	30	30	

If a spalling defect is localized on the sun gear, when the spalling defect area is engaged, an impulsive vibration can be generated. Every 31 meshing cycles (the sun gear has 31 cycles), the tooth with the spalling defect will mesh with the same planet gear one time; during this period, it will also engage with the other three planet gears one time.

The torsional acceleration responses of the sun gear under two conditions (healthy and with a spalling fault) are displayed in Figure 6. The figure clearly shows impulsive vibrations resulting from the presence of the spalling tooth. When the sun gear's spalling tooth meshes with the tooth of the planet gear, an impulsive vibration is produced. The duration of all signals given in this figure is 3*31 meshing periods T, corresponding to three revolutions of the sun gear (the carrier is considered static). Compared to the healthy condition, we see 12 spikes in the signal; these correspond to the spalling defect pair meshing with the planets four times in one revolution of the sun gear (there are four planets). The fault symptom is obvious.

Besides the responses in the time domain, the dynamic features in the frequency domain also exhibit the effects of a defect. According to [14], when the harmonic component of the meshing frequency is l=3, the motion of the torsional direction is excited. Figure 6 shows the frequency spectrum of the simulated acceleration signal corresponding to Figure 5. In this figure, sizable amplitudes appear at the triple harmonics of mesh frequency in two cases. The healthy case is characterized by the dominance of the mesh frequency and its harmonics, while in the case of a spalling fault on the sun gear tooth, a large number of sidebands appear around the meshing frequency and its harmonics. The frequency interval between sidebands is the characteristic frequency of the PGT with a spalling fault on the sun gear, or $N^*(f_s - f_c) = 56.15$Hz. In this case, we observe the defect signal's modulation of the amplitude of the mesh signal; this occurs once per revolution of a sun gear relative to the carrier.

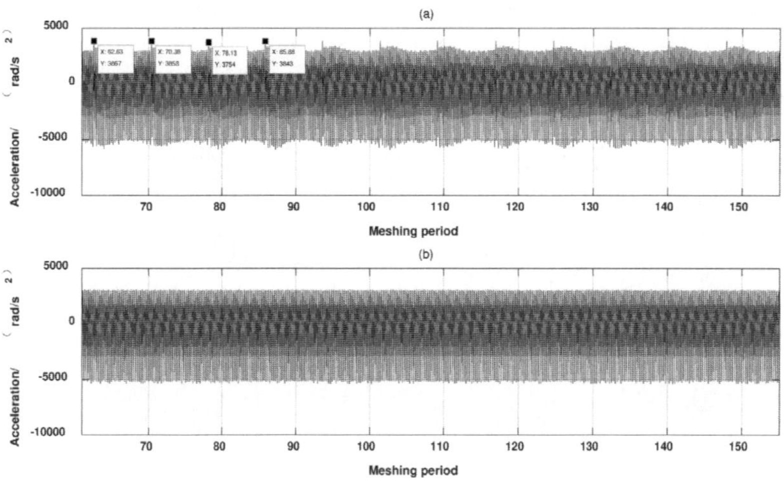

Figure 5. The dynamic response of the sun gear in torsional direction: (a) spalling fault localized on the sun gear (b) healthy planetary system

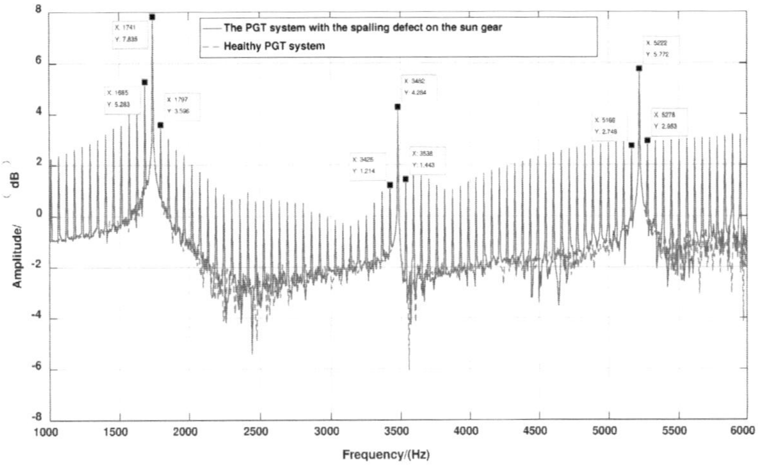

Figure 6. The frequency spectrum of the dynamic response of the sun gear in torsional direction

In addition, the trajectory of the sun gear center can reflect the effect of the spalling on the sun gear. In time duration [31**T*-31*5**T*], the trajectory of the sun gear center under two conditions (a spalling fault on the sun gear and a healthy planetary system) is shown in Figure 7. The figure indicates that the spalling defect on the sun gear will break the regularity of the trajectory of the sun gear center and result in more translational vibration of the sun gear than in the healthy case; this means a spalling defect on the sun gear has greater influence on the vibration level of the component in a translational direction.

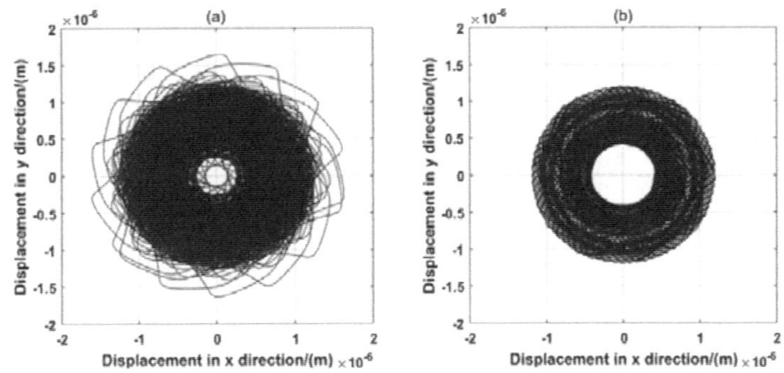

Figure 7. The trajectory of sun gear center: (a) spalling fault on the sun gear (b) healthy planetary system

As a consequence, in the case of a spalling fault localized on the sun gear, new frequency components appear at $nf_m + N\left[m\left(f_s - f_c\right)\right]$ (n, m are integers and n should obey the phase modulation law), where f_m represents mesh frequency, and f_s, f_c represent the rotational frequency of the sun gear and carrier respectively. The vibration amplitude will flatten out at higher levels. Its fluctuating amplitudes increase when a spalling defect appears.

If a spalling defect is seeded in the planet gear, every 21 meshing cycles (there are 21 teeth in a planet gear), the spalling tooth will mesh with the sun gear and ring gear one time. Figure 8 shows the acceleration of the sun gear in the torsional direction under two conditions (healthy condition and planet gear with a spalling fault). The time duration of all signals in this figure is 4*21 meshing periods *T*, corresponding to four revolutions of the planet gear (the carrier is considered static). When we compare this to the healthy condition, it is obvious that there are four spikes in the signal, with the spalling fault on the planet gear meshing with the sun gear one time in one revolution of the planet gear. The fault symptom is obvious.

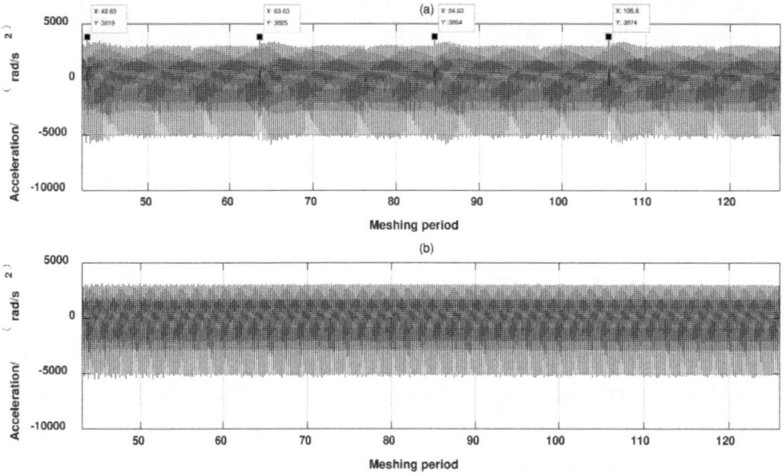

Figure 8. The dynamic response of the sun gear in torsional direction (a) spalling fault localized on planet gear (b) healthy planetary system

Figure 9 shows the frequency spectrum of the simulated acceleration signal corresponding to Figure 8. In the figure, sizable amplitudes show at the triple harmonics of mesh frequency under two conditions (healthy condition and a spalling fault on the planet gear). In the healthy condition, the gear mesh frequency and its harmonics dominate the frequency spectrum. When there is a spalling fault on the planet gear, a large number of sidebands appear around the gear mesh frequency and its harmonics. The frequency interval between sidebands is the characteristic frequency of the PGT with a spalling fault on the planet gear, or (f_p+f_c)= 20.72Hz, and the same phenomenon of amplitude modulation is observed: the amplitude of the mesh signal is modulated by the defect signal once per revolution of a planet gear relative to the carrier.

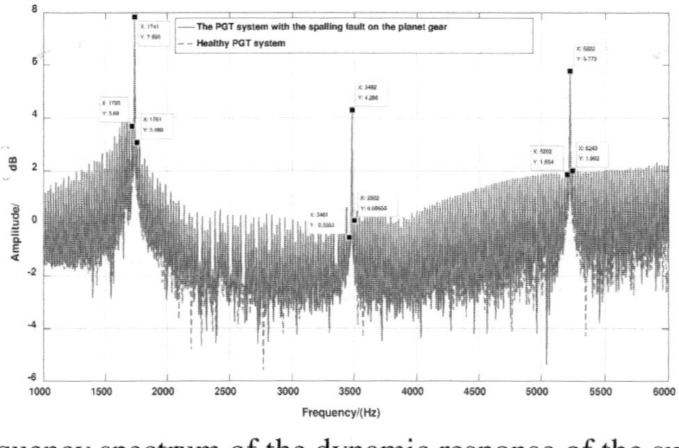

Figure 9. The frequency spectrum of the dynamic response of the sun gear in torsional direction

For the time period [21*T-21*5*T], the trajectory of the sun gear center under two conditions (a spalling fault localized on the planet gear and a healthy planetary system) is shown in Figure 10. The figure indicates that the spalling fault on the sun gear will break the regularity of the trajectory of the sun gear center and lead to a wider motion range than in the healthy condition.

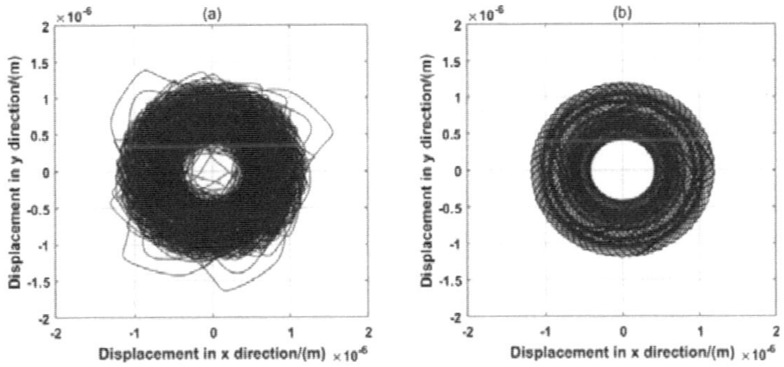

Figure 10. The trajectory of sun gear center: (a) spalling fault located on the sun gear (b) healthy planetary system

As a consequence, in the case of a spalling fault occurring on the planet gear, new frequency components appear at $nf_m + \left[m\left(f_p + f_c\right)\right]$ (n, m are integers and n should obey the phase modulation law), where f_p represents the rotational frequency of the planet gear. The vibration flattens with higher vibration levels in the torsional and translational directions.

In the case of a spalling defect localized on the ring gear, every 73 meshing cycles (there are 73 ring gears) the tooth on the ring gear with the spalling defect will mesh with the same planet gear. This means there could be a periodic impact impulse on the response of the planetary system. During this period, it also will engage with the other three planet gears one time. Figure 11 shows the torsional acceleration of the sun gear under two conditions (healthy and sun gear with spalling fault). The time duration of all signals given in this figure is 1*73 meshing periods, corresponding to one revolution of the ring gear (the carrier is considered static). When we compare this to the healthy condition, it is obvious that there are four spikes in the signal with the spalling fault on the sun gear. This corresponds to the spalling pair meshing with the planet four times in one revolution of the ring gear (there are four planets). The fault symptom is obvious.

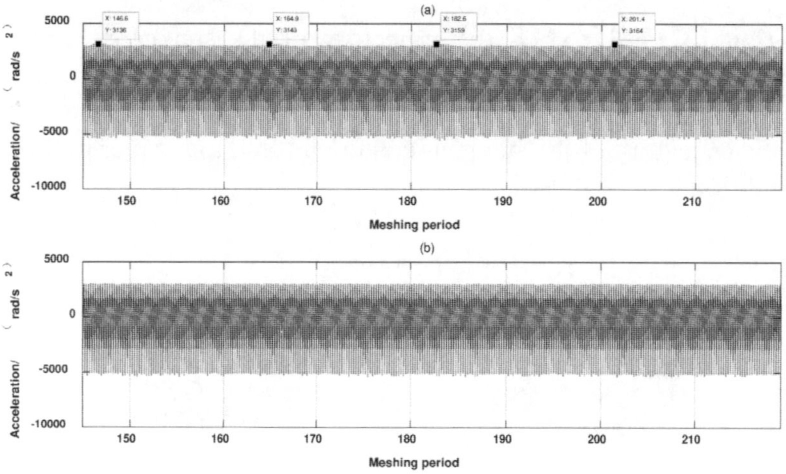

Figure 11: The dynamic response of the sun gear in torsional direction (a) spalling fault localized on the ring gear (b) healthy planetary system

Figure 12 shows the frequency spectrum of the simulated acceleration signal corresponding to Figure 11. As the figure shows, the presence of a spalling defect on the ring gear tooth can change the frequency spectrum structure, generating many sidebands around the fundamental mesh frequency and its harmonics. The frequency interval between sidebands is the characteristic frequency of the PGT with a spalling defect on the ring gear, or $N^*(f_c)$ = 23.85Hz. In this case, we observe an amplitude modulation of the mesh signal by the defect signal, once per revolution of a ring relative to the carrier.

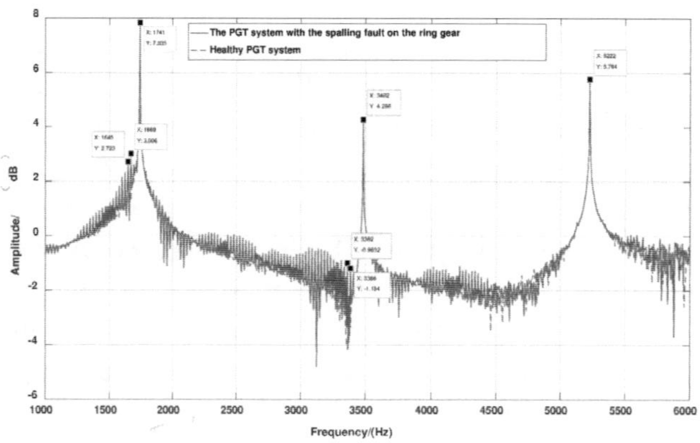

Figure 12. The frequency spectrum of the dynamic response of the sun gear in torsional direction

For the time period [2*73*T-3*73*T], the trajectory of sun gear center under two conditions (a spalling fault located on the ring gear and a healthy planetary system) is shown in Figure 13. The figure indicates that a spalling fault on the ring gear will lead to a slight disturbance in the trajectory of sun gear center compared to cases where the spalling fault is located on the sun gear or planet gear.

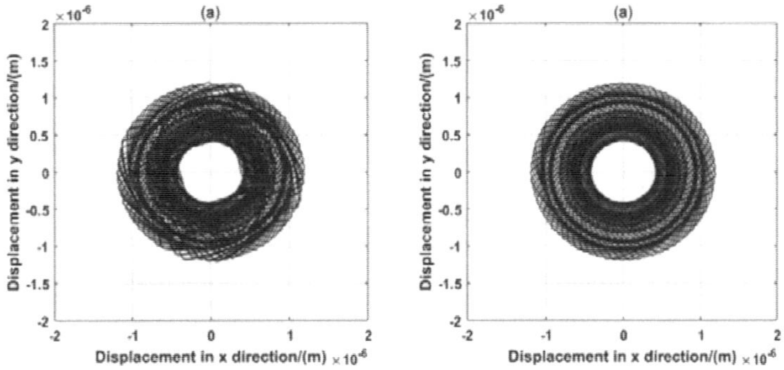

Figure 13: The trajectory of sun gear center: (a) spalling fault located on the ring gear (b) healthy planetary system

As a consequence, when a spalling defect occurs at the ring gear, the impulsive vibrations and sidebands can be observed in the dynamic response of the planetary gear set in time and frequency domains. New frequency components appear at $nf_m + Nmf_c$ (n, m are integers and n should obey the phase modulation law). The spectrum seems to be flattened at higher vibration levels, but compared to cases when a defect is localized on the sun gear or planet gear, this case (i.e., a spalling defect localized on the ring gear) has less influence on the dynamic response of the PGT system.

Conclusion:

The paper has provided an analytical formulation for mesh stiffness in the presence of a spalling defect. The mesh stiffness model, in which the spalling defect is seeded into the sun, planet, and ring gears in turn, is incorporated into a dynamic model of a PGT system to investigate the effect of tooth spalling defects. The dynamic responses computed in each of these cases show the dominance of the mesh frequency and its harmonics, with the presence of fault characteristic sidebands issued by the amplitude modulation caused by the spalling defect. The results can be applied to the condition monitoring and fault diagnosis of the planetary gear system.

Acknowledgement: This work is supported by the National Natural Science Foundation of China under Contract No.51475053 and Foundation of SKLMT-KFKT-201704 8: Diego Galar Lulea of technology wind farm condition monitoring.

References:

[1] M R Hoseini, M J Zuo. Literature review for creating and quantifying faults in planetary gearboxes. Reliability Research Lab, Mechanical Department, University of Alberta, May 2009.

[2] W Kim ,YL Ji ,J Chung. Dynamic analysis for a planetary gear with time-varying pressure angles and contact ratios. Journal of Sound & Vibration, 2012 , 331 (4) :883-901

[3] A Kahraman. Load sharing characteristics of planetary transmissions. Mechanism & Machine Theory, 1994 , 29 (8) :1151-1165

[4] A Kahraman. Planetary Gear Train Dynamics. Journal of Mechanical Design, 1994 , 116 (3) :713-720

[5] A.Kahraman, S.Vijayakar. Effect of Internal Gear Flexibility on the Quasi-Static Behavior of a Planetary Gear Set, Journal of Mechanical Design , 2001 , 123 (3) :408-415
[6] M Inalpolat, A Kahraman. A theoretical and experimental investigation of modulation sidebands of planetary gear sets. Journal of Sound & Vibration , 2009 , 323 (3) :677-696
[7] M Inalpolat, A Kahraman. A dynamic model to predict modulation sidebands of a planetary gear set having manufacturing errors. Journal of Sound & Vibration , 2010 , 329 (4) :371-393
[8] RG Parker, V Agashe, SM Vijayakar. Dynamic Response of a Planetary Gear System Using a Finite Element/Contact Mechanics Model. Journal of Mechanical Design, 2000 , 122 (3) :304
[9] T Sun, HY Hu. Nonlinear dynamics of a planetary gear system with multiple clearances. Mechanism & Machine Theory, 2003 , 38 (12) :1371-1390
[10] D.C.H. Yang, J.Y. Lin, Hertzian damping, tooth friction and bending elasticity in gear impact dynamics, J. Mech. Des. 109 (2) (1987) 189–196.
[11] X.H. Tian, Dynamic Simulation for System Response of Gearbox Including Localized Gear Faults. Master's thesis University of Alberta, 2004.
[12] Z. Chen, Y. Shao, Dynamic simulation of spur gearwith tooth root crack propagating along tooth width and crack depth, Eng. Fail. Anal. 18 (8) (2011) 2149–2164.
[13] RG Parker, J Lin. Mesh Phasing Relationships in Planetary and Epicyclic Gears. Journal of Mechanical Design , 2004 , 126 (2) :525-534
[14] RG Parker. A physical explanation for the effectiveness of planet phasing to suppress planetary gear vibration. Journal of Sound & Vibration , 2000 , 236 (4) :561-573

VIDEO MOTION AMPLIFICATION VS. OPERATING DEFLECTION SHAPES FOR MACHINERY DIAGNOSIS

by: William D. Marscher, P.E. & Maki M. Onari

Mechanical Solutions, Inc.
11 Apollo Dr
Whippany, NJ 07981 USA
(973) 326-9920
www.mechsol.com

Abstract:

Operating Deflection Shapes (ODS) has been an important tool in visualizing the vibration of the machine and its system, including piping networks. The input for ODS is the phase-linked signal set from a group of accelerometers, moved over often hundreds of test points. The data is superimposed onto a CAD model, and then scaled-up vibrations are animated at frequencies of interest. This process is time-consuming and therefore expensive each time it is applied by experts, and is error-prone. An alternative method has been developed that is based on evaluation of high resolution/ high speed videos. The method provides information equivalent to a high-sensor-count ODS, by treating each pixel as an accelerometer, using the pixel's light intensity modulation to translate information embedded in the video into vibration motion able to be observed and interpreted by human investigators.

Key Words:
Motion Amplification; video; Operating Deflection Shape; Turbomachinery

Introduction:

Turbomachinery vibration is often useful in determining whether or not a turbomachine is operating properly, and for diagnosis of problems if the operation appears improper, or if reliability issues have been experienced (e.g. fatigue cracking, or premature wear of bearings and seals). For several decades, a visual method called Operating Deflection Shapes (ODS) has been an important tool in getting a complete and simultaneous view of the vibration of the machine and its system (e.g. piping, foundation, and driver or driven machine). The input for ODS is the phase-linked signal set from a group of accelerometers, moved over often hundreds of test points while one phase-reference accelerometer is kept at a consistent location and direction. The data is superimposed onto a simplified CAD model of the machine and system, and then the exaggerated (but to-scale) vibrations are animated at frequencies where the response is sufficient to be of interest to the researcher or troubleshooter. This process takes several days for complex machinery and systems, operated at a variety of process points.

A new method has been developed that is based on evaluation of high resolution/ high speed video taken of the operating machinery. The method provides information equivalent to a high-sensor-count ODS, by treating each pixel as an accelerometer, using the pixel's light intensity modulation to determine local vibration displacement frequency spectrum. From this information, realistic magnification of slow-motion video footage permits microscopic vibration to be amplified and thereby observed and interpreted by the human investigator.

ODS Methods:

Over thirty years ago, some researchers developed the concept of "seeing" vibration. Their approach was to acquire data at many locations and directions on a vibrating structure, and then allocating those motions to a very simple CAD model. This model was then animated on a computer video screen, or the extremes of the vibration in the model were plotted by a computer printer or plotter. This technique became known as Operating Deflection Shapes. The following outline summarizes the technique:

- Based on the "natural excitation" frequency spectra of the rotating equipment
- User acquires vibration data from various locations and directions on machine (hundreds of vibration measurements
- One sensor is always kept at the same location and direction, to provide a reference signal as other probes are "roved".
- A large database is built up of amplitude vs. frequency and phase angle
- A 3-D CAD model is constructed, and assigned motion from each individual vibration data point.
- The associated software is used to amplify the results filtered at specific frequencies of interest, creating animations of the CAD model of the equipment and related piping and foundation.

As an example, consider a high-pressure "charge pump" that was experiencing excessive vibration at a US nuclear plant. This pump runs at 4800 rpm (80 Hz), driven by an 1800 rpm motor through a speed increasing gear set. The pump type is a multistage barrel. Typical vibration levels in these pumps tend to be higher than industrial pumps, and are often over 0.2 in/sec rms in a healthy pump. Depending on the specific pump type, bearing clearances, and the plant experience, trip levels may be set between 0.3 and 0.5 in/sec rms. In this case, the pump was experiencing gradual vibration increase that had not been diagnosed, and was surpassing 0.4 in/sec rms. The phase angle reading of the peak vibration versus the location of the keyphasor on the shaft was shifting, varying between 0 and 180 degrees. A combination of operation modal analysis and operating deflection shape (ODS) evaluation was performed. The testing was performed tri-directionally at 200 locations on the pump, pedestal, and foundation. Rotor displacement was also recorded from permanently mounted proximity probes.

The detailed study determined that there was significant lateral/ horizontal vibration of one of the pump non-driven-end (NDE) mounting feet versus the top mounting surface of its pedestal. This is evident in Figure 1, which will be animated during the presentation of this paper.

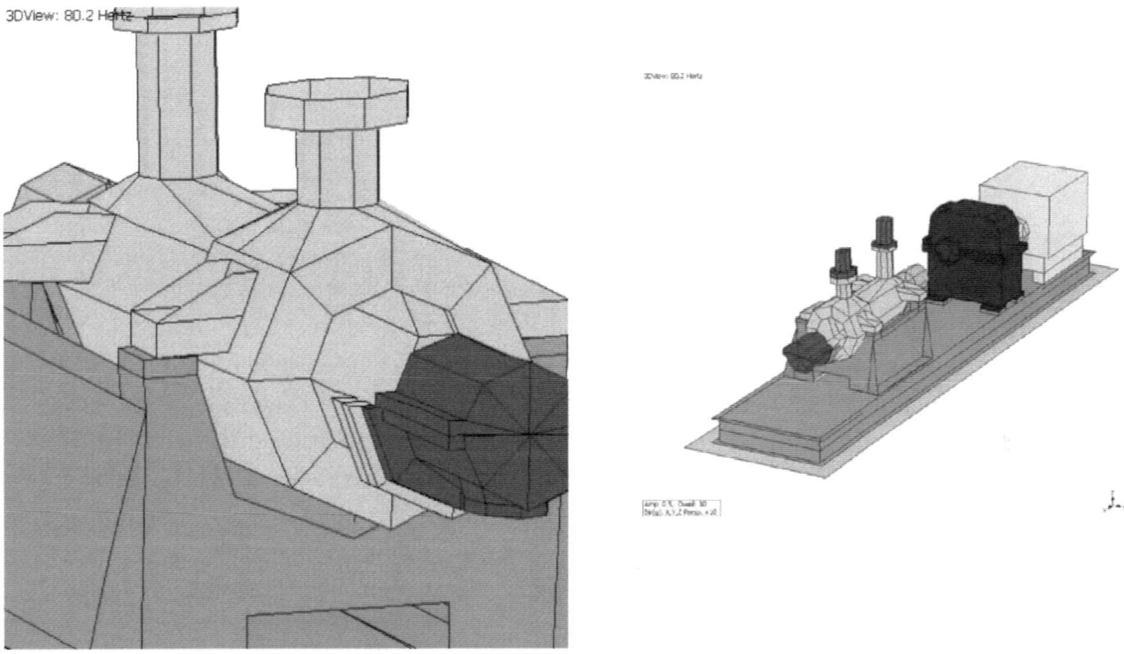

Figure 1: Soft foot evident in left center of the left-side picture. Data was taken on the entire pump/ driver/ pedestal/ foundation system, as shown in the right-side picture.

The diagnosis was that a casing (as supported by the pedestal) structural natural frequency (involving horizontal "tipping" of the casing NDE end) used to be 15% above the running speed, but was down drifting into (and sometimes beyond) the running speed of the pump, tuning in a resonance. The reason was that, when the foot was able to slide in frictional contact, it no longer supplied as stiff of a support of the casing. By tightening down on the foot attachment bolt, this "soft foot" condition was repaired, and the natural frequency involved shifted back up to order of 93 Hz, sufficiently separated from 1x running speed forces (e.g. residual imbalance) to avoid further resonance. Vibration levels decreased to below 0.1 in/sec rms.

The physical problem may seem trivial once its nature is perceived. However, some very good technical people at the plant wrestled with this problem for 6 months before a 3-D visual technique- in this case ODS- was applied. The perspective on the vibration distribution and its meaning was not there when the plant evaluated the vibration data point-by-point instead of as 'whole cloth".

Therefore, ODS was extremely useful in this case, rendering problem diagnosis and recommended solution straightforward once it was employed. However, in general ODS has its drawbacks as well. The pros and cons of ODS are summarized in the list below:

- ✓ Powerful and intuitive diagnostic tool
  - Can clearly demonstrate modes and frequencies of vibration
- ✓ Proven over decades of application
- ✗ Time consuming
  - Data acquisition – hundreds of data points
  - Post processing – compiling database to match model points
- ✗ Potential for bookkeeping error
  - Match all gathered data points to appropriate place on model
- ✗ Requires proximity – not appropriate for restricted access
  - Heat, radiation, accessibility/scaffolding requirements

This raises the thought, what if the investigator could truly see the vibration? Academic researchers, some of whom are listed in the References (Ref. 1-3) had this thought about 25 years ago, and have been gradually perfecting techniques. Basically, these techniques fall into two categories: 1) tracking of specific points, edges, or (as machine-vision scientists call them) "blobs", and 2) performing statistics, including signal vs. time as well as signal FFT frequency spectra, on the individual independent pixels. The former are called "Lagrangian Methods" by many researchers, and the later are called "Eulerian Methods". A rich literature basis exists for these methods. This paper's references emphasize the Eulerian technique

The motion-amplified video method may be summarized as follows:

- Uses high-speed, high resolution video
  - Eulerian method is equivalent of millions of accelerometers, 1 per pixel
- Analyzes/quantifies motion
  - Frequencies
  - Displacement (2-dimensional)
- Algorithms amplify motion to human visual threshold
  - Filterable by desired frequencies

Example: A High Speed Blower with Acute Noise Levels

Two small constant speed (3600 rpm) ducted fan blowers and motors installed for worker-required ventilation at a waste water facility provides an example. The vibration of each centrifugal 8-vaned blower and motor was actually quite modest, well within ISO 10816-3 specified vibration limits.

However, the room noise created by the blowers was as high as 120 dB at certain operating conditions. The authors applied both classical ODS as well as the new high-frequency high-resolution video motion amplification method to evaluate the problem. A view of the blowers and their layout in the room is shown in Figure 2:

Figure 2: Picture of the example blowers, motors, and ducting, and their layout in the machine room.

Microphone signals identified the problem frequency as 474 Hz, which was the blower rotating vane passing frequency. The question remained why, particularly given the low vibration levels. A representative ODS of one of the blowers is shown in Figure 3.

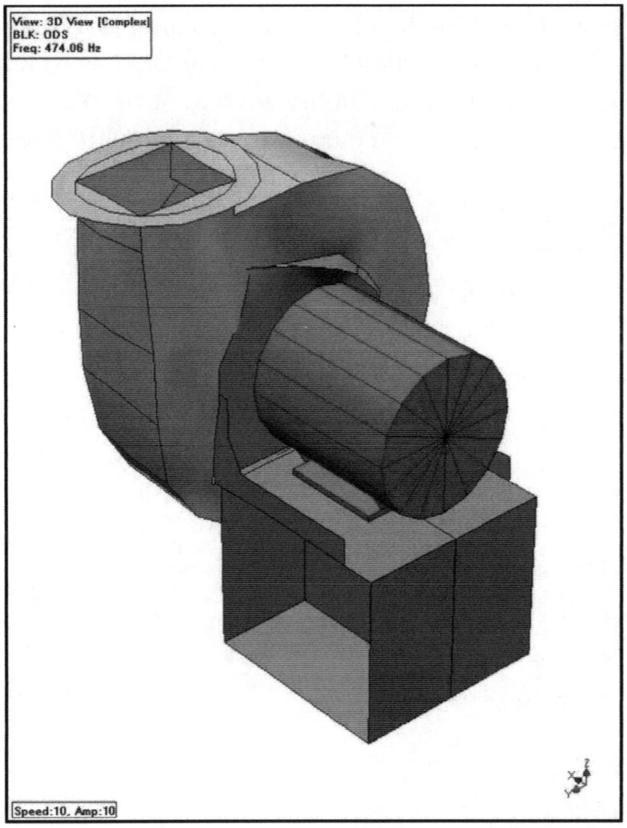

Figure 3: ODS model of the subject blower, exhibiting "undulating" axial motion of the upper flat side of the blower casing (vibration filtered to 474 Hz).

Figure 4: Motion-amplified video of the subject blower, exhibiting the same (but more easily visualized) "undulating" axial motion of the upper flat side of the blower casing.

Once the shape was identified and evaluated, it was clear that the blower was in resonance with the vane passing frequency. The upper sidewall of the casing unfortunately possessed a natural frequency very close to the vane passing frequency, and as the blower underwent changes in loading, the induction motor slip changed and the vane passing frequency drifted into and out of precise resonance with the sidewall natural frequency. When resonance was precisely tuned in, the noise level reached its 120 dB peak.

Other examples of pumps and turbines will be provided in the presentation of this paper.

Summary:

As presented, the new video-based motion amplification procedures and equipment coming onto the market can be very useful for vibration-based diagnosis of machinery. It has advantages over the classic ODS method in many (not all) instances, and typically takes much less time and "logistics" to implement.

The benefits are:

- ✓ It is a powerful and intuitive diagnostic tool
  - Realistically demonstrates modes and frequencies of vibration
  - Easy for non-experts to understand (management, etc.)
  - Comprehensive set of data points
- ✓ It does not requires "contact" – perfect for restricted areas
  - Heat, radiation, accessibility/scaffolding requirements
- ✓ It is fast
  - Millions of data points ready to evaluate in minutes, not days
- ✓ It is cost-effective
  - Helps focus effort for accelerometer-based ODS

Motion-amplified video certainly appears to represent an important addition to a vibration diagnostic tool kit.

References:

1. Timoner, SJ, and Freeman, DM, "Multi-image gradient-based algorithms for motion estimation", Optical Engineering 40.9 (2001) pp. 2003-2016
2. Liu et al, "Motion magnification", ACM Transactions of Graphics (TOG) – Proc. of ACM SIGGRAPH 2005, TOG Homepage, v. 24, Issue 3, July 2005
3. Wu, H-Y, et al, "Eulerian video magnification for revealing subtle changes in the world", ACM Trans. of Graphics (Proc. SIGGRAPH 2012), v. 31, August 2012

MFPT Society
715 Twining Road
Suite 135
Dresher PA 19025

ISBN 978-1-5108-8840-1

9 781510 888401

CURRAN ASSOCIATES INC.
proceedings.com

55th American Association of Swine Veterinarians Annual Meeting 2024

Nashville, Tennessee, USA
24 - 27 February 2024